Excel

YEAR 8

Mathematics Extension Revision & Exam Workbook

ESSENTIAL skills

AS Kalra

Reprinted 2004, 2006, 2007, 2008, 2009, 2010, 2012
Updated in 2014 for the Australian Curriculum
Reprinted 2015, 2017, 2020, 2021, 2022, 2023

Reviewed in 2024 for the NSW Curriculum and Australian Curriculum Version 9.0 changes

ISBN 978 1 74020 316 6

Pascal Press
PO Box 250
Glebe NSW 2037
(02) 9198 1748
www.pascalpress.com.au

Publisher: Vivienne Joannou
Project editor: Michael Cole-King
Edited by May McCool and Michael Cole-King
Answers checked by Peter Little and Chelsea Smithies
Additional material by Chelsea Smithies
Typeset by Precision Typesetting (Barbara Nilsson) and DiZign Pty Ltd
Cover design by DiZign Pty Ltd
Printed by Vivar Printing/Green Giant Press

Dedication

This book is dedicated to the new generation of young Australians in whose hands lies the future of our nation and who by their hard work, acquired knowledge and intelligence will take Australia successfully through the 21st century.

This book is also in the loving, living and lasting memory of my dear mum, dad and uncle, who will remain a great source of inspiration and encouragement to me for times to come.

Acknowledgements
I would especially like to express my thanks and appreciation to my dear wife and my dear son, who have helped me find the time to write this book. Without their help and support, achievement of all this work would not have been possible.

Contents

CHAPTER 12 – Probability

CHAPTER 13 – Statistics

EXAM PAPERS

ANSWERS

Introduction

There are two workbooks in this series for the Year 8 Australian Curriculum Mathematics course:

- ***Excel Essential Skills Year 8 Mathematics Revision & Exam Workbook*** and
- ***Excel Essential Skills Year 8 Mathematics Extension Revision & Exam Workbook*** (this book).

This book should be completed after the first book. It has been written specifically for the Year 8 Australian Curriculum Mathematics course and forms part of a series of eight Revision & Exam Workbooks for Years 7 to 10. Each book in the series has been specifically designed to help students **revise** their work so that they can prepare for success in their **tests** during the school year and in their **half-yearly** and **yearly** exams.

The emphasis in this book is to challenge students and extend their mathematical knowledge through extensive practice. This will ensure that students make further progress towards the Advanced Mathematics courses in senior years.

The following features will help students achieve this goal:

- This book is a **workbook**. Students write in the book, ensuring that they have all their questions and working in the same place. This is invaluable when revising for exams—no lost notes or missing pages!
- Each page is a **self-contained, carefully graded unit of work**; this means students can plan their revision effectively by completing set pages of work for each section.
- **Topics are covered in depth,** providing students with practice to enable them to focus on areas of weakness or areas for extension.
- A **Topic Test** is provided at the end of each chapter. These tests are designed to help students test their knowledge of each syllabus topic. Practising tests similar to those they will sit at school will build students' confidence and help them perform well in their actual tests.
- **Four Exam Papers** have been included to test students on the complete Year 8 Mathematics course, helping students prepare for their **half-yearly** and **yearly exams**.
- A **marking scheme** is included in both the Topic Tests and Exam Papers to give students an idea of their progress.
- A **Topic Test and Exam paper Feedback Chart**, found on the inside back cover, enables students to record their scores in all tests and exams.
- **Answers to all questions** are provided at the back of the book.
- There is a **page reference** to the ***Excel Mathematics Study Guide Year 8*** in the top right-hand corner of all pages, excluding the tests. If students need help with a specific section, they will find relevant explanations and worked examples on these pages of the study guide.

A note from the author

Mathematics is best learned if you have pen and paper with you and do every question in writing. Do not just read through the book—work through it and answer the questions, writing down all working. If this approach is coupled with a menu of motivation, realistic goal-setting and a positive attitude, it will lead to better marks in the examinations.

My best wishes are with you; I believe this book will help you achieve the best possible results. Good luck in your studies!

AS Kalra, MA, MEd, BSc, BEd

CHAPTER 1

Rational numbers

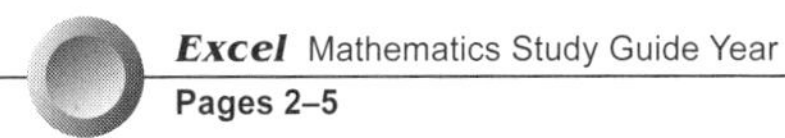
Excel Mathematics Study Guide Year 8
Pages 2–5

UNIT 1: Approximation and rounding off

QUESTION 1 Round off the following numbers to the nearest hundred.

a 56 = ________ **b** 8342 = ________ **c** 961 = ________
d 637 = ________ **e** 83 = ________ **f** 93 = ________
g 935 = ________ **h** 251 = ________ **i** 831 = ________
j 6439 = ________ **k** 6132 = ________ **l** 8538 = ________

QUESTION 2 Round off the following numbers to the nearest thousand.

a 8629 = ________ **b** 953 = ________ **c** 2364 = ________
d 5134 = ________ **e** 8309 = ________ **f** 9837 = ________
g 9635 = ________ **h** 563 = ________ **i** 5631 = ________
j 6112 = ________ **k** 998 = ________ **l** 8331 = ________

QUESTION 3 Write the following decimals correct to two decimal places.

a 8.315 = ________ **b** 0.931 = ________ **c** 72.364 = ________
d 92.612 = ________ **e** 0.2156 = ________ **f** 65.123 = ________
g 53.8135 = ________ **h** 36.213 = ________ **i** 92.987 = ________
j 98.311 = ________ **k** 48.115 = ________ **l** 63.125 = ________

QUESTION 4 Write the following decimals correct to one decimal place.

a 1.78 = ________ **b** 9.36 = ________ **c** 15.125 = ________
d 39.18 = ________ **e** 63.149 = ________ **f** 83.256 = ________
g 93.876 = ________ **h** 58.234 = ________ **i** 96.156 = ________
j 51.381 = ________ **k** 68.876 = ________ **l** 52.963 = ________

QUESTION 5 Express the following correct to 2 significant figures.

a 92.31 = ________ **b** 365 = ________ **c** 5839 = ________
d 6638 = ________ **e** 9389 = ________ **f** 56.78 = ________
g 913.68 = ________ **h** 512.64 = ________ **i** 916.86 = ________
j 1925.125 = ________ **k** 832.051 = ________ **l** 913.186 = ________

QUESTION 6 Express the following correct to 3 significant figures.

a 9364 = ________ **b** 5689 = ________ **c** 81 320 = ________
d 56.125 = ________ **e** 83.817 = ________ **f** 98.659 = ________
g 67.838 = ________ **h** 915.87 = ________ **i** 86.345 = ________
j 109.862 = ________ **k** 72.63 = ________ **l** 981.834 = ________

Rational numbers

UNIT 2: Mixed operations

Excel Mathematics Study Guide Year 8
Pages 2–5

QUESTION 1 Use your calculator to simplify the following.

a $20 - 8 + 7 =$ ______ b $56 - 18 - 6 =$ ______ c $25 + 31 + 18 - 16 =$ ______
d $56 - 13 + 8 + 9 =$ ______ e $93 - (9 + 2) =$ ______ f $(21 + 9) \times 6 =$ ______
g $3 \times 4 \times 5 =$ ______ h $93 \div 3 \times 4 \times 4 =$ ______ i $1800 \div 3 \div 2 \div 5 =$ ______
j $9 + 3 \times 2 =$ ______ k $952 \div 2 + 5 - 3 =$ ______ l $87 - 3 \times 9 =$ ______

QUESTION 2 Use your calculator to work out the following.

a $361.8 + 91.3 =$ ______ b $32.8 \times 1.5 =$ ______
c $36.4 + 42.8 - 15.2 =$ ______ d $3 \times 5.6 + 2 \times 9.1 =$ ______
e $3.2 \times 5.6 \times 3.6 =$ ______ f $91.7 + 8.9 \times 3 =$ ______
g $8.1 \times 3.2 - 5.6 =$ ______ h $8.9 \times 2.5 + 3.6 \times 1.5 =$ ______

QUESTION 3 Use your calculator to answer the following.

a $9.14 \times 2.3 - 1.5 =$ ______ b $20 \div 10 \times 8 =$ ______ c $-4.6 + 9.8 =$ ______
d $8.9 + (4.3 \times 1.2) =$ ______ e $9.24 \times 1.3 - 2.61 =$ ______ f $8.6 + 14.5 - 6.2 =$ ______
g $212.5 \times 8.3 - 150.9 =$ ______ h $9.32 \times 1.3 - 2.57 =$ ______ i $-6.3 + 8.6 - 7.5 =$ ______
j $120.5 \times 3 - 93.2 =$ ______ k $8.2 \times 7.6 \times 4.2 =$ ______ l $76 + 235 - 81 =$ ______

QUESTION 4 Use the fraction key on your calculator to calculate the following.

a $2\frac{1}{2} + 1\frac{1}{2} + \frac{2}{3} =$ ______ b $6\frac{1}{4} - 2\frac{3}{4} + 6\frac{1}{2} =$ ______ c $5\frac{3}{5} + 4\frac{2}{3} - 1\frac{3}{5} =$ ______
d $\frac{2}{15} \times \frac{3}{8} \times \frac{10}{17} =$ ______ e $9\frac{1}{2} \times 5\frac{1}{4} - 1\frac{1}{2} =$ ______ f $3\frac{2}{5} \div 1\frac{1}{5} + 3\frac{1}{2} =$ ______
g $8\frac{1}{2} \times 2\frac{1}{2} \times 1\frac{1}{4} =$ ______ h $1\frac{1}{5} \div 2\frac{1}{10} \times 4\frac{1}{5} =$ ______ i $8\frac{2}{3} \div \frac{1}{9} \times \frac{1}{6} =$ ______
j $12\frac{1}{2} \div \left(1\frac{1}{2} + 2\frac{1}{2}\right) =$ ______ k $5\frac{1}{3} \div \left(\frac{2}{3} \div 1\frac{1}{9}\right) =$ ______ l $\frac{8}{9} + 2\frac{2}{3} \div \frac{5}{6} =$ ______

QUESTION 5 Calculate the following.

a $3.2 \times 4.3 + 9.1 \times 2.3 =$ ______ b $56.7 - 8.1 \times 5.2 =$ ______ c $56 \div 7 \times 4 \div 2 =$ ______
d $86.3 + 58.9 - 30.6 =$ ______ e $(21 + 9) \times 4 =$ ______ f $9 + 8 \times 3 =$ ______
g $0.3 \times 0.6 + 6.1 \times 1.2 =$ ______ h $90 - 8 \times 5 =$ ______ i $6.9 \times 8.4 - 10.3 =$ ______

QUESTION 6 Answer the following.

a $8.5 \times 2.1 \times 1.3 =$ ______ b $231.65 \times 1.2 =$ ______ c $1.5 \times 8.6 - 3.6 =$ ______
d $9.6 \div 0.3 + 5.6 =$ ______ e $5.4 \times 3.2 \times 1.1 =$ ______ f $0.3 \times 0.8 + 5 \times 1.2 =$ ______
g $5.4 \div 9.6 \div 3.2 =$ ______ h $8.3 + 9.2 \times 2.1 =$ ______ i $632.5 \times 1.4 =$ ______

Rational numbers

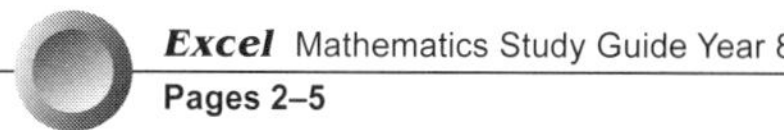

UNIT 3: Squares, cubes and powers

QUESTION 1 Use your calculator x^2 key to evaluate the following.

a $(3)^2 =$ ______ b $(7)^2 =$ ______ c $(9)^2 =$ ______

d $(5)^2 =$ ______ e $(11)^2 =$ ______ f $(13)^2 =$ ______

g $(41)^2 =$ ______ h $(48)^2 =$ ______ i $(63)^2 =$ ______

j $(58)^2 =$ ______ k $(1.7)^2 =$ ______ l $(91.3)^2 =$ ______

QUESTION 2 Use your calculator x^y or key to evaluate the following.

a $2^3 =$ ______ b $5^3 =$ ______ c $8^3 =$ ______

d $21^3 =$ ______ e $68^3 =$ ______ f $72^3 =$ ______

g $4^5 =$ ______ h $6^4 =$ ______ i $18^2 =$ ______

j $2^8 =$ ______ k $3^6 =$ ______ l $(2.5)^4 =$ ______

QUESTION 3 Answer the following.

a $(2.1)^3 =$ ______ b $(8.7)^2 =$ ______ c $(9.5)^3 =$ ______

d $(10.4)^4 =$ ______ e $(6.9)^3 =$ ______ f $(18.3)^4 =$ ______

g $(225)^2 =$ ______ h $(63.5)^2 =$ ______ i $(2.1)^6 =$ ______

j $(71.5)^2 =$ ______ k $(18.5)^3 =$ ______ l $(61.4)^2 =$ ______

QUESTION 4 Calculate the following correct to 1 decimal place.

a $(2.51)^2 =$ ______ b $(1.2)^4 =$ ______ c $(8.2)^4 =$ ______

d $(8.64)^2 =$ ______ e $(8.1)^2 =$ ______ f $(16.1)^3 =$ ______

g $(9.1)^3 =$ ______ h $(19.5)^2 =$ ______ i $(5.2)^4 =$ ______

j $(3.6)^3 =$ ______ k $(7.6)^4 =$ ______ l $(2.3)^5 =$ ______

QUESTION 5 Evaluate the following, leaving your answers in fraction form.

a $\left(2\frac{1}{2}\right)^2 =$ ______ b $\left(\frac{1}{3}\right)^3 =$ ______ c $\left(1\frac{3}{4}\right)^2 =$ ______

d $\left(4\frac{1}{3}\right)^2 =$ ______ e $\left(5\frac{3}{4}\right)^3 =$ ______ f $\left(7\frac{1}{2}\right)^4 =$ ______

g $\left(2\frac{1}{3}\right)^3 =$ ______ h $\left(3\frac{3}{5}\right)^2 =$ ______ i $\left(5\frac{5}{6}\right)^2 =$ ______

j $\left(2\frac{1}{4}\right)^3 =$ ______ k $\left(3\frac{3}{4}\right)^3 =$ ______ l $\left(4\frac{1}{4}\right)^2 =$ ______

Rational numbers

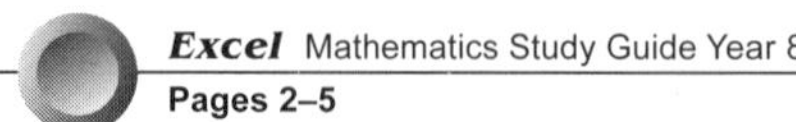

UNIT 4: Square roots and cube roots

QUESTION 1 Use your calculator square root key to simplify the following.

a $\sqrt{4} =$ ______ b $\sqrt{49} =$ ______ c $\sqrt{81} =$ ______

d $\sqrt{16} =$ ______ e $\sqrt{4} + \sqrt{64} =$ ______ f $\sqrt{225} =$ ______

g $\sqrt{121} =$ ______ h $\sqrt{25} \times \sqrt{36} =$ ______ i $\sqrt{9} \times \sqrt{16} =$ ______

j $\sqrt{169} =$ ______ k $\sqrt{625} \div \sqrt{25} =$ ______ l $\sqrt{676} - \sqrt{625} =$ ______

QUESTION 2 Use your calculator cube root key to simplify the following.

a $\sqrt[3]{27} =$ ______ b $\sqrt[3]{343} \times \sqrt[3]{8} =$ ______ c $\sqrt[3]{1000} \div \sqrt[3]{8} =$ ______

d $\sqrt[3]{8} =$ ______ e $\sqrt[3]{729} \times \sqrt[3]{27} =$ ______ f $\sqrt[3]{1331} \times \sqrt[3]{64} =$ ______

g $\sqrt[3]{64} =$ ______ h $\sqrt[3]{8} \times \sqrt[3]{729} =$ ______ i $\sqrt[3]{1.728} \times \sqrt[3]{3.43} =$ ______

j $\sqrt[3]{125} =$ ______ k $\sqrt[3]{64} \times \sqrt[3]{125} =$ ______ l $\sqrt[3]{6.4} + \sqrt[3]{8.1} =$ ______

QUESTION 3 Simplify the following correct to 2 decimal places.

a $\sqrt[3]{676} \times 8.125 =$ ______ b $\sqrt{23} \times \sqrt[3]{67} =$ ______ c $\sqrt{24.7} - \sqrt{8.35} =$ ______

d $\sqrt[3]{15.625} + 8.324 =$ ______ e $\sqrt{7} \times \sqrt[3]{5} =$ ______ f $\sqrt[3]{625} - \sqrt[3]{64.8} =$ ______

g $\sqrt{28} + \sqrt[3]{65} =$ ______ h $\sqrt{8.1} \times \sqrt[3]{5.16} =$ ______ i $\sqrt[3]{9.2} + \sqrt{8.1} =$ ______

j $\sqrt[3]{56} + \sqrt{93} =$ ______ k $\sqrt{9.6} \div \sqrt[3]{2.5} =$ ______ l $\sqrt{91.6} + \sqrt{12.6} =$ ______

QUESTION 4 Calculate the following correct to 1 decimal place.

a $\sqrt{8} + \sqrt{2} =$ ______ b $\sqrt{64} - \sqrt{9} =$ ______ c $\sqrt{169} \div \sqrt{8} =$ ______

d $\sqrt{4} - \sqrt{6} =$ ______ e $\sqrt{81} \div \sqrt{3} =$ ______ f $\sqrt{196} \times \sqrt{5} =$ ______

g $\sqrt{16} \times \sqrt{13} =$ ______ h $\sqrt{625} + \sqrt{7} =$ ______ i $\sqrt{225} - \sqrt{4} =$ ______

j $\sqrt{36} \div \sqrt{10} =$ ______ k $\sqrt{121} \times \sqrt{12} =$ ______ l $\sqrt{256} + \sqrt{11} =$ ______

QUESTION 5 Answer the following correct to 3 decimal places.

a $\sqrt{5\frac{1}{4}} \times \sqrt{2\frac{1}{2}} =$ ______ b $\sqrt{6\frac{1}{2}} - \sqrt{\frac{1}{4}} =$ ______ c $\sqrt[3]{12\frac{1}{5}} \times \sqrt{2\frac{1}{2}} =$ ______

d $\sqrt[3]{8\frac{3}{5}} \div \sqrt[3]{\frac{1}{4}} =$ ______ e $\sqrt[3]{5\frac{3}{4}} + \sqrt{\frac{3}{5}} =$ ______ f $\sqrt{18\frac{2}{3}} + \sqrt{1\frac{1}{4}} =$ ______

g $\sqrt{2\frac{1}{2}} + \sqrt{2\frac{1}{4}} =$ ______ h $\sqrt{8\frac{2}{3}} \times \sqrt{\frac{8}{9}} =$ ______ i $\sqrt[3]{15\frac{1}{5}} - \sqrt{\frac{31}{5}} =$ ______

j $\sqrt[3]{3\frac{1}{8}} - \sqrt{1\frac{1}{2}} =$ ______ k $\sqrt[3]{9\frac{3}{4}} =$ ______ l $\sqrt{19\frac{1}{2}} \div \sqrt{\frac{69}{8}} =$ ______

m $\sqrt{15\frac{5}{8}} \div \sqrt[3]{8\frac{1}{3}} =$ ______ n $\sqrt{5\frac{3}{8}} \div \sqrt{\frac{5}{7}} =$ ______ o $\sqrt[3]{16\frac{4}{5}} =$ ______

UNIT 5: Problem solving and the calculator

Question 1 Find the cost of 55 litres of petrol if 1 litre costs 85.9 cents. ______

Question 2 Find $\frac{2}{9}$ of 15.8 metres. ______

Question 3 Find the area of a square of side 8.25 m. ______

Question 4 Calculate the volume of a cube of side 3.5 cm. ______

Question 5 How much would 48 books cost if 3 books cost $25.35? ______

Question 6 Find the cost of 25 calculators at $48.95 each. ______

Question 7 Divide $1019.20 equally among 28 people. ______

Question 8 Find the average of 3.5, 8.75, 6.95, 15.6 and 11.25 ______

Question 9 Find the volume of a rectangular prism with sides 9 cm, 15.6 cm and 18.9 cm. ______

Question 10 Find the length of each side of a cube that has a volume of 24 389 cm^3. ______

Question 11 Find the sum of 25.6, 18.35, 46.286 and then subtract 19.256 ______

Question 12 Find the square root of $12\frac{1}{4}$ ______

Question 13 Find the cube root of 27.8 correct to 2 decimal places. ______

Question 14 Find the cost of 8.956 kg of cheese at $6.90 a kilogram. ______

Question 15 How many $8.95 books can be bought for $188? ______

Rational numbers

UNIT 6: Changing fractions to decimals

QUESTION **1** Write the following as decimals.

a $\frac{1}{10}=$ ____________ b $\frac{5}{10}=$ ____________ c $\frac{9}{10}=$ ____________

d $\frac{3}{100}=$ ____________ e $\frac{7}{100}=$ ____________ f $\frac{11}{100}=$ ____________

g $\frac{2}{1000}=$ ____________ h $\frac{35}{1000}=$ ____________ i $\frac{163}{1000}=$ ____________

j $\frac{834}{10\,000}=$ ____________ k $\frac{564}{100\,000}=$ ____________ l $\frac{21}{1\,000\,000}=$ ____________

QUESTION **2** Change to decimals.

a $6+\frac{5}{10}=$ ____________ b $9+\frac{7}{10}=$ ____________ c $23+\frac{18}{100}=$ ____________

d $39+\frac{7}{100}=$ ____________ e $27+\frac{3}{100}=$ ____________ f $91+\frac{81}{1000}=$ ____________

g $29+\frac{3}{1000}=$ ____________ h $52+\frac{17}{100}=$ ____________ i $63+\frac{13}{1000}=$ ____________

j $83+\frac{2}{100}=$ ____________ k $94+\frac{12}{1000}=$ ____________ l $69+\frac{35}{10\,000}=$ ____________

QUESTION **3** Change each fraction to a decimal.

a $\frac{3}{10}=$ ____________ b $\frac{12}{100}=$ ____________ c $\frac{15}{1000}=$ ____________

d $\frac{64}{100}=$ ____________ e $\frac{127}{1000}=$ ____________ f $\frac{635}{10\,000}=$ ____________

g $\frac{5}{8}=$ ____________ h $\frac{3}{5}=$ ____________ i $\frac{9}{25}=$ ____________

j $\frac{1}{8}=$ ____________ k $\frac{16}{125}=$ ____________ l $\frac{9}{8}=$ ____________

QUESTION **4** Express as decimals.

a $\frac{5}{50}=$ ____________ b $\frac{27}{100}=$ ____________ c $\frac{5}{25}=$ ____________

d $\frac{364}{100}=$ ____________ e $\frac{72}{1000}=$ ____________ f $\frac{58}{40}=$ ____________

g $\frac{85}{500}=$ ____________ h $\frac{608}{1000}=$ ____________ i $\frac{24}{50}=$ ____________

j $\frac{17}{750}=$ ____________ k $\frac{18}{125}=$ ____________ l $\frac{7}{15}=$ ____________

QUESTION **5** Write as a decimal.

a $\frac{24}{10}=$ ____________ b $\frac{638}{100}=$ ____________ c $\frac{539}{1000}=$ ____________

d $\frac{249}{10\,000}=$ ____________ e $\frac{865}{100\,000}=$ ____________ f $\frac{43}{50}=$ ____________

g $\frac{110}{200}=$ ____________ h $\frac{48}{500}=$ ____________ i $\frac{370}{5000}=$ ____________

j $\frac{19}{40}=$ ____________ k $\frac{36}{125}=$ ____________ l $\frac{563}{800}=$ ____________

Rational numbers

UNIT 7: Changing decimals to fractions

Question 1 Write the following terminating decimals as fractions in simplest form.

a 0.3 = ______ b 0.7 = ______ c 0.5 = ______
d 0.81 = ______ e 0.24 = ______ f 0.38 = ______
g 0.125 = ______ h 0.575 = ______ i 0.85 = ______
j 0.545 = ______ k 0.86 = ______ l 0.193 = ______

Question 2 Change to simple fractions.

a 0.02 = ______ b 0.04 = ______ c 0.05 = ______
d 0.003 = ______ e 0.008 = ______ f 0.025 = ______
g 0.007 = ______ h 0.005 = ______ i 0.015 = ______
j 0.805 = ______ k 0.304 = ______ l 0.708 = ______

Question 3 Change each decimal to a mixed numeral in simplest form.

a 1.2 = ______ b 3.4 = ______ c 8.9 = ______
d 2.25 = ______ e 3.64 = ______ f 5.75 = ______
g 8.3 = ______ h 9.4 = ______ i 7.2 = ______
j 6.1 = ______ k 5.42 = ______ l 8.93 = ______

Question 4 Express as a mixed numeral in simplest form.

a 6.125 = ______ b 8.342 = ______ c 9.45 = ______
d 10.325 = ______ e 29.36 = ______ f 41.175 = ______
g 18.63 = ______ h 128.48 = ______ i 526.5 = ______
j 16.801 = ______ k 19.5 = ______ l 428.55 = ______

Question 5 Write as a mixed numeral in simplest form.

a 5.64 = ______ b 9.235 = ______ c 60.724 = ______
d 83.75 = ______ e 83.42 = ______ f 81.48 = ______
g 94.25 = ______ h 49.123 = ______ i 53.231 = ______
j 51.125 = ______ k 86.514 = ______ l 32.519 = ______

Rational numbers

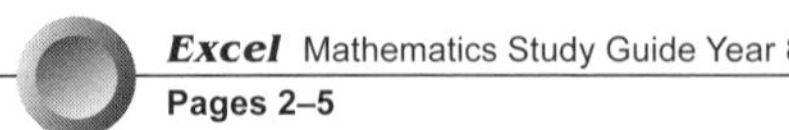

UNIT 8: Recurring and repeating decimals

Question 1 Which of the following are recurring decimals (R) and which are terminating decimals (T)?

a 1.38 = ____ **b** 0.111 = ____ **c** 3.851 = ____

d $0.\dot{3}$ = ____ **e** 8.25 = ____ **f** 0.1212= ____

g 5.8453 = ____ **h** 0.3695 = ____ **i** $0.\dot{4}\dot{3}$ = ____

j $0.\dot{7}$ = ____ **k** 5.356 = ____ **l** 0.266 66 ... = ____

Question 2 Write the following recurring decimals in a shorter form.

a 0.333 ... = ____ **b** 0.4333 ... = ____ **c** 0.135 135 ... = ____

d 0.777 ... = ____ **e** 0.807 878 ... = ____ **f** 0.233 33 ... = ____

g 0.252 525 ... = ____ **h** 0.363 636 ... = ____ **i** 0.111 11 ... = ____

j 0.505 505 ... = ____ **k** 0.277 77 ... = ____ **l** 0.7272 ... = ____

Question 3 Write the following recurring decimals in an expanded form.

a $0.\dot{2}$ = ____ **b** $5.\dot{3}$ = ____ **c** $9.\dot{5}$ = ____

d $0.\dot{2}\dot{3}$ = ____ **e** $0.86\dot{1}$ = ____ **f** $0.4\dot{9}\dot{5}$ = ____

g $0.\dot{5}\dot{4}$ = ____ **h** $0.\dot{4}\dot{5}$ = ____ **i** $0.2\dot{7}$ = ____

j $0.7\dot{3}$ = ____ **k** $0.\dot{3}7\dot{5}$ = ____ **l** $0.25\dot{9}\dot{3}$ = ____

Question 4 Write down the following as recurring decimals.

a 0.666 ... = ____ **b** 0.2626 ... = ____ **c** 1.555 ... = ____

d 0.2222 ... = ____ **e** 0.015 15 ... = ____ **f** 9.8888 ... = ____

g 5.3636 ... = ____ **h** 0.2828 ... = ____ **i** 7.9393 ... = ____

j 1.5333 ... = ____ **k** 0.111 ... = ____ **l** 0.048 88 ... = ____

Question 5 Write the following as recurring decimals.

a 0.0555 ... = ____ **b** 0.0666 ... = ____ **c** 0.2727 ... = ____

d 0.036 36 ... = ____ **e** 0.8333 ... = ____ **f** 0.0111 ... = ____

g 1.222 ... = ____ **h** 0.054 54 ... = ____ **i** 2.777 ... = ____

j 0.727 272 ... = ____ **k** 0.048 48 ... = ____ **l** 0.006 262 ... = ____

Question 6 Write the following fractions as recurring decimals.

a $\frac{2}{11}$ = ____ **b** $\frac{3}{9}$ = ____ **c** $\frac{5}{6}$ = ____

d $\frac{7}{9}$ = ____ **e** $\frac{8}{15}$ = ____ **f** $\frac{5}{27}$ = ____

g $\frac{8}{55}$ = ____ **h** $\frac{7}{66}$ = ____ **i** $\frac{11}{9}$ = ____

j $\frac{25}{7}$ = ____ **k** $\frac{63}{13}$ = ____ **l** $\frac{6}{7}$ = ____

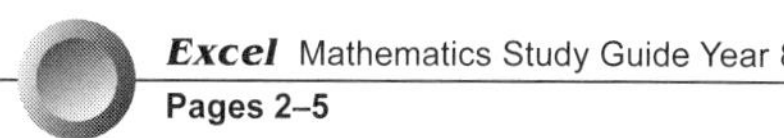

UNIT 9: Irrational numbers

Question 1 Write rational or irrational for each number.

a $\frac{2}{9}$ ______ **b** 0.3 ______ **c** $\sqrt{7}$ ______

d $\sqrt{11}$ ______ **e** π ______ **f** $6\frac{1}{4}$ ______

g 19 ______ **h** $\sqrt{64}$ ______ **i** $8 + \sqrt{2}$ ______

j $\sqrt{9} + \sqrt{36}$ ______ **k** $0.\dot{3}$ ______ **l** $\sqrt[3]{17}$ ______

Question 2 Write rational or irrational or neither for each number.

a $\sqrt{49}$ ______ **b** $\sqrt{9^2}$ ______ **c** $\sqrt[3]{27}$ ______

d $\sqrt{-36}$ ______ **e** $\frac{8}{19}$ ______ **f** $\sqrt{0.9}$ ______

g $\sqrt{1.69}$ ______ **h** $\frac{1}{\sqrt[3]{-64}}$ ______ **i** $\sqrt[3]{11}$ ______

j $\sqrt{\frac{1}{9}}$ ______ **k** $\frac{22}{7}$ ______ **l** $5 + \sqrt{19}$ ______

Question 3 Which of the following are irrational?

$\sqrt{13}$ 26% $\sqrt{0.9}$ $\frac{22}{7}$ $\sqrt{196}$ π

______

Question 4 Between which two consecutive integers does each of the following lie?

a $\sqrt{10}$ ______ **b** $\sqrt{29}$ ______ **c** $\sqrt{84}$ ______

d $\sqrt{216}$ ______ **e** $\sqrt{30}$ ______ **f** $\sqrt{105}$ ______

Rational numbers

UNIT 10: Equivalent ratios

QUESTION 1 Express the following ratios in simplest form.

a $4:8 =$ ______ **b** $5:15 =$ ______ **c** $8:24 =$ ______
d $6:36 =$ ______ **e** $5:35 =$ ______ **f** $4:48 =$ ______
g $2:18 =$ ______ **h** $3:24 =$ ______ **i** $8:72 =$ ______
j $9:36 =$ ______ **k** $12:60 =$ ______ **l** $11:99 =$ ______

QUESTION 2 Simplify the following ratios.

a $25:65 =$ ______ **b** $24:72 =$ ______ **c** $18:54 =$ ______
d $10:95 =$ ______ **e** $5:65 =$ ______ **f** $8:84 =$ ______
g $14:63 =$ ______ **h** $7:91 =$ ______ **i** $6:28 =$ ______
j $4:22 =$ ______ **k** $8:60 =$ ______ **l** $16:36 =$ ______

QUESTION 3 Express as a ratio in its simplest form.

a $5:15:25 =$ ______ **b** $\frac{1}{3}:\frac{1}{9} =$ ______ **c** $2\frac{1}{4}:4 =$ ______
d $1\frac{1}{2}:2 =$ ______ **e** $\frac{1}{4}:\frac{5}{4} =$ ______ **f** $3:\frac{3}{5} =$ ______
g $\frac{1}{7}:\frac{3}{7} =$ ______ **h** $0.5:0.7 =$ ______ **i** $1.2:2.6 =$ ______
j $2.5:3 =$ ______ **k** $2:\frac{1}{6} =$ ______ **l** $1.2:6 =$ ______

QUESTION 4 Express the following ratios in simplest form.

a \$8 : \$42 = ______ **b** 80 cents : \$3 = ______
c 12 h : 3 days = ______ **d** 800 g : 2 kg = ______
e 60 cents : 80 cents = ______ **f** 5 days : 5 weeks = ______
g 80 cm : 2.5 m = ______ **h** 40 s : 1 min = ______
i 4 kg : 1600 g = ______ **j** 800 mL : 1 L = ______
k 5 km : 500 m = ______ **l** 10 cm : 80 mm = ______

QUESTION 5 Simplify the following ratios.

a 60 cm : $2\frac{1}{2}$ m = ______ **b** 600 g : 9 kg = ______
c 50 min : 3 h = ______ **d** 2 h : 40 min = ______
e $3\frac{1}{2}:9 =$ ______ **f** $1:\frac{1}{8} =$ ______
g $2.5:3.5 =$ ______ **h** $3.5:4 =$ ______

Rational numbers

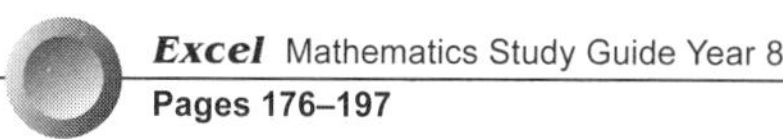

UNIT 11: Using ratios

Question 1

a Divide 180 kg in the ratio 1 : 8

b Divide $45 in the ratio 2 : 7

c Divide $840 in the ratio 1 : 4

d $500 is divided in the ratio 3 : 2. Find the larger part.

e $2800 is divided in the ratio 2 : 5. Find the smaller part.

Question 2

a Two sides of a rectangle are in the ratio 2 : 5. If the shorter side is 10 cm, what is the length of the rectangle?

b The ratio of girls to boys is 4 : 5. If there are 12 girls, how many boys are there?

c The three angles of a triangle are in the ratio 2 : 3 : 4. Find the size of each angle.

d Find the ratio of the areas of two squares whose sides are 4 cm and 5 cm respectively.

e The ratio of children to adults is 8 : 3. If there are 24 children, how many adults are there?

Question 3

a A piece of rope is cut into two lengths in the ratio 2 : 5. If the shorter length is 8 m, find the length of the original rope.

b Divide $96 between two girls in the ratio of their ages, 6 years and 10 years.

c The dimensions of a rectangle are 8 cm and 12 cm. What is the ratio of its length to its perimeter?

d Divide $4800 in the ratio 3 : 4 : 5

Rational numbers

UNIT 12: Rates

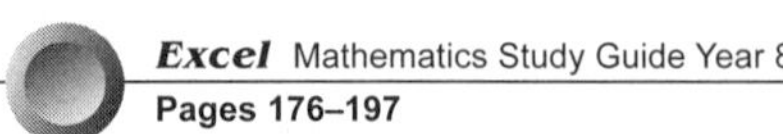
Excel Mathematics Study Guide Year 8
Pages 176–197

QUESTION 1 Complete the following sentences.

a 480 km in 6 hours is a rate of ____________ per hour.

b 64 books bought for $1280 is at a rate of ____________ per book.

c If 1600 litres of water flows through a tap in 5 hours, it flows at a rate of ____________ per minute.

d 5 kg of meat cost $39.75 which equals ____________ per kg.

QUESTION 2 Complete the equivalent rates.

a 60 km/h = ________ km/min

b 40 L/h = ________ L/day

c 60 m/min = ________ m/h

d $5/min = ________ $/h

e 30 mL/min = ________ mL/h

f 36°/min = ________ °/s

QUESTION 3

a Michelle drives 240 km in 3 hours. Find her average speed.

____

b A car travels at the speed of 30 m/s. How many kilometres does it travel in 1 hour?

____

c A car uses petrol at a rate of 10.8 L/100 km. How many litres would be used to travel 450 km?

____

QUESTION 4

a Change 150 km/h to km/min

____

b Change 100 km/h to km/s

____

c Andrew delivered 680 bottles of milk every evening between 4 pm and 9 pm Find his hourly rate of delivery.

____

QUESTION 5

a A car travels 900 km and covers this distance in 5 hours 20 minutes. Calculate the average speed per hour.

____

b A tree grows to a height of 16.8 metres over a period of $6\frac{1}{2}$ years. What is the average annual growth rate in metres per year?

____

c In a cricket match runs were scored at a rate of 4 runs per over. How many overs did it take to score 148 runs?

____

Rational numbers

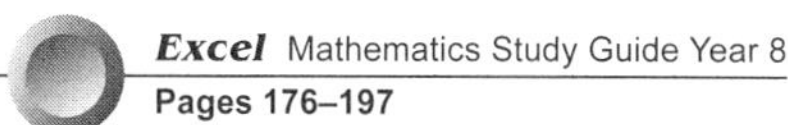

UNIT 13: Conversion of rates

QUESTION **1** Complete the equivalent rates.

a 80 mL/min = ________________ mL/h

b 1000 mm/s = ________________ cm/s

c 24.5 t/day = ________________ kg/h

d 60 m/min = ________________ m/h

e 275 L/min = ________________ kL/min

f 100 L/h = ________________ L/day

g 120 km/h = ________________ km/min

h $12.50/h = ________________ $/day

i 27°/min = ________________ °/s

j 90 km/h = ________________ km/day

k 54 ha/day = ________________ m^2/day

l 2.35 m/s = ________________ cm/min

QUESTION **2** A speed of 58 metres per second is how many:

a metres per minute? ________________

b metres per hour? ________________

c kilometres per hour? ________________

d kilometres per day? ________________

QUESTION **3** A car is travelling at 110 km/h. How many metres does it travel in one minute?

____

QUESTION **4** A flow rate of 45 mL per second is how many:

a mL per minute? ________________

b mL per hour? ________________

c L per hour? ________________

d kL per hour? ________________

QUESTION **5** Change the following.

a 96 km/h to m/s ________________

b 30 m/s to km/h ________________

c 120 km/h to m/s ________________

d 72 km/h to m/min ________________

Rational numbers

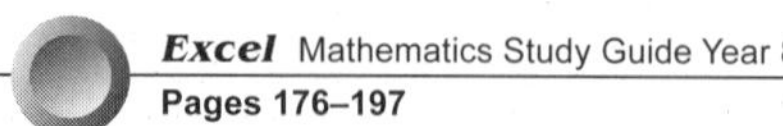

UNIT 14: Scale drawing

QUESTION 1 Write each of the following scales in ratio form.

a 2 mm to 3 m ______

b 1 cm to 2 m ______

c 1 cm to 300 m ______

d 20 cm to 1 km ______

e 6 mm to 1 m ______

f 5 cm to 2 m ______

g 1 mm to 30 m ______

h 30 cm to 1 m ______

i 1 mm to 8 m ______

QUESTION 2 Using a scale of 1 : 100, what length is represented by each of the following?

a 1 cm ______

b 4 cm ______

c 6 cm ______

d 5 mm ______

e 7 mm ______

f 20 m ______

g 6 mm ______

h 8 mm ______

i 15 m ______

QUESTION 3 Using a scale of 1 : 1000, what real length is represented by each of the following?

a 4 mm ______

b 10 cm ______

c 8 m ______

d 7.5 cm ______

e 9.3 m ______

f 23.45 m ______

g 9 mm ______

h 15.2 m ______

i 48.25 m ______

QUESTION 4 The real distance between two points is given below. Find the following distances between the two points in a scale drawing of scale 1 cm to 100 m.

a 600 m ______

b 700 m ______

c 1380 m ______

d 50 m ______

e 5000 m ______

f 3965 m ______

QUESTION 5

a The plan of a house is drawn to a scale of 1 : 100. If a room measures 48 mm by 56 mm on the plan, how big is the room in real life?

______

b A drawing has a scale of 1 : 100. convert the real distance of 58 m to a scaled distance.

______

Rational numbers

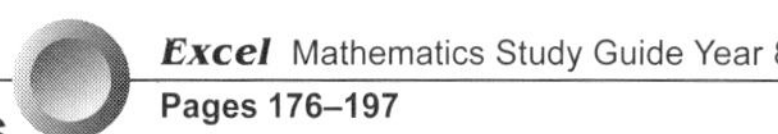

UNIT 15: Problem solving with ratios and rates

QUESTION **1** In a class of 35 students there are 15 boys. What is the ratio of:

a boys to girls? ______

b girls to the total number of students? ______

QUESTION **2** Tom buys a camera for \$80 and sells it for \$100. What is the ratio of:

a cost price to selling price? ______

b selling price to cost price? ______

QUESTION **3** John drives 560 km in 7 hours. Find his average speed.

______

QUESTION **4** Divide \$560 in the ratio 3 : 4

______

QUESTION **5** Two angles of a triangle are 30° and 70°. What is the ratio of:

a the third angle to the sum of the angles of the triangle?

______

b the smallest angle to the largest angle?

______

QUESTION **6** A car travels 840 km in 7 hours. What is the average speed of the car per hour?

______

QUESTION **7** If a 40 kg bag of potatoes costs \$12 calculate the price of 1 kg of potatoes.

______

QUESTION **8** The areas of two squares are 9 m^2 and 16 m^2. What is the ratio of the smaller side to the larger side?

______

QUESTION **9** The ratio of tax to income is 1 : 8. If the tax is \$500, what is the income?

______

QUESTION **10** The complementary angles x and y are in the ratio 1 : 2. What is the size of each angle?

______

QUESTION **11** A silo is being filled with wheat at the rate of 1.5 t/min. Express this rate in kg/s.

______

Rational numbers

TOPIC TEST — PART A

Instructions
- This part consists of 10 multiple-choice questions.
- Fill in only ONE CIRCLE for each question.
- Each question is worth 1 mark.
- Calculators are allowed.

Time allowed: 10 minutes **Total marks: 10**

Marks

1 Which of these sets of numbers is arranged in ascending order? 1

(A) 62%, 0.603, $\frac{5}{8}$ (B) 62%, $\frac{5}{8}$, 0.603 (C) 0.603, 62%, $\frac{5}{8}$ (D) $\frac{5}{8}$, 0.603, 62%

2 Which is the smallest number? 1

(A) 0.04 (B) 0.3 (C) 0.32 (D) 0.041

3 147.658 31 correct to two decimal places is 1

(A) 140.00 (B) 147.65 (C) 147.66 (D) 150.65

4 When a tank is $\frac{2}{5}$ full there are 12 000 L in it. The total capacity of the tank is 1

(A) 4800 L (B) 6800 L (C) 18 000 L (D) 30 000 L

5 A car travels 24 km between 9:30 am and 10:15 am What is its average speed in km/h? 1

(A) 18 km/h (B) 30 km/h (C) 32 km/h (D) 36 km/h

6 Ore is removed from a mine at the rate of 270 tonnes per hour. Express this in kg/s. 1

(A) 7.5 kg/s (B) 75 kg/s (C) 450 kg/s (D) 4500 kg/s

7 An amount is divided in the ratio 2 : 3. If the smaller part is $30, the original amount was 1

(A) $45 (B) $50 (C) $75 (D) $78

8 If you divide $65 in the ratio 6 : 7, what is the smaller part? 1

(A) $13 (B) $26 (C) $30 (D) $35

9 Whilst travelling 100 km a car uses 20 L of petrol. At this rate, how many litres of petrol are needed to travel 1 km? 1

(A) 0.02 L (B) 0.2 L (C) 0.5 L (D) 5 L

10 The ratio of 3.5 kg to 56 kg is 1

(A) 8 : 1 (B) 1 : 16 (C) 16 : 1 (D) 28 : 3

Total marks achieved for PART A

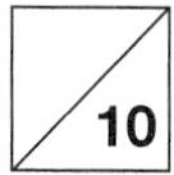

Rational numbers

TOPIC TEST — PART B

Instructions
- This part consists of 15 questions.
- Each question is worth 1 mark.
- Write only the answer in the answer column.

Time allowed: 15 minutes **Total marks: 15**

	Questions	Answers	Marks
1	Find the cost of 10 books if 2 books cost \$6.95		1
2	Simplify $6^2 + \sqrt{64} + \sqrt[3]{343}$		1
3	Find the smallest number which is greater than 2000 and is divisible by 6		1
4	Find the value of $\frac{1}{\sqrt{23.5} + 3.5^2}$ to 2 decimal places.		1
5	Express the following as a ratio in simplest form. 20 minutes : 2 hours		1
6	\$125 is divided in the ratio 12 : 13. What is the larger part?		1
7	A car is travelling at 45 km/h. How many metres does it travel in 4 seconds?		1
8	Write $3\frac{3}{5} : 2\frac{1}{10}$ in its simplest form.		1
9	Change 120 km/h to m/s.		1
10	How much time will it take to travel 1200 km at 90 km/h?		1
11	Divide 640 m in the ratio 2 : 3		1
12	A 30 litre container full of water is leaking at a rate of 100 mL per hour. How much will be left in the container after 3 days?		1
13	The three angles of a triangle are in the ratio 2 : 3 : 4. Find the size of the smallest angle.		1
14	There are 70 girls at a party. If the ratio of boys to girls is 7 : 5, how many boys are at the party?		1
15	$\sqrt{15}$ lies between which two consecutive integers?		1

Total marks achieved for PART B /15

Chapter 2

Integers

UNIT 1: Plotting on number lines

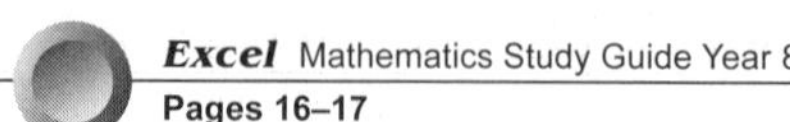
Excel Mathematics Study Guide Year 8
Pages 16–17

Question 1 Giving *distance* and *direction*, write where you would be from your starting point (0) if you travelled on the east-west line.

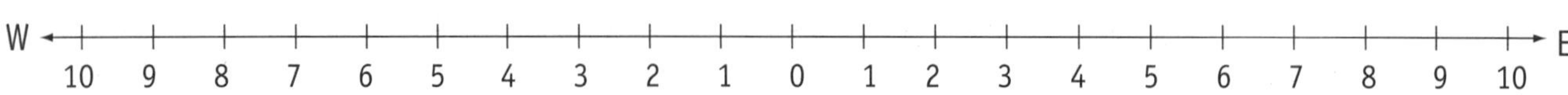

a 4 km west and then 7 km east ____________________

b 6 km west and then 11 km east ____________________

c 2 km west and then 8 km east ____________________

d 6 km east and then 7 km west ____________________

e 2 km east and then 5 km east ____________________

f 3 km west and then 6 km west ____________________

Question 2 Plot the following sets of points on the number line.

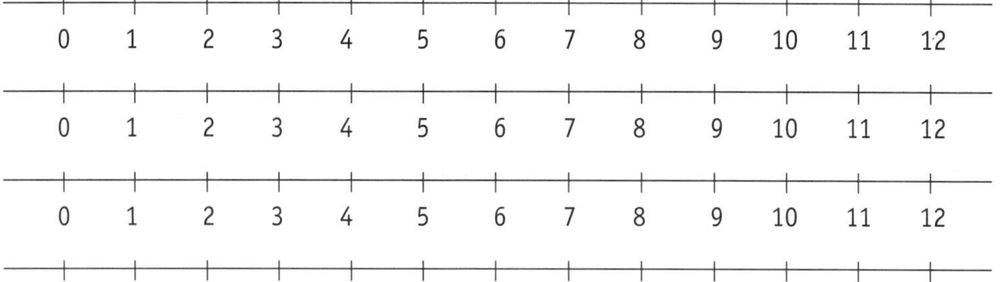

a {1, 3, 4, 6} 0 1 2 3 4 5 6 7 8 9 10 11 12

b {0, 2, 5, 7} 0 1 2 3 4 5 6 7 8 9 10 11 12

c {0, 2, 4, 6} 0 1 2 3 4 5 6 7 8 9 10 11 12

d {1, 3, 5, 7} 0 1 2 3 4 5 6 7 8 9 10 11 12

e {0, 1, 2, 3, 4} 0 1 2 3 4 5 6 7 8 9 10 11 12

f {1, 5, 6, 7, 9} 0 1 2 3 4 5 6 7 8 9 10 11 12

Question 3 Graph each set of points on a separate number line.

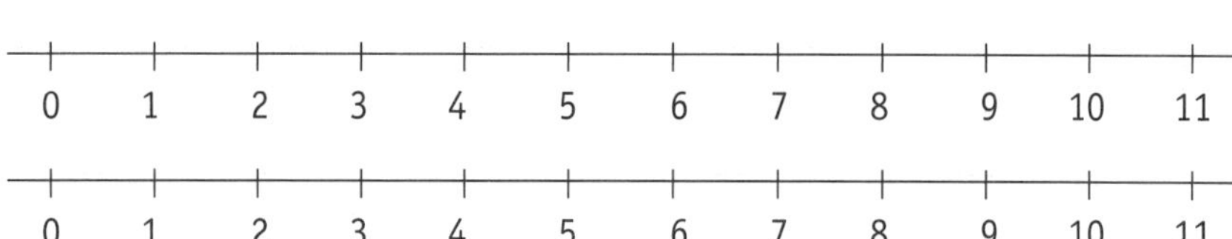

a the numbers from 3 to 7 0 1 2 3 4 5 6 7 8 9 10 11

b the numbers 1, 2, 5, 6, 8, 9 0 1 2 3 4 5 6 7 8 9 10 11

c odd numbers less than 10 0 1 2 3 4 5 6 7 8 9 10 11

d even numbers less than 10 0 1 2 3 4 5 6 7 8 9 10 11

e prime numbers less than 10 0 1 2 3 4 5 6 7 8 9 10 11

f the numbers 2, 3, 4, 8, 9, 10 0 1 2 3 4 5 6 7 8 9 10 11

Integers

UNIT 2: Extending the number line

Use the number line to answer the following questions.

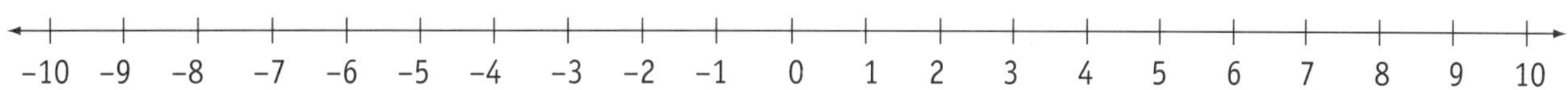

QUESTION 1 Represent each of the following trips on the number line using a directed number (a positive or negative number).

a 0 to 7 ______ **b** 1 to 8 ______
c −8 to 3 ______ **d** 3 to 9 ______
e −2 to 5 ______ **f** −1 to 10 ______
g 6 to 10 ______ **h** −8 to −1 ______
i −7 to −3 ______ **j** −9 to 5 ______

QUESTION 2 Which is the smaller number?

a 10 or 6 ______ **b** 8 or −1 ______ **c** −8 or −2 ______ **d** −2 or 2 ______
e 6 or 1 ______ **f** −7 or −5 ______ **g** −3 or 7 ______ **h** 10 or 8 ______
i −6 or −1 ______ **j** −4 or −2 ______ **k** −6 or 7 ______ **l** 8 or 9 ______
m −8 or −1 ______ **n** −7 or 2 ______ **o** 2 or 8 ______

QUESTION 3 Use the number line to complete the following.

a 5 − 8 = ______ **b** 4 + 7 = ______ **c** 6 + 3 = ______
d 3 + 4 = ______ **e** 9 − 2 = ______ **f** 8 − 2 = ______
g −2 + 7 = ______ **h** 7 − 3 = ______ **i** 5 − 9 = ______
j 4 − 8 = ______ **k** 8 − 10 = ______ **l** −3 − 4 = ______

QUESTION 4 Arrange in descending order.

a 2, 4, −2, −7, −9, −1 ______
b −2, −8, −6, −9, 3, 5 ______
c 8, 9, 3, 1, 2, −4 ______
d −3, −5, 2, 6, 3, −1 ______
e 2, 5, 8 −4, −2, −6 ______

QUESTION 5 Write > or < to make the following statements true.

a 8 ______ −5 **b** 4 ______ −8 **c** 3 ______ −5
d 6 ______ 9 **e** −5 ______ 5 **f** 4 ______ −9
g −6 ______ 2 **h** −2 ______ 2 **i** 8 ______ 10
j 9 ______ −4 **k** −7 ______ 3 **l** −2 ______ −7
m −8 ______ −6 **n** −3 ______ 2 **o** −1 ______ 6

Integers

UNIT 3: Addition of integers

Excel Mathematics Study Guide Year 8
Pages 16–17

Question 1 Find the sum of the following, using the number line if necessary.

a $4 + 9 =$ ______ b $1 + 2 =$ ______ c $1 + 7 =$ ______
d $6 + 2 =$ ______ e $3 + 6 =$ ______ f $2 + 9 =$ ______
g $8 + 3 =$ ______ h $2 + 4 =$ ______ i $4 + 7 =$ ______

Question 2 Find each sum.

a $1 + -1 =$ ______ b $3 + -8 =$ ______ c $5 + -9 =$ ______
d $2 + -3 =$ ______ e $9 + -5 =$ ______ f $8 + -6 =$ ______
g $4 + -2 =$ ______ h $10 + -4 =$ ______ i $7 + -7 =$ ______

Question 3 Add the following.

a $-9 + 1 =$ ______ b $-18 + 6 =$ ______ c $-15 + 11 =$ ______
d $-10 + 3 =$ ______ e $-24 + 7 =$ ______ f $-17 + 4 =$ ______
g $-12 + 5 =$ ______ h $-28 + 10 =$ ______ i $-7 + 15 =$ ______

Question 4 Find the answers to these additions.

a $-2 + -3 =$ ______ b $-7 + -1 =$ ______ c $-1 + -7 =$ ______
d $-5 + -9 =$ ______ e $-3 + -2 =$ ______ f $-10 + -5 =$ ______
g $-9 + -4 =$ ______ h $-8 + -10 =$ ______ i $-4 + -8 =$ ______
j $-12 + -17 =$ ______ k $-6 + -14 =$ ______ l $-11 + -6 =$ ______

Question 5 Find the values of the following.

a $6 + 2 + 5 =$ ______ b $17 + -3 + -2 =$ ______ c $10 + 2 + -3 =$ ______
d $-18 + 8 =$ ______ e $-8 + -6 =$ ______ f $-20 + -5 =$ ______
g $7 + -4 + -1 =$ ______ h $-14 + 3 + 8 =$ ______ i $-9 + -6 =$ ______
j $10 + -5 + -3 =$ ______ k $15 + -8 + -3 =$ ______ l $-8 + 4 =$ ______

Question 6 Find the missing number.

a $4 +$ ______ $= 2$ b ______ $+ -5 = 0$ c $10 + -3 =$ ______
d $8 +$ ______ $= 6$ e $5 + -9 + -2 =$ ______ f $-7 + -8 =$ ______
g $9 + -6 =$ ______ h $8 + 4 =$ ______ i $-4 +$ ______ $= -13$
j $-10 +$ ______ $= -21$ k $-3 + -4 + -2 =$ ______ l $15 + 6 =$ ______
m $15 + -3 + -4 =$ ______ n $-5 +$ ______ $= -14$ o $-4 +$ ______ $= -18$

Integers

UNIT 4: Subtraction of integers

QUESTION **1** Find the answers to the following questions.

a 8 – 5 = ________ **b** 10 – 7 = ________ **c** 25 – 18 = ________
d 6 – 4 = ________ **e** 14 – 2 = ________ **f** 35 – 8 = ________
g 5 – 2 = ________ **h** 18 – 3 = ________ **i** 12 – 1 = ________
j 9 – 3 = ________ **k** 31 – 6 = ________ **l** 18 – 8 = ________

QUESTION **2** Complete these subtractions. Use the number line if necessary.

a 6 – –2 = ________ **b** 18 – –6 = ________ **c** 27 – –10 = ________
d 10 – –3 = ________ **e** 25 – –12 = ________ **f** 14 – –7 = ________
g 11 – –4 = ________ **h** 9 – –11 = ________ **i** 19 – –9 = ________
j 15 – –5 = ________ **k** 8 – –8 = ________ **l** 32 – –13 = ________

QUESTION **3** Find each difference.

a –20 – 2 = ________ **b** –8 – 6 = ________ **c** –7 – 12 = ________
d –10 – 8 = ________ **e** –16 – 9 = ________ **f** –17 – 4 = ________
g –15 – 10 = ________ **h** –12 – 3 = ________ **i** –27 – 7 = ________
j –5 – 5 = ________ **k** –24 – 13 = ________ **l** –38 – 11 = ________

QUESTION **4** Find the answers to the following subtractions.

a –3 – –1 = ________ **b** –2 – –5 = ________ **c** –13 – –8 = ________
d –5 – –6 = ________ **e** –8 – –2 = ________ **f** –6 – –3 = ________
g –7 – –10 = ________ **h** –12 – –7 = ________ **i** –15 – –9 = ________
j –9 – –12 = ________ **k** –11 – –4 = ________ **l** –4 – –11 = ________

QUESTION **5** Find the values of the following.

a –1 – –3 = ________ **b** –34 – 23 = ________ **c** –88 – 35 = ________
d –12 – –7 = ________ **e** 45 – –19 = ________ **f** –91 – –31 = ________
g –23 – 11 = ________ **h** –56 – –27 = ________ **i** 14 – –39 = ________
j –36 – –15 = ________ **k** 67 – 43 = ________ **l** –19 – –47 = ________

QUESTION **6** Simplify the following.

a 9 – 4 = ________ **b** –23 – 10 = ________ **c** 17 – 9 = ________
d –7 – (–6) = ________ **e** 11 – (–14) = ________ **f** 5 – (–12) = ________
g –11 – 8 = ________ **h** 36 – (–6) = ________ **i** 28 – 11 = ________
j –15 – (–9) = ________ **k** 3 – 8 = ________ **l** 2 – 17 = ________

Integers

UNIT 5: Multiplication of integers

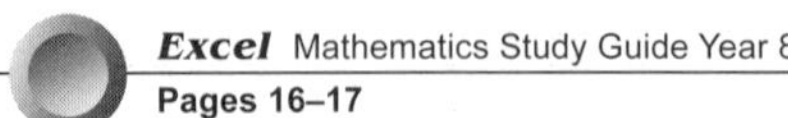
Excel Mathematics Study Guide Year 8
Pages 16–17

QUESTION 1 Multiply the following.

a $5 \times 1 =$ ______ b $15 \times 8 =$ ______ c $14 \times 9 =$ ______

d $12 \times 6 =$ ______ e $7 \times 2 =$ ______ f $18 \times 3 =$ ______

g $11 \times 11 =$ ______ h $13 \times 5 =$ ______ i $9 \times 10 =$ ______

j $16 \times 21 =$ ______ k $17 \times 7 =$ ______ l $21 \times 4 =$ ______

QUESTION 2 Work out the following.

a $4 \times -2 =$ ______ b $1 \times -3 =$ ______ c $3 \times -15 =$ ______

d $2 \times -9 =$ ______ e $5 \times -12 =$ ______ f $18 \times -5 =$ ______

g $7 \times -6 =$ ______ h $8 \times -4 =$ ______ i $6 \times -8 =$ ______

j $15 \times -7 =$ ______ k $10 \times -11 =$ ______ l $21 \times -5 =$ ______

QUESTION 3 Find the answers to the following.

a $-3 \times 10 =$ ______ b $-6 \times 8 =$ ______ c $-7 \times 9 =$ ______

d $-10 \times 15 =$ ______ e $-2 \times 11 =$ ______ f $-1 \times 12 =$ ______

g $-5 \times 13 =$ ______ h $-11 \times 14 =$ ______ i $-8 \times 15 =$ ______

j $-9 \times 6 =$ ______ k $-4 \times 5 =$ ______ l $-12 \times 7 =$ ______

QUESTION 4 Simplify.

a $-7 \times -10 =$ ______ b $-1 \times -4 =$ ______ c $-8 \times -14 =$ ______

d $-2 \times -11 =$ ______ e $-9 \times -12 =$ ______ f $-3 \times -13 =$ ______

g $-10 \times -2 =$ ______ h $-4 \times -3 =$ ______ i $-11 \times -5 =$ ______

j $-5 \times -6 =$ ______ k $-12 \times -1 =$ ______ l $-6 \times -7 =$ ______

QUESTION 5 Find each product.

a $4 \times -3 =$ ______ b $-10 \times -10 =$ ______ c $-9 \times 6 =$ ______

d $-3 \times 5 =$ ______ e $12 \times 12 =$ ______ f $-15 \times 5 =$ ______

g $8 \times -6 =$ ______ h $-11 \times -14 =$ ______ i $18 \times -4 =$ ______

j $-6 \times 8 =$ ______ k $8 \times -9 =$ ______ l $-25 \times -7 =$ ______

QUESTION 6 Complete the following tables.

a

×	8	−6	10
6			
−8			
3			

b

×	−8	9	−5
4			
−2			
5			

c

×	4	−6	8
−7			
12			
−15			

Integers

UNIT 6: Division of integers

Question 1 Simplify the following.

a $-15 \div 3 =$ ______ b $-36 \div 9 =$ ______ c $-42 \div 6 =$ ______
d $-21 \div 7 =$ ______ e $-48 \div 8 =$ ______ f $-35 \div 5 =$ ______
g $-24 \div 8 =$ ______ h $-64 \div 4 =$ ______ i $-26 \div 2 =$ ______
j $-30 \div 5 \div 2 =$ ______ k $-84 \div 2 \div 7 =$ ______ l $-36 \div 9 \div 2 =$ ______

Question 2 Work out the following divisions.

a $6 \div -2 =$ ______ b $18 \div -9 =$ ______ c $27 \div -9 =$ ______
d $9 \div -3 =$ ______ e $39 \div -3 =$ ______ f $21 \div -7 =$ ______
g $12 \div -4 =$ ______ h $24 \div -8 =$ ______ i $33 \div -11 =$ ______
j $15 \div -5 =$ ______ k $30 \div -15 =$ ______ l $36 \div -4 =$ ______

Question 3 Simplify the following.

a $-10 \div -2 =$ ______ b $-25 \div -5 =$ ______ c $-14 \div -7 =$ ______
d $-12 \div -4 =$ ______ e $-27 \div -3 =$ ______ f $-8 \div -4 =$ ______
g $-36 \div -6 =$ ______ h $-38 \div -2 =$ ______ i $-9 \div -3 =$ ______
j $-54 \div 3 \div -9 =$ ______ k $-64 \div -4 \div 2 =$ ______ l $-78 \div 2 \div -3 =$ ______

Question 4 If $a = 5$, $b = -3$, $c = -6$ and $d = 10$, find the value of each of the following.

a $a \div b =$ ______ b $c^2 \div d =$ ______ c $ac \div d =$ ______
d $c \div d =$ ______ e $a^2 \div b^2 =$ ______ f $c^2 \div b^2 =$ ______
g $a^2 \div b =$ ______ h $d \div ab =$ ______ i $d \div c =$ ______
j $d^2 \div a =$ ______ k $bd \div ac =$ ______ l $cd \div ab =$ ______

Question 5 Fill in the missing number.

a $-60 \div$ ______ $= -6$ b $-144 \div 12 =$ ______ c $-12 \div 4 =$ ______
d ______ $\div -4 = 9$ e $48 \div$ ______ $= -6$ f ______ $\div -7 = 7$
g $-64 \div -16 =$ ______ h $-64 \div -4 =$ ______ i $-33 \div$ ______ $= -3$
j ______ $\div -7 = -6$ k $-63 \div$ ______ $= 7$ l $28 \div$ ______ $= -7$

Question 6 Simplify the following.

a $-76 \div 2 \div -2 =$ ______ b $48 \div -8 =$ ______ c $-81 \div -3 \div -3 =$ ______
d $-46 \div 2 =$ ______ e $-25 \div -5 =$ ______ f $72 \div -2 \div -9 =$ ______
g $68 \div 4 =$ ______ h $-70 \div -2 \div 5 =$ ______ i $68 \div -4 =$ ______
j $-72 \div -3 \div 3 =$ ______ k $66 \div 3 \div -11 =$ ______ l $-56 \div 7 =$ ______

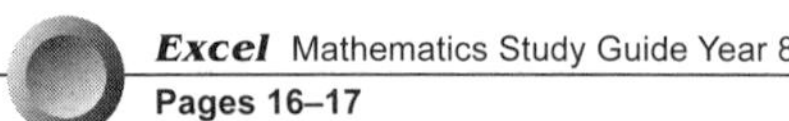

UNIT 7: The four operations with integers

QUESTION 1 Find the answers to the following.

a $7 + -8 =$ ______ b $36 + 3 =$ ______ c $18 - -4 =$ ______
d $-9 - 5 =$ ______ e $45 + -1 =$ ______ f $-24 - -6 =$ ______
g $10 + -7 =$ ______ h $-54 - 2 =$ ______ i $36 + 7 =$ ______
j $-21 - 6 =$ ______ k $-9 - -3 =$ ______ l $-48 + 8 =$ ______

QUESTION 2 Find the value of the following.

a $-28 \div -4 =$ ______ b $36 \div -4 =$ ______ c $-72 \div -4 =$ ______
d $-15 + -3 =$ ______ e $25 - 6 =$ ______ f $-42 - -12 =$ ______
g $19 - -8 =$ ______ h $28 + -10 =$ ______ i $-45 + 14 =$ ______
j $-29 + 7 =$ ______ k $-32 - 9 =$ ______ l $-64 - -15 =$ ______

QUESTION 3 Work out the following.

a $63 \div -9 =$ ______ b $-32 \times 2 =$ ______ c $-42 \div -7 =$ ______
d $-56 \div -7 =$ ______ e $-48 \div -8 =$ ______ f $20 \times 5 =$ ______
g $28 \div 4 =$ ______ h $81 \div -9 =$ ______ i $-10 \times -6 =$ ______
j $-24 \times -3 =$ ______ k $96 \div -6 =$ ______ l $-8 \times 9 =$ ______

QUESTION 4 Simplify the following.

a $(5 \times 4) \times -2 =$ ______ b $10 - 4 \times -5 =$ ______ c $-6 + -5 + -8 =$ ______
d $(-3 \times -2) \times -5 =$ ______ e $-64 \div -8 \times 3 =$ ______ f $16 \div (-6 + 2) =$ ______
g $24 \div (-3 \times -2) =$ ______ h $(-6 - 6) \times -5 =$ ______ i $30 \times (15 \div 5) =$ ______
j $(8 \times -3) \div -4 =$ ______ k $-300 \div -50 \times 2 =$ ______ l $(-8 - 4) \div 6 =$ ______

QUESTION 5 Fill in the missing number.

a $72 \div -9 =$ ______ b ______ $\div 2 = -12$ c $24 -$ ______ $= 18$
d $9 \times -6 =$ ______ e ______ $\div 9 = -6$ f $-10 -$ ______ $= -7$
g $8 + -7 =$ ______ h $15 + -6 =$ ______ i $8 \times$ ______ $= -80$
j $-6 - -3 =$ ______ k $-18 - -5 =$ ______ l $-9 +$ ______ $= -15$

QUESTION 6 Complete the tables.

a

×	6	−8	10	−12	−3	5
−3						
5						
−9						

b

×	−5	−3	2	6	−8	9
4						
−8						
7						

Integers

UNIT 8: Indices

Excel Mathematics Study Guide Year 8
Pages 16–17

QUESTION 1 Evaluate the following.

a $2^3 =$ ______ b $5^2 =$ ______ c $-3^3 =$ ______ d $3^2 =$ ______

e $4^2 =$ ______ f $(-1)^2 =$ ______ g $5^3 =$ ______ h $8^2 =$ ______

i $(-1)^3 =$ ______ j $4^3 =$ ______ k $(-2)^3 =$ ______ l $(-1)^4 =$ ______

m $2^5 =$ ______ n $(-3)^4 =$ ______ o $(-2)^4 =$ ______ p $3^4 =$ ______

QUESTION 2 Evaluate the following.

a $(-1)^{10} =$ ______ b $(-5)^4 =$ ______ c $(-3)^5 =$ ______ d $(-1)^{15} =$ ______

e $(-2)^5 =$ ______ f $2 \times (-4)^2 =$ ______ g $(-1)^{99} =$ ______ h $(-10)^3 =$ ______

i $-(8)^2 =$ ______ j $(-6)^3 =$ ______ k $(-2)^6 =$ ______ l $-4 \times (-3)^3 =$ ______

m $9^2 =$ ______ n $7^3 =$ ______ o $(-2)^2 \times (-5)^2 =$ ______ p $(-5)^3 =$ ______

QUESTION 3 Calculate the following.

a $(-2)^3 \times (-3)^2 =$ ______ b $(-9)^2 - (-4) =$ ______ c $(-4)^3 + (-1)^2 =$ ______

d $4^2 + (-5)^2 =$ ______ e $(-5)^2 + (-10)^2 =$ ______ f $(-8)^2 \div (2)^5 =$ ______

g $(-1)^5 + (-3)^3 =$ ______ h $(-7)^2 \times (-1)^3 =$ ______ i $(-5)^3 + -5^2 =$ ______

j $8^2 \div (-2)^4 =$ ______ k $(-3)^4 - (-1)^5 =$ ______ l $(-6)^2 \times (-2)^3 =$ ______

m $7^2 \times (-3)^2 =$ ______ n $6^3 + (-2)^3 =$ ______ o $(-10)^4 \div 10^2 =$ ______

p $(-2)^2 + (-2)^3 =$ ______ q $(-2)^4 + (-3)^3 =$ ______ r $8^3 \div (-2)^5 =$ ______

QUESTION 4 Evaluate the following.

a $9^2 \div 3 - 4^2 \times 2 =$ ______

b $(-4)^3 + -4^3 =$ ______

c $(-3)^2 \times -3^2 =$ ______

d $4 \times (-2)^2 + 5 \times (-1)^4 =$ ______

e $(-10)^2 \times 5^2 =$ ______

f $2^5 \div 2^2 =$ ______

g $2 \times 10^3 + (-10)^2 =$ ______

h $(-5)^2 + 4 \times -10 =$ ______

i $(-10)^2 \times (-10)^3 =$ ______

j $-6 \times (-10)^3 \times (-1)^5 =$ ______

k $(-8)^2 + 2 \times -20 =$ ______

l $(-5)^3 + (-3)^2 =$ ______

m $(-9)^2 + 3 \times -6 =$ ______

n $3 \times (-4)^2 + 8 \times (-2)^4 =$ ______

o $(-6)^3 - (-5)^2 =$ ______

p $(-3)^2 + -3^2 + (-1)^2 =$ ______

q $(-10)^3 + (-10)^2 - (-2)^6 =$ ______

r $-5 \times (-10)^2 \times -10 =$ ______

Excel Mathematics Study Guide Year 8
Pages 16–17

UNIT 9: Order of operations

QUESTION **1** Find the answers to the following.

a $9 - 6 \times 5 =$ ______ **b** $8 + 7 - 16 =$ ______ **c** $-200 \div 25 \times 3 =$ ______

d $12 - 5 \times 5 =$ ______ **e** $12 - 14 + 16 =$ ______ **f** $(-2 \times -6) \times -4 =$ ______

g $-7 + 3 \times 3 =$ ______ **h** $5 \times -8 \div 4 =$ ______ **i** $12 \div (-5 + 8) =$ ______

j $-4 - 5 - 6 =$ ______ **k** $-80 \div 5 \times 3 =$ ______ **l** $(-4 + 6) - (3 - 5) =$ ______

QUESTION **2** Find the value of the following.

a $24 \div (3 \times -2) =$ ______ **b** $(-6 - 6) \div (4 - 7) =$ ______ **c** $-32 \div (4 \times 8) =$ ______

d $12 - 2 \times 4 - 9 =$ ______ **e** $12 - 16 \div 8 - 5 =$ ______ **f** $10 - (-2 - 14) \div 4 =$ ______

g $-5 + 12 \times -9 =$ ______ **h** $7 \times -2 + 10 =$ ______ **i** $25 \div 5 - 5 =$ ______

j $-6 + 14 \times -2 =$ ______ **k** $12 \times -3 + 18 =$ ______ **l** $-6 + 18 \div 6 =$ ______

QUESTION **3** Work out the following.

a $(-9 + 18) \div 3 =$ ______ **b** $10 - 70 \div 7 =$ ______ **c** $-12 - 6 + 32 =$ ______

d $8 - 50 \div 5 =$ ______ **e** $-8 - 4 \times 5 =$ ______ **f** $-8 - (6 + 12) =$ ______

g $(-7 - 5) \times 8 =$ ______ **h** $-12 \times -4 + 18 =$ ______ **i** $\dfrac{-9 - 7}{-16} =$ ______

j $\dfrac{15 \times -6}{10} =$ ______ **k** $84 \div 21 + 68 =$ ______ **l** $-48 \div 6 - 4 =$ ______

QUESTION **4** Simplify the following.

a $(15 \times 4) \div 3 =$ ______ **b** $10 - 2 \times -5 =$ ______ **c** $-6 + -5 \times -8 =$ ______

d $(-3 \times -2) \times -8 =$ ______ **e** $-64 \div -8 \times 8 =$ ______ **f** $216 \div (-6 + 2) =$ ______

g $324 \div (-3 \times -2) =$ ______ **h** $(-6 - 16) \times -5 =$ ______ **i** $80 \times (15 \div 5) =$ ______

j $(28 \times -3) \div -4 =$ ______ **k** $-30 \div -5 \times 2 =$ ______ **l** $(-8 - 4) \times 6 =$ ______

QUESTION **5** Find the value of the following.

a $72 \div -9 \times 5 =$ ______ **b** $25 - 5 - 42 =$ ______ **c** $-5 + 12 \times -3 + 8 =$ ______

d $9 \times -6 + 12 =$ ______ **e** $60 \div -10 + 20 =$ ______ **f** $(-4 + 8) \times [-5 - (-4)] =$ ______

g $(8 + -7) \times 32 =$ ______ **h** $[-18 - (-18)] \div 18 =$ ______ **i** $(-7 - 3) \times 11 =$ ______

j $(-6 - -3) \div 3 =$ ______ **k** $[20 + (-16)] \times 15 =$ ______ **l** $-22 \times (-4) + 9 =$ ______

QUESTION **6** Work out the following.

a $(-96 - 4) \times -12 =$ ______ **b** $14 \times (5 + 7 + 8) =$ ______ **c** $80 - 10 + 7 - 16 =$ ______

d $220 + 75 \times 2 + 5 =$ ______ **e** $\dfrac{-8 \times 12 + 24}{(2 \times 4)} =$ ______ **f** $\dfrac{20 \times 6 - 24}{6 + 3 \times 6} =$ ______

g $\dfrac{8 - 16 \times 3}{10 + (10 \times -2)} =$ ______ **h** $28 - 16 \div 4 - 14 =$ ______ **i** $16 - (-8 - 12) \div 5 =$ ______

j $-2 \times [-14 \div (-3 + 10)] =$ ______ **k** $18 - [16 - (-4 - 5)] =$ ______ **l** $84 \div [(-40 + 19)] \div -4 =$ ______

Integers

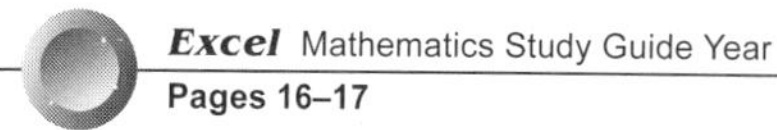

UNIT 10: Problem solving with integers

QUESTION 1 Where would I be after a trip of 5 km south, followed by a trip of 9 km north?

QUESTION 2 What would be the net result of a deposit of $700 in my account followed by a withdrawal of $900?

QUESTION 3 Two numbers have a product of 56 and their sum of −15. What are the numbers?

QUESTION 4 If the sum of two numbers is −2 and their product is −24, find the numbers.

QUESTION 5 I am a negative number divisible by 7 and greater than −9. What number am I?

QUESTION 6 Find the sum of a profit of $9000 and a loss of $21 000.

QUESTION 7 The temperature fell 5°C during the day and fell another 5°C during the night. What was the total change in temperature?

QUESTION 8 What is the combined effect of a gain of 8 kg and a loss of 7 kg?

QUESTION 9 Find the two integers whose sum is −11 and product is 30.

QUESTION 10 Start with the integer 5, add 7 and multiply the result by −3. What is the answer?

QUESTION 11 If 6 more than −4 is added to the product of 5 and 9, what is the result?

QUESTION 12 Divide −72 by −9 and multiply the result by −5. What is the answer?

QUESTION 13 Which of the integers −9, −5, −2, −1, 3, 7 is closest to zero?

QUESTION 14 From the integers −7, −5, −3, 4, 9, find the three integers whose sum is −1

QUESTION 15 Multiply 8 by −12 and add 6 to the result. What is the answer?

Integers

TOPIC TEST — PART A

Instructions
- This part consists of 10 multiple-choice questions.
- Fill in only ONE CIRCLE for each question.
- Each question is worth 1 mark.
- Calculators are NOT allowed.

Time allowed: 15 minutes **Total marks: 10**

	Question					Marks
1	$-16 + 4$ equals	Ⓐ -12	Ⓑ -20	Ⓒ 12	Ⓓ 20	1
2	$-9 + -2 + -1$ equals	Ⓐ -12	Ⓑ 12	Ⓒ -11	Ⓓ -10	1
3	$(-3)^2$ equals	Ⓐ -9	Ⓑ 9	Ⓒ -6	Ⓓ 6	1
4	$(-7 \times 5) + 9$ equals	Ⓐ -26	Ⓑ 26	Ⓒ -18	Ⓓ 18	1
5	$-7 - (8 - -2)$ equals	Ⓐ 3	Ⓑ -3	Ⓒ -17	Ⓓ 17	1
6	$-10 + -4 \times -5$ equals	Ⓐ -10	Ⓑ 10	Ⓒ -30	Ⓓ 30	1
7	If $x = 3$ and $y = -8$ then $xy + 7$ equals	Ⓐ 17	Ⓑ -17	Ⓒ 31	Ⓓ -31	1
8	$-9 \times -6 \div -2$ equals	Ⓐ -27	Ⓑ 27	Ⓒ -54	Ⓓ 54	1
9	$-64 \div 16 - -3$ equals	Ⓐ -7	Ⓑ 1	Ⓒ -1	Ⓓ 7	1
10	$(-8 \times -4) \div -16$ equals	Ⓐ 2	Ⓑ -2	Ⓒ 16	Ⓓ -16	1

Total marks achieved for PART A

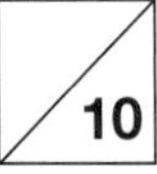

Integers

TOPIC TEST

PART B

Instructions
- This part consists of 15 questions.
- Each question is worth 1 mark.
- Write only the answer in the answer column.

Time allowed: 15 minutes **Total marks: 15**

	Questions	Answers	Marks
1	$18 \div -6 =$		1
2	$5 - -12 =$		1
3	$(-7)^2 =$		1
4	$-8 - 10 =$		1
5	$15 - 8 \times 4 =$		1
6	$-8 \times -5 \times -2 =$		1
7	$-8 \times (5 - 9) =$		1
8	$5 - 8\frac{1}{2} =$		1
9	$(-6 - -15) \times (8 - 12) =$		1
10	$-36 \div 9 \div -4 =$		1
11	$(-3)^2 \times (5 - 12) =$		1
12	$\frac{-6 - (-18)}{5 - 7} =$		1
13	$8 \times -2 - (6 - 11) =$		1
14	$1 + -15 =$		1
15	$5 - -12 + 8 \times -3 =$		1

Total marks achieved for PART B /15

CHAPTER 3

Indices

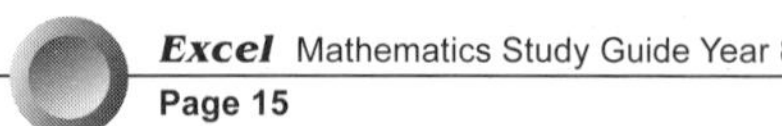

UNIT 1: Index notation

QUESTION 1 Write the following in expanded form.

a $4^3 =$ ______ **b** $8^6 =$ ______ **c** $5^3 =$ ______

d $6^4 =$ ______ **e** $2^4 =$ ______ **f** $8^2 =$ ______

g $3^5 =$ ______ **h** $3^6 =$ ______ **i** $9^3 =$ ______

QUESTION 2 Write the following without indices (in expanded form).

a $(1.5)^3 =$ ______ **b** $3^2 \times 5^5 =$ ______

c $2^4 \times 3^6 =$ ______ **d** $3^4 \times 4^2 =$ ______

e $7^5 =$ ______ **f** $\left(\frac{3}{4}\right)^5 =$ ______

g $5^2 \times 8^6 =$ ______ **h** $(9)^5 =$ ______

i $(-5)^3 =$ ______ **j** $\left(-\frac{2}{3}\right)^5 =$ ______

QUESTION 3 Change the following to expanded form.

a $6^3 =$ ______ **b** $2^6 =$ ______

c $7^4 =$ ______ **d** $3^2 \times 4^3 =$ ______

e $2^5 \times 3^4 =$ ______ **f** $5^2 \times 3^3 \times 4^4 =$ ______

g $2^3 \times 3^4 \times 4^5 =$ ______ **h** $5^3 \times 6^5 =$ ______

i $-3 \times 2^3 \times 5^5 =$ ______ **j** $17 \times 3^3 \times 5^9 =$ ______

QUESTION 4 Write the following in index form.

a $3 \times 3 \times 3 \times 3 \times 3 \times 3 =$ ______ **b** $5 \times 5 \times 5 \times 5 =$ ______

c $2 \times 2 \times 2 \times 2 \times 2 \times 2 \times 2 \times 2 =$ ______ **d** $9 \times 9 \times 9 \times 9 \times 9 =$ ______

e $7 \times 7 \times 7 \times 7 \times 7 \times 7 =$ ______ **f** $4 \times 4 \times 4 =$ ______

g $-3 \times -3 \times -3 \times -3 =$ ______ **h** $\times \quad \times \quad \times \quad =$ ______

i $1.8 \times 1.8 \times 1.8 \times 1.8 \times 1.8 =$ ______ **j** $3.5 \times 3.5 \times 3.5 \times 3.5 =$ ______

QUESTION 5 Write the following using index notation.

a $2 \times 2 \times 2 \times 3 \times 3 \times 3 =$ ______ **b** $5 \times 5 \times 5 \times 6 \times 6 =$ ______

c $4 \times 4 \times 4 \times 4 \times 6 \times 6 \times 6 =$ ______ **d** $2 \times 2 \times 2 \times 2 \times 8 \times 8 =$ ______

e $4 \times 4 \times 4 \times 4 \times 4 \times 5 \times 5 =$ ______ **f** $2 \times 2 \times 2 \times 2 \times 2 \times 4 \times 4 =$ ______

g $7 \times 7 \times 7 \times 7 \times 7 \times 3 \times 3 \times 3 =$ ______ **h** $7 \times 7 \times 7 \times 5 \times 5 \times 5 =$ ______

i $5 \times 5 \times 5 \times 2 \times 2 \times 2 =$ ______ **j** $4 \times 4 \times 4 \times 5 \times 5 \times 5 \times 5 =$ ______

QUESTION 6 Evaluate the following.

a $6^5 =$ ______ **b** $3^4 =$ ______ **c** $(1.1)^3 =$ ______

d $5^4 =$ ______ **e** $7^3 =$ ______ **f** $9^4 =$ ______

g $2^3 =$ ______ **h** $8^3 =$ ______ **i** $10^7 =$ ______

Indices

UNIT 2: Index laws—multiplication with indices

Question 1 Simplify the following, writing your answers in index form.

a $3^4 \times 3^8 =$ ______ **b** $6^3 \times 6^7 =$ ______ **c** $7^2 \times 7^6 =$ ______
d $5^7 \times 5^5 =$ ______ **e** $2^4 \times 2^6 =$ ______ **f** $6^8 \times 6^2 =$ ______
g $10^3 \times 10^9 =$ ______ **h** $11^5 \times 11^4 =$ ______ **i** $7^7 \times 7^6 =$ ______
j $6^8 \times 6^{10} =$ ______ **k** $10^9 \times 10^{12} =$ ______ **l** $13^{17} \times 13^{12} =$ ______

Question 2 Simplify the following, writing your answers in index form.

a $5 \times 5^2 =$ ______ **b** $2^3 \times 2^2 =$ ______ **c** $6 \times 6^3 =$ ______
d $7^2 \times 7 =$ ______ **e** $6^3 \times 6 =$ ______ **f** $3^3 \times 3^4 =$ ______
g $11^2 \times 11^2 =$ ______ **h** $10^2 \times 10 =$ ______ **i** $5 \times 5^4 =$ ______
j $3^4 \times 3^2 =$ ______ **k** $5^2 \times 5^5 =$ ______ **l** $2^3 \times 2^5 =$ ______

Question 3 Simplify the following.

a $5^3 \times 5^2 =$ ______ **b** $6^5 \times 6^7 =$ ______ **c** $6^2 \times 6^9 =$ ______
d $3^8 \times 3^6 =$ ______ **e** $5^3 \times 5^6 =$ ______ **f** $10^9 \times 10^3 =$ ______
g $2^7 \times 2^5 =$ ______ **h** $11^4 \times 11^8 =$ ______ **i** $3^3 \times 3^4 \times 3^7 =$ ______
j $7^5 \times 7^2 \times 7^3 =$ ______ **k** $3^9 \times 3^{12} =$ ______ **l** $2^5 \times 2^{17} =$ ______

Question 4 Simplify the following.

a $13^4 \times 13^3 =$ ______ **b** $2^5 \times 2^4 =$ ______ **c** $2^8 \times 2^3 =$ ______
d $6^9 \times 6^2 =$ ______ **e** $3^7 \times 3^4 =$ ______ **f** $6^3 \times 6^7 =$ ______
g $3^3 5^5 \times 3^2 5^2 =$ ______ **h** $11^5 \times 11^5 =$ ______ **i** $7^3 \times 7^2 =$ ______
j $7^2 \times 7^3 =$ ______ **k** $5^5 \times 5^3 =$ ______ **l** $10^4 \times 10^6 =$ ______

Question 5 Simplify the following.

a $6 \times 6^4 =$ ______ **b** $10^7 \times 10 =$ ______ **c** $2^2 \times 2^3 =$ ______
d $5^6 \times 5 =$ ______ **e** $6^3 \times 6^2 =$ ______ **f** $7^4 \times 7^8 =$ ______
g $2^5 \times 2^9 =$ ______ **h** $3^7 \times 3^8 =$ ______ **i** $6^5 \times 6^3 \times 6^2 =$ ______
j $7 \times 7^3 =$ ______ **k** $5^4 \times 5^3 \times 5^2 =$ ______ **l** $11^9 \times 11^6 =$ ______

Question 6 Find the missing term in each of the following.

a $7^7 \times \square = 7^{10}$ **b** $\square \times 6^6 = 6^9$ **c** $5^4 \times \square = 5^{11}$
d $9^3 \times \square = 9^5$ **e** $3^5 \times \square = 3^8$ **f** $4^4 \times \square = 4^5$
g $8^2 \times \square = 8^9$ **h** $\square \times 2^2 = 2^3$ **i** $\square \times 3^2 = 3^6$
j $4^2 \times \square = 4^8$ **k** $\square \times 9^4 = 9^7$ **l** $\square \times 2^2 = 2^4$

Indices

UNIT 3: Index laws—division with indices

QUESTION 1 Simplify the following, writing your answers in index form.

a $10^9 \div 10^5 =$ ______ b $6^8 \div 6^5 =$ ______ c $11^{16} \div 11^3 =$ ______

d $3^7 \div 3^4 =$ ______ e $5^{19} \div 5^6 =$ ______ f $13^{10} \div 13^5 =$ ______

g $2^{29} \div 2^6 =$ ______ h $3^{21} \div 3^6 =$ ______ i $7^{15} \div 7^4 =$ ______

j $5^{10} \div 5^3 =$ ______ k $2^{18} \div 2^{12} =$ ______ l $5^{14} \div 5^5 =$ ______

QUESTION 2 Simplify the following, leaving your answers in index form.

a $3^8 \div 3^3 =$ ______ b $11^9 \div 11^4 =$ ______ c $7^7 \div 7^3 =$ ______

d $6^{18} \div 6^{16} =$ ______ e $6^{12} \div 6^8 =$ ______ f $3^9 \div 3^2 =$ ______

g $5^{21} \div 5^{17} =$ ______ h $10^5 \div 10^2 =$ ______ i $5^9 \div 5^6 =$ ______

j $13^{15} \div 13^{13} =$ ______ k $5^{18} \div 5^{13} =$ ______ l $2^8 \div 2^4 =$ ______

QUESTION 3 Simplify the following, writing your answers in index form.

a $\frac{3^{15}}{3^8} =$ ______ b $\frac{10^{10}}{10^4} =$ ______ c $\frac{6^9}{6^5} =$ ______

d $\frac{5^{12}}{5^8} =$ ______ e $\frac{3^{40}}{3^7} =$ ______ f $\frac{7^{12}}{7^8} =$ ______

g $\frac{2^{43}}{2^{11}} =$ ______ h $\frac{2^9}{2^6} =$ ______ i $\frac{5^{32}}{5^9} =$ ______

j $\frac{7^{47}}{7^{13}} =$ ______ k $\frac{11^{18}}{11^6} =$ ______ l $\frac{3^{16}}{3^{13}} =$ ______

QUESTION 4 Simplify the following.

a $5^{10} \div 5^2 =$ ______ b $2^{11} \div 2^5 =$ ______ c $6^{21} \div 6^{16} =$ ______

d $6^8 \div 6^2 =$ ______ e $5^{12} \div 5^3 =$ ______ f $11^9 \div 11^3 =$ ______

g $10^{15} \div 10^7 =$ ______ h $3^{23} \div 3^8 =$ ______ i $13^{18} \div 13^6 =$ ______

j $3^{21} \div 3^{14} =$ ______ k $7^{32} \div 7^{18} =$ ______ l $2^{54} \div 2^{28} =$ ______

QUESTION 5 Simplify the following.

a $11^7 \div 11^4 =$ ______ b $6^9 \div 6^2 =$ ______ c $6^7 \div 6^6 =$ ______

d $6^8 \div 6^3 =$ ______ e $7^5 \div 7^2 =$ ______ f $10^8 \div 10^7 =$ ______

g $3^7 \div 3^4 =$ ______ h $2^9 \div 2^4 =$ ______ i $3^{10}5^8 \div 3^7 5^2 =$ ______

j $2^{10}5^7 \div 2^8 5^4 =$ ______ k $4^5 5^9 \div 4^3 5^4 =$ ______ l $2^9 7^6 \div 2^4 7^5 =$ ______

QUESTION 6 Find the missing term in each of the following.

a $6^9 \div \square = 6^4$ b $6^{14} \div \square = 6^8$ c $9^{12} \div \square = 9^7$

d $\square \div 8 = 8^5$ e $9^9 \div 9^6 = \square$ f $8^8 \div 8^5 = \square$

g $3^6 \div \square = 3^2$ h $\frac{\square}{4^2} = 4^7$ i $\frac{3^8}{\square} = 3^6$

j $3^4 4^4 \div \square = 3^3 4^2$ k $\square \div 2^2 = 2^6$ l $\frac{4^8 5^{12}}{\square} = 4^5 5^6$

UNIT 4: Index laws—powers of powers

Question 1 Simplify the following, leaving your answers in the simplest index form.

a $(11^5)^5 =$ ______ b $(2^4)^2 =$ ______ c $(6^8)^8 =$ ______
d $(7^2)^8 =$ ______ e $(5^6)^4 =$ ______ f $(3^6)^9 =$ ______
g $(10^4)^6 =$ ______ h $(6^7)^8 =$ ______ i $(11^8)^5 =$ ______
j $(3^4)^6 =$ ______ k $(2^6)^9 =$ ______ l $(7^4)^9 =$ ______

Question 2 Simplify the following, writing your answers in index form.

a $(2^8)^2 =$ ______ b $(10^2)^8 =$ ______ c $(6^2)^2 =$ ______
d $(5^8)^1 =$ ______ e $(6^2)^2 =$ ______ f $(3^2)^2 =$ ______
g $(13^2)^3 =$ ______ h $(9^1)^8 =$ ______ i $(6^1)^3 =$ ______
j $(2^6)^2 =$ ______ k $(2^8)^8 =$ ______ l $(6^2)^8 =$ ______

Question 3 Simplify the following, leaving your answers in index form.

a $(2^6)^7 =$ ______ b $(3^7)^8 =$ ______ c $(6^8)^6 =$ ______
d $(5^6)^9 =$ ______ e $(7^4)^6 =$ ______ f $(10^8)^8 =$ ______
g $(6^6)^{15} =$ ______ h $(11^4)^{10} =$ ______ i $(2^8)^5 =$ ______
j $(3^9)^9 =$ ______ k $(6^9)^2 =$ ______ l $(7^8)^2 =$ ______

Question 4 Simplify the following, leaving your answers in index form.

a $(7^5)^8 =$ ______ b $(8^9)^5 =$ ______ c $(3^{10})^4 =$ ______
d $(6^{18})^8 =$ ______ e $(2^{15})^6 =$ ______ f $(6^{12})^9 =$ ______
g $(5^9)^2 =$ ______ h $(7^{10})^8 =$ ______ i $(2^{11})^6 =$ ______
j $(11^{16})^8 =$ ______ k $(3^9)^7 =$ ______ l $(10^{14})^3 =$ ______

Question 5 By first changing the base to simplest index form, simplify the following.

a $(36^2)^2 =$ ______ b $(8^3)^8 =$ ______ c $(81^2)^4 =$ ______
d $(125^2)^8 =$ ______ e $(49^9)^2 =$ ______ f $(36^2)^9 =$ ______
g $(81^6)^4 =$ ______ h $(216^2)^3 =$ ______ i $(121^4)^2 =$ ______
j $(64^9)^2 =$ ______ k $(64^2)^8 =$ ______ l $(169^{12})^2 =$ ______

Question 6 By first changing the base to simplest index form if necessary, simplify the following.

a $(4^2)^8 =$ ______ b $(8^3)^4 =$ ______ c $(9^4)^6 =$ ______
d $(6^{16})^2 =$ ______ e $(9^6)^8 =$ ______ f $(6^5)^8 =$ ______
g $(2^8)^2 =$ ______ h $(8^5)^3 =$ ______ i $(4^2)^9 =$ ______
j $(9^7)^6 =$ ______ k $(12^8)^8 =$ ______ l $(9^9)^4 =$ ______

Indices

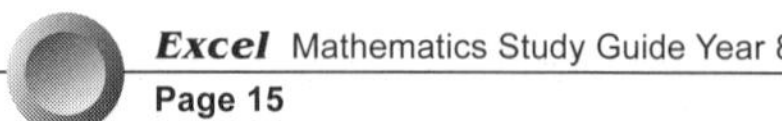

UNIT 5: Index laws—the zero index

QUESTION 1 Simplify the following.

a $(48)^0 =$ ______ b $8^0 \times 3^0 =$ ______ c $x^0y^0 =$ ______
d $(86)^0 =$ ______ e $(xy)^0 =$ ______ f $9y^0 =$ ______
g $x^0y^0 =$ ______ h $8^4p^0 =$ ______ i $(4^3)^0 =$ ______
j $(6xy)^0 =$ ______ k $(9ab)^0 =$ ______ l $16a^0 =$ ______

QUESTION 2 Use your calculator to verify true or false for the following.

a $9^0 = 1$ ______ b $93^0 = 1$ ______ c $\left(\frac{5}{8}\right)^0 = 1$ ______
d $(2.3)^0 = 1$ ______ e $-(6)^0 = -1$ ______ f $(6 \times 12)^0 = 1$ ______
g $-5 \times 7^0 = -5$ ______ h $-61^0 - 3^0 - 5^0 = -3$ ______ i $8 \times 6^0 = 8$ ______
j $21 \times (-5)^0 = 21$ ______ k $6 \times 4^0 \times (-9)^0 = 6$ ______ l $12 \div 4^0 = 12$ ______

QUESTION 3 Simplify the following.

a $6 \times 2y^0 =$ ______ b $(8a^0)^2 =$ ______ c $7^0 + 8m^0 =$ ______
d $6 \times (6a)^0 =$ ______ e $9 \times 4x^0 =$ ______ f $(-15)^0 + 8 =$ ______
g $(ab)^0 \times 9 =$ ______ h $16^0 + 9^0 =$ ______ i $-9x^0 + 12 =$ ______
j $(8 + 8)^0 =$ ______ k $(5a^2)^0 + (3b^2)^0 =$ ______ l $9(y^2)^0 \times 8(x^7)^0 =$ ______

QUESTION 4 Simplify the following.

a $6x^0 + (6x)^0 =$ ______ b $\frac{10y^0}{(10y)^0} =$ ______ c $\frac{(6t)^0}{6t^0} =$ ______
d $-8^0 - (-8)^0 =$ ______ e $9(2a - 3b)^0 =$ ______ f $12a^4b^0 =$ ______
g $28x^0y^4 =$ ______ h $2ab^0c^0 =$ ______ i $(8a^4)^0 =$ ______
j $(\frac{2}{3} \times 8)^0 =$ ______ k $(6xyz)^0 =$ ______ l $14 \times 7^0 =$ ______

QUESTION 5 Simplify the following, leaving your answers in index form.

a $2^8 \times 2^0 =$ ______ b $3^0 \times 3^6 =$ ______ c $4^8 \times 4^0 =$ ______
d $10^0 \times 10^8 =$ ______ e $8^8 \times 8^0 =$ ______ f $5^6 \times 5^0 =$ ______
g $8^0 \times 8^9 =$ ______ h $5^9 \times 5^0 =$ ______ i $6^9 \times 6^0 =$ ______
j $3^6 \times 3^4 \times 3^0 =$ ______ k $7^2 \times 7^0 \times 7^8 =$ ______ l $(-3)^0 \times -3^0 \times -3^0 =$ ______

QUESTION 6 Simplify the following.

a $(6y)^0 \times 6y^0 =$ ______ b $\frac{9b^0}{8a^0} =$ ______ c $\frac{(6x^2)^0 \times (xy)^0}{26x^0} =$ ______
d $\frac{6^4 \times 6^5}{6^9} =$ ______ e $\frac{5a^0 \times (2a^3)^0}{20a^0} =$ ______ f $(6y)^0 + 6y^0 =$ ______
g $\frac{9 \times (5x)^0}{8x^0} =$ ______ h $\frac{16p^0}{8n^0} =$ ______ i $\frac{(4p^3)^0 \times 2p^0}{32p^0} =$ ______
j $8y^0 \times (8y)^0 \div 4y^0 =$ ______ k $\frac{2a^9 \times (a^3b^2)^0}{(2ab)^0} =$ ______ l $6x^0 \times (6x)^0 =$ ______

Indices

UNIT 6: Index laws—miscellaneous questions

QUESTION 1 Simplify the following leaving answers in index form.

a $9^4 \times 9^6 =$ ______ **b** $3^2 \times 3^6 =$ ______ **c** $9^6 \times 9^2 =$ ______

d $4^6 \times 4^6 =$ ______ **e** $8^2 \times 8^7 =$ ______ **f** $8^2 \times 8 \times 8^3 =$ ______

g $8^2 \times 8^8 =$ ______ **h** $5^8 \times 5^2 =$ ______ **i** $6^4 \times 6^2 =$ ______

QUESTION 2 Simplify, giving answers in index form.

a $4^9 \div 4^6 =$ ______ **b** $2^8 \div 2^6 =$ ______ **c** $6^7 \div 6^6 =$ ______

d $6^6 \div 6^3 =$ ______ **e** $4^9 \div 4^4 =$ ______ **f** $8^8 \div 8^6 =$ ______

g $3^6 \div 3^2 =$ ______ **h** $6^6 \div 6^4 =$ ______ **i** $4^6 \div 4^3 =$ ______

QUESTION 3 Give the answers in simplest index form.

a $(3^2)^8 =$ ______ **b** $(2^8)^2 =$ ______ **c** $(9^8)^2 =$ ______

d $(4^8)^4 =$ ______ **e** $(6^3)^2 \times 6^4 =$ ______ **f** $(8^9)^2 \div 8^6 =$ ______

g $(5^4)^8 \times (5^6)^2 =$ ______ **h** $(3^6)^6 =$ ______ **i** $(5^8)^4 =$ ______

QUESTION 4 Simplify the following.

a $4m^0 =$ ______ **b** $(6m)^0 =$ ______ **c** $8n^0 =$ ______

d $6m^0 \times (8m)^0 =$ ______ **e** $2^6b^0 =$ ______ **f** $(x^2)^0 =$ ______

g $(5y^8)^0 =$ ______ **h** $5^9 \div 5^9 =$ ______ **i** $9^6 \div 9^4 =$ ______

QUESTION 5 Simplify the following, giving the answers in index form.

a $8^9 \times 8^2 =$ ______ **b** $6^9 \times 6^2 =$ ______ **c** $(2^6)^4 =$ ______

d $(5^6)^2 =$ ______ **e** $5^8 \div 5^3 =$ ______ **f** $6^2 \times 6^2 =$ ______

g $2^2 6^8 \times 2^4 6^8 =$ ______ **h** $4^8 3^6 \div 4^2 3^4 =$ ______ **i** $8^2 \times 8 =$ ______

j $9^2 \div 9^2 =$ ______ **k** $7^9 4^9 \div 7^6 4^9 =$ ______ **l** $(9^2)^2 =$ ______

Indices

TOPIC TEST **PART A**

Instructions
- This part consists of 10 multiple-choice questions.
- Fill in only ONE CIRCLE for each question.
- Each question is worth 1 mark.
- Calculators are NOT allowed.

Time allowed: 15 minutes **Total marks: 10**

Marks

1 $3^6 \times 3^6$ equals

Ⓐ 9^6 Ⓑ 3^{36} Ⓒ 3^{12} Ⓓ 9^{12} **1**

2 $8^8 \div 8^4$ equals

Ⓐ 8^4 Ⓑ 8^2 Ⓒ 1^4 Ⓓ 1^2 **1**

3 $\dfrac{6^8 8^1}{6^4 8^4}$ equals

Ⓐ $\dfrac{6^2}{8^3}$ Ⓑ $\dfrac{6^4}{8^3}$ Ⓒ $6^2 8^3$ Ⓓ $6^4 8^3$ **1**

4 $5^4 \times (5^{10} \div 5^2)$ equals

Ⓐ 5^{20} Ⓑ 5^{12} Ⓒ 5^{10} Ⓓ 5^6 **1**

5 $2a^0 \times 8^0$ equals

Ⓐ 16 Ⓑ 8 Ⓒ 1 Ⓓ 2 **1**

6 $(6^2)^3$ equals

Ⓐ 6^5 Ⓑ 6^6 Ⓒ 6^{23} Ⓓ 6^{32} **1**

7 $(9^2)^0$ equals

Ⓐ 9^2 Ⓑ 9 Ⓒ 0 Ⓓ 1 **1**

8 $\dfrac{5^4 \times 5^8}{5^2}$ equals

Ⓐ 5^6 Ⓑ 5^{10} Ⓒ 5^{14} Ⓓ 5^8 **1**

9 $(9^{10} \div 9^2)^2$ equals

Ⓐ 9^{25} Ⓑ 9^8 Ⓒ 9^{64} Ⓓ 9^{16} **1**

10 $(2^3 7^4)^2$ equals

Ⓐ $2^5 7^6$ Ⓑ $2^6 7^8$ Ⓒ 14^{24} Ⓓ 14^9 **1**

Total marks achieved for PART A 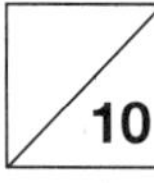

Indices

TOPIC TEST — PART B

Instructions
- This part consists of 3 questions.
- Each question part is worth 1 mark.
- Write only the answer in the answer column.

Time allowed: 15 minutes — **Total marks: 15**

Questions	Answers	Marks
1 Simplify the following.		
a $4^0 \times (4^2)^3 =$	______	1
b $6^7 \div 6^5 =$	______	1
c $(3^3)^4 =$	______	1
d $8t^0 + (9t)^0 + (3t^3)^0 =$	______	1
e $8^{20} \div (8^4)^2 =$	______	1
2 Simplify the following.		
a $8^9 \times 8^5 =$	______	1
b $3^8 \div 3^4 =$	______	1
c $\frac{2^6 3^4}{2^4 3^2} =$	______	1
d $\frac{8^0 \times (1)^9}{(3)^3} =$	______	1
e $(5^2)^2 \div 5^3 =$	______	1
3 Simplify the following, writing your answers in the simplest index form.		
a $12^5 \times 12^4 \times 12^9 =$	______	1
b $7^0 \div 7^2 =$	______	1
c $(1000a^{12}b^{15})^0 =$	______	1
d $9(a^3)^0 + (9a)^0 + 9^0 =$	______	1
e $25^2 \times 25^3 \times 64^4 \times 64^5 =$	______	1

Total marks achieved for PART B

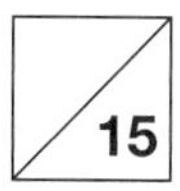

CHAPTER 4

Percentages

Excel Mathematics Study Guide Year 8
Pages 26–41

UNIT 1: The meaning of percentage

QUESTION 1 Complete the following sentences.

a 63% means 63 out of ______________.

b 95% means 95 out of ______________.

c 29% means 29 out of ______________.

d A percentage is actually a fraction with a denominator of ______________.

e The symbol of per cent is ___ and it means out of ______________.

QUESTION 2 In the diagram given opposite:

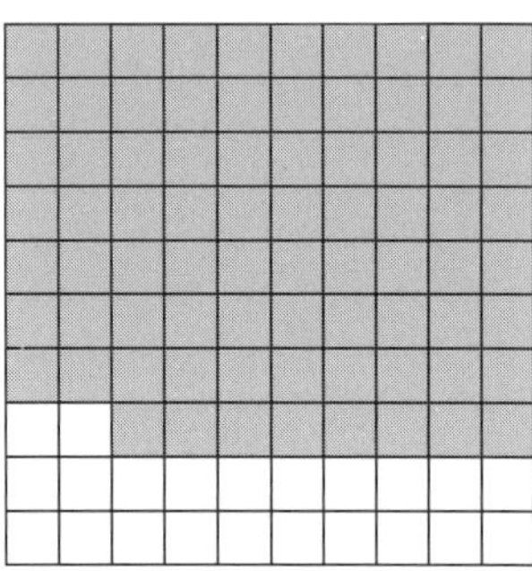

a what percentage is shaded? ______________

b what percentage is unshaded? ______________

c if 63% of a diagram is shaded, how much is unshaded? ______________

d if 81% of a diagram is shaded, how much is unshaded? ______________

e if 13% of a diagram is shaded, how much is unshaded? ______________

QUESTION 3 In a grid of 100 squares, what percentage is unshaded when the following are shaded?

a 27% ______________ **b** 66% ______________ **c** 52% ______________

d 69% ______________ **e** 39% ______________ **f** 18% ______________

g 34% ______________ **h** 87% ______________ **i** 84% ______________

QUESTION 4 Work out the following subtractions.

a 100% – 82% = ___________ **b** 100% – 39% = ___________ **c** 100% – 54% = ___________

d 100% – 19% = ___________ **e** 100% – 67% = ___________ **f** 100% – 47% = ___________

QUESTION 5 Add the following percentages.

a 30% + 27% = ______________ **b** 30% + 6% = ______________

c 10% + 20% + 7% = ______________ **d** 12% + 23% + 32% = ______________

QUESTION 6

a If 22% of students are away from a class, what percentage of students are present? ______________

b In a class spelling test, if James got 89% of the words correct, what percentage of the words did he spell incorrectly? ______________

Percentages

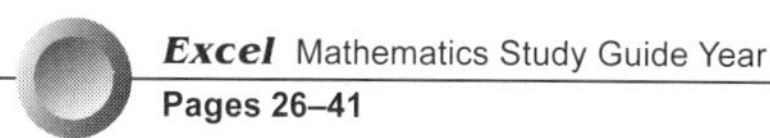

UNIT 2: Changing percentages to fractions and fractions to percentages

QUESTION 1 Change the following percentages to fractions in simplest form.

a 20% = ________ b 30% = ________ c 40% = ________

d 15% = ________ e 25% = ________ f 35% = ________

g 37% = ________ h 88% = ________ i 93% = ________

j 18% = ________ k 23% = ________ l 47% = ________

QUESTION 2 Convert the following percentages to fractions in simplest form.

a 1.2% = ________ b 3.6% = ________ c 12.8% = ________

d 8.2% = ________ e 3.7% = ________ f 5.9% = ________

g $\frac{1}{8}$% = ________ h $\frac{2}{5}$% = ________ i $\frac{3}{20}$% = ________

j $\frac{1}{5}$% = ________ k $\frac{7}{10}$% = ________ l $50\frac{3}{4}$% = ________

QUESTION 3 Express the following percentages as fractions in simplest form.

a 39% = ________ b $12\frac{1}{2}$% = ________ c $40\frac{1}{4}$% = ________

d $3\frac{3}{4}$% = ________ e $9\frac{1}{2}$% = ________ f 85% = ________

g $38\frac{1}{2}$% = ________ h 42% = ________ i $33\frac{1}{3}$% = ________

QUESTION 4 Change the following fractions to percentages.

a $\frac{7}{100}$ = ________ b $\frac{9}{100}$ = ________ c $\frac{3}{100}$ = ________

d $\frac{21}{100}$ = ________ e $\frac{33}{100}$ = ________ f $\frac{47}{100}$ = ________

g $\frac{16}{75}$ = ________ h $\frac{24}{150}$ = ________ i $\frac{21}{60}$ = ________

j $\frac{9}{120}$ = ________ k $\frac{15}{80}$ = ________ l $\frac{17}{90}$ = ________

QUESTION 5 Convert the following fractions to percentages.

a $\frac{2}{5}$ = ________ b $\frac{6}{65}$ = ________ c $\frac{5}{40}$ = ________

d $\frac{3}{7}$ = ________ e $\frac{1}{3}$ = ________ f $\frac{1}{4}$ = ________

Percentages

UNIT 3: Changing percentages to decimals and decimals to percentages

QUESTION 1 Change the following percentages to decimals.

a 10% = ______ **b** 30% = ______ **c** 50% = ______

d 90% = ______ **e** 80% = ______ **f** 60% = ______

g 85% = ______ **h** 65% = ______ **i** 45% = ______

j 95% = ______ **k** 115% = ______ **l** 35% = ______

QUESTION 2 Express the following percentages as decimals.

a 8% = ______ **b** 72% = ______ **c** 64% = ______

d 64.8% = ______ **e** 78.5% = ______ **f** 39.4% = ______

g 73.6% = ______ **h** 55.7% = ______ **i** 76.4% = ______

j 93.2% = ______ **k** 78.5% = ______ **l** 32.9% = ______

QUESTION 3 Convert the following percentages to decimals.

a $2\frac{1}{2}\%$ = ______ **b** $3\frac{1}{4}\%$ = ______ **c** $15\frac{1}{5}\%$ = ______

d $12\frac{1}{4}\%$ = ______ **e** $5\frac{1}{2}\%$ = ______ **f** $38\frac{1}{4}\%$ = ______

QUESTION 4 Convert the following decimals to percentages.

a 0.80 = ______ **b** 0.70 – ______ **c** 0.90 – ______

d 0.19 = ______ **e** 0.38 = ______ **f** 0.65 = ______

g 0.325 = ______ **h** 0.138 = ______ **i** 0.257 = ______

j 0.256 = ______ **k** 0.406 = ______ **l** 0.548 = ______

QUESTION 5 Express the following decimals as percentages.

a 0.001 = ______ **b** 0.007 = ______ **c** 0.003 = ______

d 0.123 = ______ **e** 0.246 = ______ **f** 0.876 = ______

g 0.004 = ______ **h** 0.005 = ______ **i** 0.008 = ______

j 0.023 = ______ **k** 0.034 = ______ **l** 0.078 = ______

QUESTION 6 Convert the following decimals to percentages.

a 1.25 = ______ **b** 2.87 = ______ **c** 3.91 = ______

d 2.56 = ______ **e** 8.67 = ______ **f** 5.227 = ______

g 7.98 = ______ **h** 4.67 = ______ **i** 9.42 = ______

Percentages

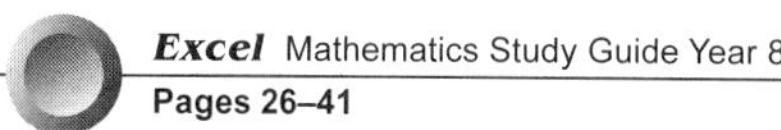

UNIT 4: Finding the percentage of a quantity

QUESTION 1 Calculate the following.

a 40% of 2000 = ______ b 80% of 700 = ______ c 20% of 800 = ______
d 170% of 400 = ______ e 60% of 300 = ______ f 130% of 900 = ______
g 140% of 800 = ______ h 40% of 600 = ______ i 60% of 500 = ______
j 110% of 1200 = ______ k 30% of 1300 = ______ l 30% of 1000 = ______

QUESTION 2 Find the following.

a 15% of 1100 = ______ b 45% of 350 = ______ c 105% of 960 = ______
d 65% of 2500 = ______ e 95% of 800 = ______ f 125% of 450 = ______
g 85% of 1500 = ______ h 45% of 225 = ______ i 15% of 675 = ______
j 5% of 300 = ______ k 95% of 840 = ______ l 65% of 825 = ______

QUESTION 3 Evaluate the following.

a 8% of 130 = ______ b 21% of 560 = ______ c 64% of 1256 = ______
d 4% of 245 = ______ e 46% of 1024 = ______ f 92% of 670 = ______
g 85% of 368 = ______ h 88% of 492 = ______ i 98% of 830 = ______
j 84% of 962 = ______ k 68% of 1132 = ______ l 68% of 1360 = ______

QUESTION 4 Work out the following.

a 2.5% of 1300 = ______ b 9.3% of 9600 = ______ c 4.4% of 780 = ______
d 6.7% of 4100 = ______ e 5.6% of 2600 = ______ f 9.9% of 1290 = ______
g 4.9% of 5800 = ______ h 3.8% of 6700 = ______ i 6.5% of 3900 = ______
j 8.5% of 164 = ______ k 8.3% of 10 360 = ______ l 9.2% of 1450 = ______

QUESTION 5 Find the answers to the following.

a $2\frac{1}{2}$% of 6000 = ______ b $33\frac{1}{3}$% of 100 = ______ c $45\frac{3}{4}$% of 90 000 = ______
d $58\frac{1}{2}$% of 11 000 = ______ e $75\frac{1}{2}$% of 5000 = ______ f $12\frac{1}{2}$% of 30 000 = ______
g $25\frac{1}{4}$% of 4000 = ______ h $1\frac{1}{2}$% of 7000 = ______ i $90\frac{1}{2}$% of 80 000 = ______

QUESTION 6 Evaluate the following.

a 8% of 35 000 = ______ b 8.4% of 100 = ______ c $12\frac{1}{2}$% of 86 400 = ______
d $6\frac{1}{4}$% of 3000 = ______ e 50% of 71 000 = ______ f 20.6% of 53 700 = ______
g 3.5% of 87 500 = ______ h $16\frac{3}{4}$% of 4000 = ______ i 60% of 20 000 = ______

Percentages

Excel Mathematics Study Guide Year 8
Pages 26–41

UNIT 5: Increasing or decreasing by a given percentage

QUESTION **1** Increase the following by the given percentage.

a $90 by 70% = ______________________
b 272 h by 30% = ______________________
c $450 by 20% = ______________________
d 2800 ha by 10% = ______________________
e $600 by 60% = ______________________
f 356 min by 40% = ______________________
g 1500 g by 80% = ______________________
h 1850 L by 90% = ______________________
i 2620 kg by 50% = ______________________
j 460 t by 100% = ______________________
k $840 by 65% = ______________________
l 5870 cm by 45% = ______________________

QUESTION **2** Decrease the following by the given percentage.

a $285 by 40% = ______________________
b $96 by 70% = ______________________
c 216 m by 30% = ______________________
d 1580 L by 60% = ______________________
e $930 by 20% = ______________________
f $4864 by 35% = ______________________
g $235 by 10% = ______________________
h $1525 by 65% = ______________________
i $680.50 by 15% = ______________________
j $550 by 80% = ______________________
k $240 by 80% = ______________________
l $660 by 90% = ______________________

QUESTION **3** Find the answers to the following.

a Increase $300 by 40% = ______________________
b Increase 1500 L by 70% = ______________________
c Decrease $550 by 30% = ______________________
d Decrease 240 t by 45% = ______________________
e Decrease 28 m by 65% = ______________________
f Decrease 600 by 55% = ______________________
g Increase $860 by 20% = ______________________
h Increase $800 by 60% = ______________________
i Increase $1024 by 60% = ______________________
j Decrease $2000 by 34% = ______________________
k Decrease $230 by 5% = ______________________
l Increase $2500 by 85% = ______________________

QUESTION **4**

a The price of a new car is $60 000 and after 1 year it decreases by 25%. Find the new value of the car.

____

Find the simple interest earned on:

b $3500 at 15% p.a. for 1 year ______________________

____

c $80 000 at 8% p.a. for 3 years ______________________

____

d $16 540 at 12% p.a. for 6 years ______________________

____

Percentages

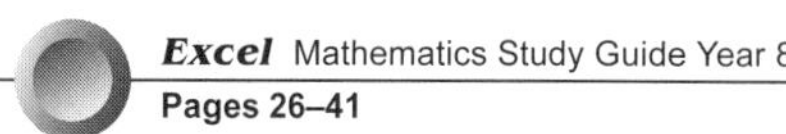

UNIT 6: Expressing quantities as percentages

QUESTION 1 What percentage is:

a 80 of 800? = ______________________
b 20 cents of $30? = ______________________
c 20 of 1600? = ______________________
d 60 of 240? = ______________________
e 24 of 80? = ______________________
f 18 of 90? = ______________________
g 16 of 64? = ______________________
h 30 of 75? = ______________________
i 70 of 280? = ______________________
j 50 cents of $88? = ______________________
k 6 h of 1 day? = ______________________
l 24 cm of 6 m? = ______________________

QUESTION 2 What percentage is the first quantity of the second quantity? Write your answer as a fraction where necessary.

a 12 L, 240 L = ______________________
b 240 g, 8 kg = ______________________
c 12 m, 72 m = ______________________
d 100 cm, 8 m = ______________________
e 20 kg, 180 kg = ______________________
f $1.25, $15 = ______________________
g 50 mL, 300 mL = ______________________
h 375 mL, 15 L = ______________________
i 20 s, 12 min = ______________________
j $2.50, $5 = ______________________
k 6 h, 5 days = ______________________
l $31, 200 cents = ______________________

QUESTION 3 What percentage is:

a 25 cents of $5? = ______________________
b 200 g of 10 kg? = ______________________
c 10 min of 8 h? = ______________________
d $10 of 1500 cents? = ______________________
e 250 m of 6 km? = ______________________
f 64 kg of 5 t? = ______________________
g 20 cents of $5? = ______________________
h 12 min of 9 h? = ______________________
i 15 mm of 90 cm? = ______________________
j 12 mm of 3 m? = ______________________
k 18 cm of 360 cm = ______________________
l 200 g of 4 kg? = ______________________

QUESTION 4 Work out the following.

a What percentage is 20 seconds of 3 minutes?

____

b What percentage is 80 of 248?

____

c Express 16 grams as a percentage of 640 grams.

____

d Interest charges are $320 on a loan of $6400. What percentage is the interest on the loan?

____

UNIT 7: Unitary method and percentages

Excel Mathematics Study Guide Year 8
Pages 26–41

QUESTION **1** Work out the following.

a If 10% of a number is 68, what is the number? __________

b If 20% of a number is 230, what is the number? __________

c Find the total cost if 40% of it is $720. __________

d If 60% of a number is 800, what is the number? __________

e If 50% of a number is 80, what is the number? __________

QUESTION **2** Use the unitary method to solve the following.

a 70% of a number is 54. What is the number? __________

b 60% of a number is 360. What is the number? __________

c 35% of a number is 28. What is the number? __________

d 48% of a number is 152. What is the number? __________

e 85% of a number is 69. What is the number? __________

QUESTION **3** Find 100% of these numbers if percentage parts are given below.

a 2% is 20 __________

b 15% is 98 __________

c 5% is 40 __________

d 35% is 80 __________

e 30% is 80 __________

f $12\frac{1}{2}$% is 960 __________

g 35% is 98 __________

h $33\frac{1}{3}$% is 1920 __________

i 40% is 300 __________

j 55% is 680 __________

k 10% is 140 __________

l 20% is 75 __________

QUESTION **4** Find answers to the following.

a John spends $110 per week on food. If he spends 30% of his income on food, what is his weekly income?

____

b I spent 30% of my allowance on a movie which cost me $25.50. How much is my allowance?

____

c If $66\frac{2}{3}$% of a number is 83, find the number.

____

d Lisa's income increased by 8%. If her income rose by $880, find her previous income.

____

Percentages

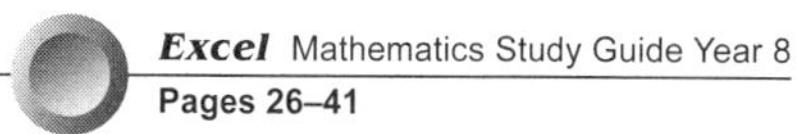

UNIT 8: Commission

QUESTION 1 Michael works for a company and gets a commission of 7% on all sales. How much commission will he receive in a week in which his sales total \$65 000?

QUESTION 2 Janelle is a car sales representative and is paid a retainer (basic wage) of \$800 per week and a commission of 9% on all sales made. Find her weekly income in a week in which she sells a car to the value of:

a \$65 000

b \$83 000

QUESTION 3 Beatrice is a sales person and earns \$850 a week plus $8\frac{1}{2}$% commission on sales. Her weekly sales total is \$98 000. Find:

a her commission

b her total earnings for the week

QUESTION 4 Guy is a real estate agent and receives 3% commission on the first \$300 000 and 2% on the value thereafter. Find his commission for selling a property worth \$850 000.

QUESTION 5 Charlotte sells cosmetics and is paid 10.5% commission on all sales.

a What is her income if she sells \$9570 worth of cosmetics in a week?

b How much would she need to sell (to the nearest dollar) to have an income of \$2350? ____________

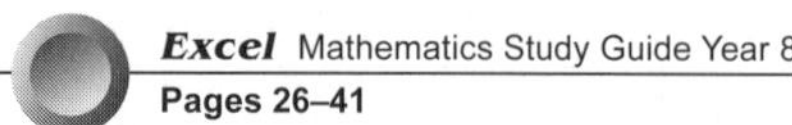

UNIT 9: Discounts

QUESTION 1 Complete the following table.

	Name of the item bought	Marked price	Percentage discount	Discount in dollars	Sale price
a	Chair	$890	25%		
b	Calculator	$38	15%		
c	Text book	$74	5%		
d	Pens	$325	10%		
e	Computer	$1890	20%		
f	Study lamp	$65	12%		

QUESTION 2 A study desk with a marked price of $965 is discounted by 15%, find:

a the discount

b the sale price

QUESTION 3 Claudia spent $865 after getting a discount of $115 on a new watch.

a What was the price of the watch before the discount?

b What percentage discount was given?

QUESTION 4 A track suit that generally costs $430 is reduced by 20%. Find:

a the discount

b the sales price

QUESTION 5 Steph paid $225 after receiving a discount of 18% on a pair of shoes.

a How much did the shoes cost before the discount?

b How much discount did she get?

QUESTION 6 A dinner set was discounted by 55% and sold for $1237.

a How much did the set cost before the discount?

b How much was the discount?

Percentages

UNIT 10: Profit and loss

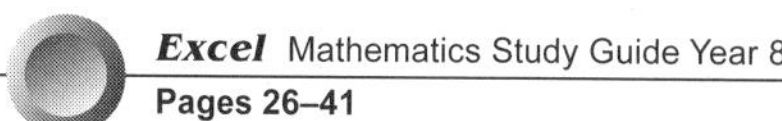

QUESTION 1 James bought a computer for $1400 and 2 months later sold it for $900. Find:

a his loss on the sale

b his loss as a percentage of the cost price

c his loss as a percentage of the selling price

QUESTION 2 Annalese bought a car for $30 000 and sold it to Robert at a loss of 45% after it was involved in an accident. Robert repaired it and then sold it, making a profit of 50% on his purchase price.

a How much did Robert pay for the car?

b For how much did Robert sell the car?

QUESTION 3 A book is sold for $75 including 10% GST.

a How much GST is charged?

b What is the selling price of the book without GST?

QUESTION 4 A bicycle is on sale for $680. This is 65% of the regular price. What is the regular price?

Percentages

UNIT 11: Problem solving with percentages

QUESTION 1 There are 40 students in a class and 30% are absent. How many are present?

QUESTION 2 Find 8% of 2 kg, giving the answer in grams.

QUESTION 3 If 7% of a number is 133, what is the number?

QUESTION 4 Sarah pays 20% of her total weekly income in tax. If she receives $420 income per week, find:

a Sarah's total weekly tax. ______

b how much money Sarah receives after the tax has been taken out. ______

QUESTION 5 Increase $800 by 8% and then decrease the result by 8%.

QUESTION 6 Find the simple interest on $350 for 2 years at a rate of 9% p.a.

QUESTION 7 A book was bought for $2.25 and sold for $2.40. What is the percentage gain?

QUESTION 8 Add 125% of $84 to 21% of $200.

QUESTION 9 21 out of 35 students play tennis. What percentage of the class play tennis?

QUESTION 10 I spend 20% of my money to buy a car for $8600. How much did I have before buying the car?

QUESTION 11 A block of land bought for $20 600 increased in value 12%. What is the new value?

QUESTION 12 If after a $12\frac{1}{2}$% pay rise I receive $480, how much did I earn before the pay rise?

QUESTION 13 If 40% of a number is 120, find the number.

QUESTION 14 Find the simple interest on $3000 for 2 years at $6\frac{1}{4}$% p.a.

Percentages

TOPIC TEST

PART A

Instructions
- This part consists of 10 multiple-choice questions.
- Fill in only ONE CIRCLE for each question.
- Each question is worth 1 mark.
- Calculators are allowed.

Time allowed: 15 minutes **Total marks: 10**

Marks

1 As a decimal $4\frac{1}{2}\%$ equals

Ⓐ 0.42 Ⓑ 0.45 Ⓒ 0.042 Ⓓ 0.045 [1]

2 15% of 1.5 tonnnes equals

Ⓐ 22.5 kg Ⓑ 100 kg Ⓒ 225 kg Ⓓ 1000 kg [1]

3 A toy was reduced in price from $40 to $36. What is the percentage discount?

Ⓐ 4% Ⓑ 10% Ⓒ 11% Ⓓ 96% [1]

4 What would John pay for a CD player that sells for $550 but is discounted by 15%?

Ⓐ $82.50 Ⓑ $467.50 Ⓒ $535 Ⓓ $632.50 [1]

5 To increase 80 by 13% you evaluate

Ⓐ $80 + 1.13$ Ⓑ 80×0.13 Ⓒ 80×1.13 Ⓓ $80 + \frac{13}{100}$ [1]

6 Express 7% as a decimal.

Ⓐ 7.0 Ⓑ 0.7 Ⓒ 0.07 Ⓓ none of these [1]

7 In a test a student gets 10 questions wrong out of 25. What percentage of questions are correct?

Ⓐ 15% Ⓑ 40% Ⓒ 60% Ⓓ 90% [1]

8 After 20% of a bill is paid, $60 still remains to be paid. How much was the total bill?

Ⓐ $12 Ⓑ $63 Ⓒ $72 Ⓓ $75 [1]

9 $\frac{8}{20}$ equals

Ⓐ 8% Ⓑ 16% Ⓒ 40% Ⓓ 80% [1]

10 A concrete mixture contains 1 part cement, 2 parts sand and 2 parts gravel. What percentage of the mixture is sand?

Ⓐ 20% Ⓑ $33\frac{1}{3}\%$ Ⓒ 40% Ⓓ $66\frac{2}{3}\%$ [1]

Total marks achieved for PART A /10

Percentages

TOPIC TEST PART B

Instructions
- This part consists of 15 questions.
- Each question is worth 1 mark.
- Write only the answer in the answer column.

Time allowed: 15 minutes **Total marks: 15**

	Questions	Answers	Marks
1	Express 910% as a fraction.		1
2	What percentage is 5 of 30?		1
3	If 150 is 20% of a number, what is the number?		1
4	Find 31% of 300.		1
5	Increase 180 kg by 10%.		1
6	Decrease 860 by 5%.		1
7	Write $\frac{3}{5}$ as a percentage.		1
8	Write 0.74 as a percentage.		1
9	If 4% of an amount is 15, what is the amount?		1
10	Increase 100 by 8% and decrease the result by 8%.		1
11	If 5% = 5 then 12% =		1
12	Write the simplest fraction for 88%.		1
13	Find $33\frac{1}{3}$% of 270.		1
14	Change 66% to a fraction.		1
15	Find 100% given that 12% is 108.		1

Total marks achieved for PART B /15

Chapter 5

Basic algebra

UNIT 1: Algebraic expressions

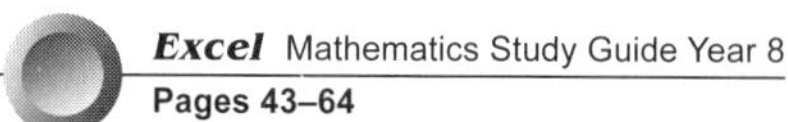
Excel Mathematics Study Guide Year 8
Pages 43–64

Question 1 Write an expression for the following.

a The sum of $5a$ and $7b$ ______

b 12 less than $2x$ ______

c The double of $5x$ ______

d The difference between x and y ______

e Two thirds of $3y$ ______

Question 2 Write an algebraic expression for the following.

a Four times the number y minus 6 ______

b The sum of $3a$ and $2b$ ______

c The sum of $2x$, y and z ______

d The product of $8a$ and $2b$ ______

e The average of $9a$ and $11a$ ______

Question 3 If x represents any number, write an algebraic expression for the following.

a Five times the number plus 11 ______

b One fifth of the number plus 9 ______

c The sum of the number and 35 ______

d The difference between the number and 7 ______

e Nine more than the number ______

Question 4 Show the sum of each of the following.

a 16 and x ______

b $2a$ and $8b$ ______

c $5a$, $6b$ and $9c$ ______

d $8x$, $12y$ and $5z$ ______

e $8m$, $12n$ and $11p$ ______

Question 5 Write algebraic expressions for the following.

a To the sum of $8a$ and $7b$, add $6y$ ______

b From the product of $2x$ and $8y$, take away 18 ______

c Divide $5a$ by $7b$ and then add 19 ______

d Divide the sum of x and y by 20 ______

e 16 plus x, all divided by 8 ______

Basic algebra

UNIT 2: Simplifying algebraic expressions

QUESTION 1 Write down the coefficient for each of the following terms.

a $6x$ ______ **b** $49m$ ______ **c** $116t$ ______

d $124x$ ______ **e** $114y$ ______ **f** $154n$ ______

g $96q$ ______ **h** $85y$ ______ **i** $82a$ ______

j $141p$ ______ **k** $45l$ ______ **l** $8m$ ______

QUESTION 2 Circle the like terms.

a $8a, 4b, 16a$ ______ **b** $6x, 6y, 9x$ ______ **c** $7m, 8n, 8x$ ______

d $6c, 8d, 9c$ ______ **e** $14y, 8a, 6y$ ______ **f** $5l, 4m, 6l$ ______

g $8a, 4c, 6c$ ______ **h** $4x, 5a, 6a$ ______ **i** $18y, y, 13x$ ______

QUESTION 3 Simplify the following.

a $15a + 7a =$ ______ **b** $y + y + y + y + y =$ ______ **c** $m + m + m + m + m =$ ______

d $x + x + x + x + x =$ ______ **e** $a + a + a + a + a =$ ______ **f** $d + d + d + d =$ ______

g $20d + 6d =$ ______ **h** $9k + 6k + k =$ ______ **i** $14a + 15a =$ ______

QUESTION 4 Simplify the following expressions.

a $16x - 8x =$ ______ **b** $28b - 19b =$ ______ **c** $85x - 12x =$ ______

d $20a - 8a =$ ______ **e** $49x - 6x =$ ______ **f** $18a - 6a =$ ______

g $12y - 6y =$ ______ **h** $14c - 8c =$ ______ **i** $21y - 2y =$ ______

QUESTION 5 Simplify the following.

a $19 \times a =$ ______ **b** $6 \times a \times b =$ ______ **c** $x \times y \times z =$ ______

d $16 \times x \times y =$ ______ **e** $42 \times m \times n \times l =$ ______ **f** $8 \times m \times n =$ ______

g $a \times b \times 16 =$ ______ **h** $4 \times 6 \times m \times n =$ ______ **i** $44 \times a \times d \times e =$ ______

QUESTION 6 Write the following expressions in expanded form.

a $28ab =$ ______ **b** $8abc =$ ______

c $9xy =$ ______ **d** $71mn =$ ______

e $19abc =$ ______ **f** $48m^2 =$ ______

g $12xyz =$ ______ **h** $19a^2b =$ ______

i $44mnp =$ ______ **j** $8a^3 =$ ______

k $6ace =$ ______ **l** $47a^2b^2 =$ ______

m $46x =$ ______ **n** $28x^2y^3 =$ ______

Basic algebra

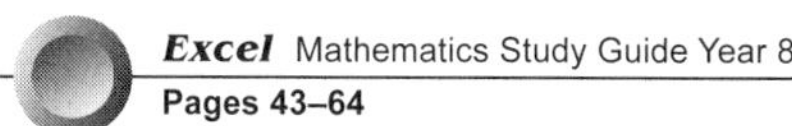

UNIT 3: Collecting like terms

Question 1 Collect like terms to simplify the following.

a $a + a + a + a + a =$ __________

b $x + x + x + y + y + y + y + y + y =$ __________

c $x + x + x + x + y + y + y =$ __________

d $l + l + l + l + l + m + m + m =$ __________

e $m + m + m + n + n + n + n =$ __________

f $k + k + k + m + m + m + m =$ __________

Question 2 Simplify the following by collecting like terms.

a $5x + 7y + 9x + 6x + 2y =$ __________

b $14a + 17 + 5a + 12 + 8 =$ __________

c $2a + 5b + 3a + 7b + 6a + 3b =$ __________

d $7x + 3y + 8y + 9x =$ __________

e $6m + 4n + 8m + 3n =$ __________

f $3a + 4b + 8a + 11b + 5a + 6b =$ __________

Question 3 Simplify the following.

a $46x - 9x - 3x + 2y =$ __________

b $40x + 9y - 8x + 7y + 3y =$ __________

c $19p + 3l + 8p - l + 2p =$ __________

d $9m - 4n + 12m + 6n + 8m =$ __________

e $15y + 31x + 4y + 8x + 2y =$ __________

f $18xyz - 14xyz + 8x - 9y =$ __________

g $20a - 9a + 6b + 10b + b + 6a =$ __________

h $19a + 8b + 15a - 5b - 2b =$ __________

Question 4 Simplify the following.

a $10m - 2m - m + 5n - 3n =$ __________

b $8p - 5p + 3p - 4p + 3q =$ __________

c $9y^2 + 7y^2 - 8y^2 + x^2 + 5x^2 =$ __________

d $12x - 7x - 3x - 2y + 7y =$ __________

e $9a - 7b + 3a - a - 6b =$ __________

f $y + 5y - 2y + 8x - 3x =$ __________

g $10x + 5x + 2x - 7x + 9y =$ __________

h $11d - d - 5d - 3d + 8c - 9c =$ __________

Question 5 Simplify the following.

a $9x - x - 4x - 10x =$ __________

b $a^2 - 7a - a + 15 + 18 - 3a^2 =$ __________

c $7y - 3x - y - 4y =$ __________

d $p^3 + 3p^2 - p^2 - 10 + 4p^3 + 25 =$ __________

e $15a + 2b - b - 7a - 9a + 7b =$ __________

f $5x + 10y + 9x - 2y - 9y =$ __________

g $ab + 4b + 3ab + 9b =$ __________

h $2xy + 4x - xy + 2x - 9x =$ __________

Question 6 Collect like terms and simplify.

a $xy + yz - xy - yz + 5xz =$ __________

b $15m^2n - 6mn^2 + 7mn^2 + 8m^2n =$ __________

c $5a^2 + 15a^2 - 4a^2 + 8a^2 =$ __________

d $18xyz + 8x + 5xyz - 4x^2 =$ __________

e $20xyz + 12xyz - 5xyz + 15x - 7x =$ __________

f $25x^2 + 17y^2 - 19x^2 =$ __________

g $9x^2y^2 + 8x^2 + 2x^2y^2 + 9x^2 =$ __________

h $9a^3 + 7a^3 - 4b^3 + 8b^3 - 6b^3 - 3a^3 =$ __________

Basic algebra

UNIT 4: Algebraic abbreviations

Question 1

a Write $18 \times y$ in a shorter way. ______

b Write $12 \times 4 \times a$ in a shorter way. ______

c Write $15 \times m \times m \times m$ in another way. ______

d Write $6x^2y$ showing multiplication signs. ______

e What is the difference in the meaning of $24y$ and $2 \times 4 \times y$? ______

Question 2 Write the following expressions without multiplication or division signs.

a $23 \times 5 \times a =$ ______
b $15 \times n \times 9 =$ ______
c $2a + 7 \times b =$ ______
d $14 \times (x + 16) =$ ______
e $13x \div 5 =$ ______
f $a \times b \times c =$ ______
g $27 \times a \times 6 =$ ______
h $18a \times b \times b =$ ______
i $14 \times m \times 5 \times n =$ ______
j $12 \times m + 5 =$ ______
k $9 + 14 \times x =$ ______
l $13 \times a + 4 \times b =$ ______
m $23 \div xy =$ ______
n $6a \div 7b =$ ______
o $3m \div 10 =$ ______
p $(m + n) \div 6 =$ ______
q $15x \div (3a + 2) =$ ______
r $(8x + 3) \div 2a =$ ______

Question 3 Write the following expressions by showing all multiplication or division signs.

a $19a =$ ______
b $13y =$ ______
c $42a =$ ______
d $14x - 5 =$ ______
e $8m + 32 =$ ______
f $60 - 8x =$ ______
g $ab + 36 =$ ______
h $12x - 3y =$ ______
i $a^3b^2c^2 =$ ______
j $14xyz =$ ______
k $17m^2 + 42 =$ ______
l $x^2 - y^2 =$ ______

Question 4 Write the following in a shortened form.

a $6 \times a \times a \times a \times b =$ ______
b $42 \times a \times a \times b \times b =$ ______
c $5 \times (x + 9) =$ ______
d $15 \times (m - 5) =$ ______
e $7 \times a \times (c + 2) =$ ______
f $4 \times 7 \times (x - 2) =$ ______
g $m \times n \times (p + 3) =$ ______
h $6 \times (5 \times a + 9) =$ ______
i $18 + (4 \times 5 + 4 \times m) =$ ______
j $8 \times (6 \times a - 2 \times b) =$ ______
k $(3a + 2) \times (a + 5) =$ ______
l $(4a + 3) \times (2a - 5) =$ ______
m $n \times n \times 7 \times n \times n =$ ______
n $6 \times a \times a \times a \times 3 =$ ______
o $x \times x \times x \times y \times y \times y =$ ______
p $(3 \times m + 2) \times (5 \times m - 3) =$ ______
q $(a + 2) \times (5a + 7) =$ ______
r $(a - 3) \times (4a - 7) =$ ______

Basic algebra

UNIT 5: Addition and subtraction of like terms

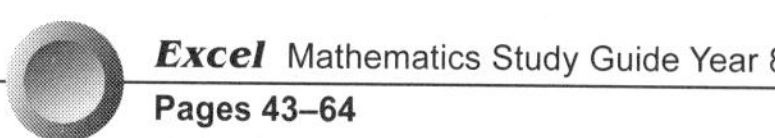

QUESTION 1 Add the following expressions.

a $3x + 8x =$ ______

b $5x + 11x =$ ______

c $9x + 7x =$ ______

d $18x + 9x =$ ______

e $16x + 25x =$ ______

f $30x + 14x + x =$ ______

g $6xy + 8xy =$ ______

h $9mn + 11nm =$ ______

i $15a + 16a + 7a =$ ______

j $8x + 3x + 7x =$ ______

k $15a^2 + 26a^2 + a^2 =$ ______

l $7p + 6p + 5p + p =$ ______

QUESTION 2 Subtract the following expressions.

a $6a - 2a =$ ______

b $15a - 7a =$ ______

c $8a - 3a =$ ______

d $17x - 2x =$ ______

e $14m - 6m - m =$ ______

f $8y - 3y - y =$ ______

g $15x - 7x - 2x =$ ______

h $6xy - 3xy - xy =$ ______

i $25x^2 + 7x^2 - 4x^2 - 6x^2 =$ ______

j $14y - 3y - y - 2y =$ ______

k $8x - 2x - x =$ ______

l $19p - 3p - 5p =$ ______

QUESTION 3 Simplify the following expressions by adding or subtracting.

a $8a + 7a - 3a =$ ______

b $5x + 6x - x - 2x =$ ______

c $5a + 15a - 6a =$ ______

d $8a + 5a + 3a - 2a =$ ______

e $19xy + 2xy - xy =$ ______

f $7mn + 3nm - 2mn =$ ______

g $9x^2 + 8x^2 - 3x^2 - 5x^2 =$ ______

h $9a^2 + 7a^2 - a^2 - 2a^2 =$ ______

i $5x - 2x + 3x - x =$ ______

j $8t + 6t + 3t - t =$ ______

k $6m + 7m - 2m - m =$ ______

l $8a + 16a - 5a - 7a =$ ______

QUESTION 4 Simplify the following.

a $9x + 3y - 6x + 2y =$ ______

b $20a + 3b + 8b - 11a =$ ______

c $6m + 3n - n - m =$ ______

d $6x^2 - 4x^2 - x^2 + 3x^2 =$ ______

e $8a + 3b + 2b - 6a =$ ______

f $10a + 4a - 5b - b =$ ______

g $7y + 8y - 3x - 4x =$ ______

h $6xy + 3yx - xy - 2yx =$ ______

QUESTION 5 Simplify the following expressions.

a $15 - 8x + 7 - 2x =$ ______

b $8m + 3n - 5m =$ ______

c $18x - 3y + y + x =$ ______

d $2x^2 + 9x^2 - 7y^2 =$ ______

e $10m + 3n - 2m - n =$ ______

f $19xy + 6yz - 13yx =$ ______

g $12a + 6b - 3a =$ ______

h $6a + 9b - 3b - 3a =$ ______

i $9t + 18 - 3t - 15 =$ ______

j $6ab + 3ba + 9ab =$ ______

Basic algebra

UNIT 6: Multiplication of pronumerals

Excel Mathematics Study Guide Year 8
Pages 43–64

QUESTION **1** Multiply the following expressions.

a $2 \times 5a =$ ______________

b $3 \times 6b =$ ______________

c $8 \times 9y =$ ______________

d $7 \times 3y =$ ______________

e $(-6) \times ab =$ ______________

f $(-3) \times (-2a) =$ ______________

g $3a \times (-b) =$ ______________

h $(-2a) \times (-3b) =$ ______________

i $(-6x) \times (-3) =$ ______________

j $(9ab) \times (-5a) =$ ______________

k $(-2a) \times (-8a) =$ ______________

l $(-6x) \times (-2x) =$ ______________

QUESTION **2** Find the products of the following.

a $6a \times 3b =$ ______________

b $(-6x) \times (-x) =$ ______________

c $5x \times 8y =$ ______________

d $(-8y) \times (-y) \times 3 =$ ______________

e $9a \times 3b =$ ______________

f $6a \times (-a) \times 5 =$ ______________

g $6y \times 8y =$ ______________

h $t \times (-3t) \times (-2) =$ ______________

i $9x \times 3y \times 2x =$ ______________

j $5a \times (-6ab) =$ ______________

QUESTION **3** Work out the following products.

a $2a \times ab =$ ______________

b $3 \times 6a \times 5 =$ ______________

c $5a \times 6bc \times a =$ ______________

d $ab \times bc \times ca =$ ______________

e $9ab \times (-6bc) =$ ______________

f $6a \times 7ap =$ ______________

g $(-8x) \times 5 =$ ______________

h $3x \times 2y \times y =$ ______________

QUESTION **4** Find the following products.

a $8a \times 5b =$ ______________

b $(-7k) \times (-2k) =$ ______________

c $9x \times (-8y) \times xy =$ ______________

d $8ax \times a \times ax =$ ______________

e $(-6m) \times (-6m) =$ ______________

f $-6t \times 4m \times (-m) =$ ______________

g $y \times -y \times -3y =$ ______________

h $(-5x) \times (-2) \times (-3x) =$ ______________

QUESTION **5** Simplify the following expressions.

a $8ab \times (-ab) =$ ______________

b $7 \times 3a \times (-4a) =$ ______________

c $9x \times y \times (-2x) =$ ______________

d $2a \times 3b \times 5c =$ ______________

e $(-a) \times (-2a) \times (-2b) =$ ______________

f $(-x) \times (-y) \times (-x) =$ ______________

g $(-5p) \times (-5p) =$ ______________

h $6a \times (-3a) \times 2 =$ ______________

UNIT 7: Division of pronumerals

QUESTION 1 Work out the following divisions.

a $\frac{18a}{6} =$ ______ b $\frac{9pq}{3} =$ ______ c $\frac{12m}{15m^2} =$ ______

d $\frac{15x}{5} =$ ______ e $\frac{21xy}{7x} =$ ______ f $\frac{18a^2}{3a} =$ ______

g $\frac{12a}{6} =$ ______ h $\frac{36x^2}{4x} =$ ______ i $\frac{30y}{10y^2} =$ ______

QUESTION 2 Find the following divisions.

a $15x \div 5x =$ ______ b $-36xy \div -9x =$ ______

c $12mn \div 4m =$ ______ d $64m^2 \div 8m =$ ______

e $5x \div -5 =$ ______ f $(-48ab) \div (-6a) =$ ______

g $8x \div 2x =$ ______ h $72a^2 \div 9a =$ ______

i $32xy \div 8x =$ ______ j $15ab \div (-5a) =$ ______

k $27ab \div 9a =$ ______ l $16x^2y \div 4xy =$ ______

QUESTION 3 Divide the following algebraic expressions.

a $(-24)x \div 8 =$ ______ b $(-6ab) \div 3a =$ ______

c $38x^2y \div 19x^2 =$ ______ d $(-15x^2) \div (-5x) =$ ______

e $27x^2y^2 \div 9x^2y =$ ______ f $(-16k) \div (-4k) =$ ______

g $(-52x^2) \div (-13) =$ ______ h $(-36y) \div (-3) =$ ______

i $20pq \div -4p =$ ______ j $50mn \div 5m =$ ______

QUESTION 4 Simplify the following.

a $28x \div -7x =$ ______ b $68xy \div 4xy =$ ______

c $32x^2 \div (-4x) =$ ______ d $(-18x) \div 9x =$ ______

e $(-30x) \div (-15) =$ ______ f $(-48m^3) \div (-8m^2) =$ ______

g $(-25y^2) \div (-5y) =$ ______ h $(-64ab) \div (-8a) =$ ______

QUESTION 5 Find the answers to the following.

a $24ab \div (-6a) =$ ______ b $(-24x^3y^2) \div (-6x^2y^2) =$ ______

c $15ab \div -15ab =$ ______ d $36xy \div 4x =$ ______

e $mn^2 \div mn \div n =$ ______ f $15a^2b^2 \div 15abc =$ ______

g $3abc \div bc \div ba =$ ______ h $(-16x^2y) \div (-8xy) =$ ______

Basic algebra

UNIT 8: Multiplication and division of pronumerals

Question 1 Multiply the following.

a $8a \times 2a =$ ______

b $5a \times 4a \times 3c =$ ______

c $9x \times (-3x) =$ ______

d $6a \times 3a \times (-2a) =$ ______

e $15xy \times 3x^2 =$ ______

f $5y \times (-3y) \times (-y) =$ ______

g $8a \times 3b \times 4a =$ ______

h $8 \times 9a \times a^2 =$ ______

Question 2 Divide the following expressions.

a $15a \div 3a =$ ______

b $(-72a) \div (-8a) =$ ______

c $8pq \div (-4p) =$ ______

d $35x^2 \div -5x =$ ______

e $16ab \div (-4a) =$ ______

f $28a^2b \div (-7ab) =$ ______

g $20p \div 5 =$ ______

h $64x^2y^2z^2 \div 16x^2yz =$ ______

Question 3 Work out the following divisions.

a $(-27x^2) \div (3x^2) =$ ______

b $15ab \div (-3a^2) =$ ______

c $16xy \div (-16xy) =$ ______

d $35a^2b^2c^2 \div 5abc =$ ______

e $24ab \div 4a \div 2b =$ ______

f $56t^3 \div 8t^2 =$ ______

g $48x^2y^2 \div 4xy \div 3x =$ ______

h $10ab \div -2a =$ ______

i $15xy \div 6xy =$ ______

j $15abc \div -3bc =$ ______

Question 4 Simplify the following expressions.

a $5a \times -3a =$ ______

b $4a \times 5b \times 2c =$ ______

c $8x \times (-7x) =$ ______

d $-8a \times 3a \times -2a =$ ______

e $15a \times (-2ab) =$ ______

f $10x \times 2y \times (-2x) =$ ______

g $18a^2 \times 2ab^2 =$ ______

h $ab \times a^2b^2 \times -ab =$ ______

i $8x \times y \times y^2 =$ ______

j $-5x \times 3xy =$ ______

Question 5 Simplify the following.

a $\dfrac{8x \times 9y}{12xy} =$ ______

______

b $\dfrac{15a \times 8b}{10ab} =$ ______

______

c $\dfrac{8a \times 5b}{3a \times 4b} =$ ______

______

d $\dfrac{(6ab)^2 \times ab}{9a^3b} =$ ______

______

e $\dfrac{(3a)^2 \times (4b)^2}{24ab} =$ ______

______

f $\dfrac{(-2x) \times (-3) \times (-5x)}{-6x \times -5} =$ ______

______

Basic algebra

UNIT 9: Algebraic expressions and the order of operations

QUESTION **1** Simplify the following.

a $12x \times 5x \times 2$

b $6a \times 12ab \div a^2$

c $6mn \times 5m \div n$

d $3a \times 12a \times 5b$

e $-4a \times 8a \times -3$

f $18xyz \div 6xy \div z$

g $42x \times 7 \div 14x$

h $20 \times 8m \div 15$

i $5 \times 8mn \div 4n$

QUESTION **2** Find answers to the following.

a $(6x)^2 \div 3x \times 2$

b $(-5) \times (-8x) \times 8$

c $18ab \div 9a \div 8b$

d $35ab \div -35ab$

e $(8ab)^2 \div 16a^2b^2$

f $9m \times -5 \times (-2m)$

g $-5a \times 8b \times -4ab$

h $56xy \div 7x \div 4y$

i $20a \times 3a \div 5a$

QUESTION **3** Use the rules for the order of operations to simplify the following.

a $24y \div 6 - 4y$

b $10x + 6 \times 3x$

c $-12n - 6n \times 2$

d $15 \times 8p - 35p \div 7$

e $42mn \div 6m - 8n$

f $-6x \times 8 - (4x + 8x)$

g $35t - 8 \times 4t - 6t$

h $16a \div 4 - 28a \div 7$

i $15a \div (8a - 3a)$

QUESTION **4** Simplify the following.

a $\dfrac{6x + 18x}{2x \times 3}$

b $\dfrac{96a \div 16}{28a^2 \div 7a}$

c $\dfrac{42x^2 \div 7x}{2x + 4x}$

d $\dfrac{9y \times 6y}{12y^2 \div 2y}$

e $\dfrac{18n \times 4n}{(2n)^2}$

f $\dfrac{48a - -16a}{3a + 5a}$

g $5x + (9 \times 5x) - 4x$

h $15a - (15a \div 5) - 2a$

i $(14y + 2 \times 8y) \div 5y$

Basic algebra

UNIT 10: Grouping symbols

QUESTION 1 Expand the following expressions.

a $3(x+y)$ ______ b $5(2a+3b)$ ______ c $4(2a+3)$ ______

d $6(8a-7b)$ ______ e $8(m+3p)$ ______ f $3(5a-4m)$ ______

g $5x(x+8)$ ______ h $7(x-5)$ ______ i $9a(a-2b)$ ______

QUESTION 2 Write these expressions without grouping symbols.

a $-3(x+2)$ ______ b $-5x(2x+1)$ ______ c $-2(2x-7)$ ______

d $-2x(8x-3y)$ ______ e $-3(4x-5)$ ______ f $-y(x+y^2)$ ______

g $-8(8x-9)$ ______ h $-m(3m+8)$ ______ i $-2x(x+1)$ ______

QUESTION 3 Expand the following expressions.

a $5(2x+7)$ ______ b $7(3x-11)$ ______ c $-3(4x-9)$ ______

d $-4(2x+5)$ ______ e $8(6y+3)$ ______ f $-3a(5a+5)$ ______

g $-x(2x+7)$ ______ h $-m(8m-5)$ ______ i $-2t(3t+5)$ ______

QUESTION 4 Expand and simplify.

a $2(x+3)+5x$ b $4(y+3)+7y+5$ c $5(xy-7)-3xy$

d $9(p-7)-8p-7$ e $4y(y+3)-2y^2$ f $8x-3(x-2)$

g $9p-3(8-p)$ h $5m+4(m-4)$ i $9m(2m+1)-10m^2$

QUESTION 5 Expand and simplify.

a $4(m+3)+2(m+1)$ b $5(x+3)+3(x-1)$ c $8(2x+7)+3(x+2)$

d $8(2x+7)-4(2x-5)$ e $5(y+7)-2(y-1)$ f $9(3x+1)-5(2x-3)$

g $7(n-4)-2(n+2)$ h $x(2x+3)-3(x+1)$ i $6(x+8)-3(x-2)$

Basic algebra

UNIT 11: Substitution (1)

QUESTION 1 If $a = 5$, find the value of each expression.

a $7a =$ ______
b $9a + 1 =$ ______
c $9a - 1 =$ ______
d $a^2 =$ ______
e $5a^2 =$ ______
f $(a + 1)(a - 1) =$ ______
g $a^3 =$ ______
h $a^2 - 2a =$ ______
i $(2a + 7) \times 3a =$ ______
j $7a - 12 =$ ______
k $180 \div 6a =$ ______
l $(a - 2) \div 15 =$ ______

QUESTION 2 If $a = 2$, $b = 4$, $c = 5$ and $d = 6$, find the value of each of the following expressions.

a $a + b =$ ______
b $a + b + c + d =$ ______
c $abc - d^2 =$ ______
d $abc =$ ______
e $2a + 3b + 4c =$ ______
f $3b + 5c - d =$ ______
g $16bc =$ ______
h $8c - d =$ ______
i $8d \div b =$ ______
j $5a - a^2 =$ ______
k $(c + d) - a^2 =$ ______
l $b^2 + d^2 - abc =$ ______

QUESTION 3 Complete the following.

a $b = 2a + 3$

a	0	1	2	3	4
b					

b $y = 4x - 5$

x	1	2	3	4
y				

c $m = 6l - 3$

l	0	1	2	3	4
m					

d $y = 4x - 1$

x	1	2	3	4	5	6
y						

e $p = 3q + 3$

q	0	1	2	3	4
p					

f $v = 3u - 7$

u	1	2	3	4	5	6
v						

QUESTION 4 Complete the table.

	x	y	$x + y$	$x - y$	xy	x^2	y^2	$4x + 5y$
a	2	3						
b	4	6						
c	5	4						
d	3	8						
e	6	7						
f	5	9						
g	7	5						
h	6	8						
i	3	10						

UNIT 12: Substitution (2)

Question 1 If $m = 7$, $n = 4$ and $p = 12$, evaluate the following.

a $mn =$ ______ **b** $mnp =$ ______

c $mn \div p =$ ______ **d** $m^2n^2 =$ ______

e $mnp^2 =$ ______ **f** $m^2 + n^2 + p^2 =$ ______

g $m + n + p =$ ______ **h** $7m - p =$ ______

i $m^2 - 5p =$ ______ **j** $m^2 + 10 + p =$ ______

k $mn^2p - mn =$ ______ **l** $8m^2 - 12 =$ ______

Question 2 If $a = 4$, find the value of the following expressions.

a $5a + 7 =$ ______ **b** $(a - 5)^2 =$ ______

c $a^2 - 9 =$ ______ **d** $85 - 4a =$ ______

e $3a(a + 4) =$ ______ **f** $5a^2 + 8 =$ ______

g $5(a + 6) =$ ______ **h** $a^3 =$ ______

i $\sqrt{25 - a^2} =$ ______ **j** $3(2a + 7) =$ ______

k $3a^2 =$ ______ **l** $(a + 7)(a - 7) =$ ______

Question 3 If $x = 7.2$, $y = 3.5$ and $z = 6.4$, evaluate each expression correct to 1 decimal place.

a $x + y =$ ______ **b** $x + y + z =$ ______

c $2x + 2y + 2z =$ ______ **d** $x^2 + y^2 - z^2 =$ ______

e $(x + y)^2 =$ ______ **f** $3xyz =$ ______

g $x^3 + x^2 =$ ______ **h** $\sqrt{xy} =$ ______

i $(x - z) \div y =$ ______ **j** $\sqrt{x + y + z} =$ ______

k $(x - z) + 8 =$ ______ **l** $\frac{xy}{z} =$ ______

Question 4 If $a = 3$, $b = 4$, $c = 5$ and $d = 6$, find the value of each expression.

a $6c^2 - ab =$ ______ **b** $a^2 + b^2 + c^2 =$ ______

c $abcd =$ ______ **d** $ab - cd =$ ______

e $b^2 + c^2 - a^2 =$ ______ **f** $b^2 - 5 =$ ______

g $6ab + 7bc =$ ______ **h** $a^2 + b^2 - cd =$ ______

i $3abc \div d =$ ______ **j** $(a + b)^2 - cd =$ ______

Basic algebra

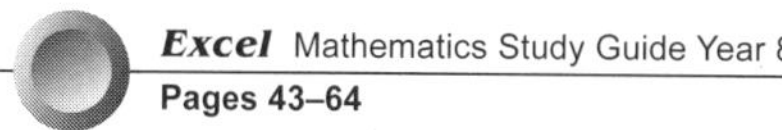
Excel Mathematics Study Guide Year 8
Pages 43–64

UNIT 13: Common factors

QUESTION 1 Factorise the following by taking out the highest common factor.

a $7x + 7 =$ ______ **b** $10a - 2 =$ ______ **c** $7y + 14 =$ ______
d $8x - 4 =$ ______ **e** $3a + 15 =$ ______ **f** $4x - 12y =$ ______
g $3y + 3 =$ ______ **h** $5m + 30 =$ ______ **i** $32x - 28 =$ ______
j $5a - 5 =$ ______ **k** $6a - 2b =$ ______ **l** $5a + 35 =$ ______

QUESTION 2 Factorise the following.

a $y^2 + 3y =$ ______ **b** $6y^3 - 3y^2 =$ ______ **c** $6x^3 - 3x^2 =$ ______
d $2m^2 + 6m =$ ______ **e** $8a^2 - 4a =$ ______ **f** $m^2 - 9m =$ ______
g $8a^2 + 24a =$ ______ **h** $15y^2 - 6y =$ ______ **i** $16a^2 - 8a =$ ______
j $x^2 + 9x =$ ______ **k** $9a^3 - 6a^2 =$ ______ **l** $28x^2 - 7x =$ ______

QUESTION 3 Factorise the following by taking out the highest common factor.

a $-3x - 3 =$ ______ **b** $-4y^2 + 16 =$ ______ **c** $-6n - 3 =$ ______
d $-5x - 40 =$ ______ **e** $-3x + 27 =$ ______ **f** $-7t - 14 =$ ______
g $-2m^2 - 4 =$ ______ **h** $-5m^2 + 35 =$ ______ **i** $-11x^2 - 22 =$ ______
j $-6a^2 + 12 =$ ______ **k** $-8a^2 - 32 =$ ______ **l** $-13y + 39 =$ ______

QUESTION 4 Factorise.

a $-6x^2 - 12x =$ ______ **b** $-16x^2 - 4x =$ ______ **c** $-6x - 9 =$ ______
d $-8xy - 4y =$ ______ **e** $-9x^3 + 27x^2 =$ ______ **f** $-15x^2 - 5xy =$ ______
g $-6mn - 3 =$ ______ **h** $-5x^2y^2 + 25xy =$ ______ **i** $-6mn + 15m^2 =$ ______
j $-8xyz + 4xy =$ ______ **k** $-9x^2 - 15x =$ ______ **l** $-x - 3yx =$ ______

QUESTION 5 Factorise the following.

a $8a + 16b =$ ______ **b** $5a - 25 =$ ______ **c** $8x - 4y =$ ______
d $9a - 9b =$ ______ **e** $8 - 8t =$ ______ **f** $m^3 - 4m^2 =$ ______
g $16ab - a^2 =$ ______ **h** $14m - 7 =$ ______ **i** $16ab - 8a^2b =$ ______
j $15x^2y^2 - 5xy =$ ______ **k** $-6m - 36 =$ ______ **l** $-4m - 12n =$ ______

QUESTION 6 Factorise the following.

a $3a + 3b + 3c =$ ______ **b** $16pq - 7p =$ ______ **c** $-x^2 - 7x =$ ______
d $8x + 8y - 16z =$ ______ **e** $tx - x =$ ______ **f** $15a + 5b - 10c =$ ______
g $4x - 8y =$ ______ **h** $y^3 - y^2 =$ ______ **i** $18x - 9y + 6 =$ ______
j $16x - 32y =$ ______ **k** $15mn - 3n =$ ______ **l** $6a^2 + 3a + 9 =$ ______
m $m^2 - 2mn =$ ______ **n** $8pq - q^2 =$ ______ **o** $-6x - 6xy - 12 =$ ______

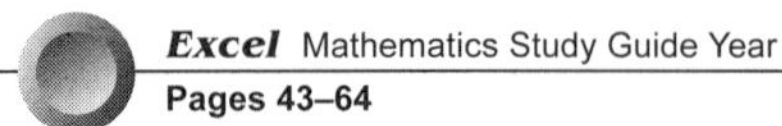

UNIT 14: Problem solving and algebra

QUESTION 1 Find expressions for the perimeter of each of the following shapes.

a

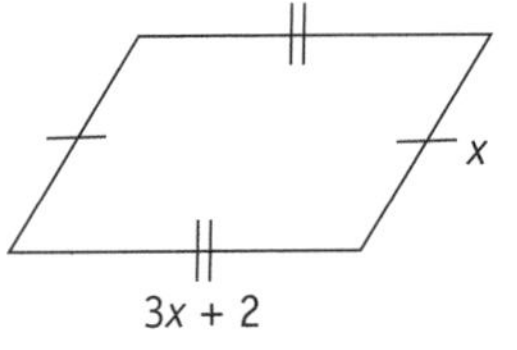

b

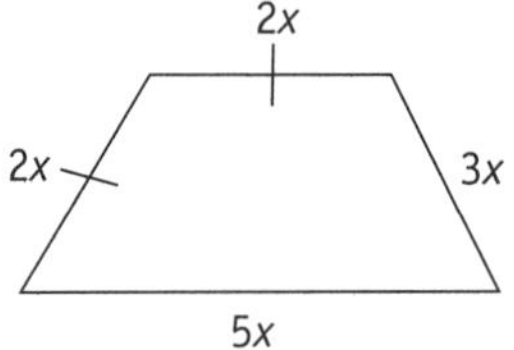

c

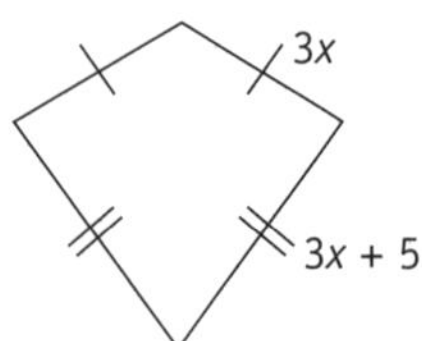

QUESTION 2 Find expressions for the area of each of the following shapes.

a

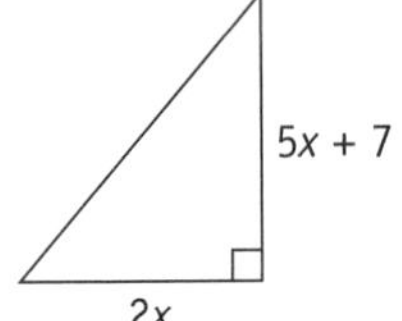

b

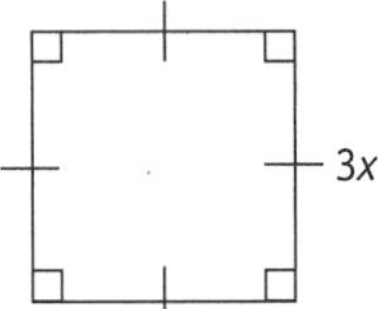

c

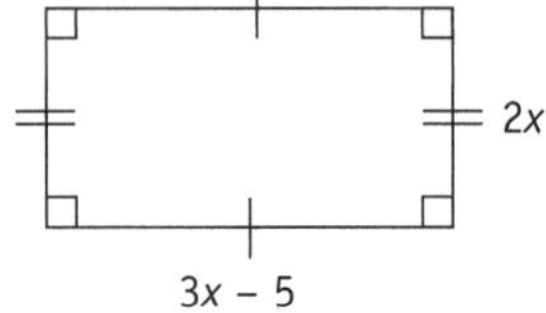

QUESTION 3 Find the sum of $3x$ and $5y$.

QUESTION 4 What is the product of $8m$ and $5n$?

QUESTION 5 Find the average of $6x$, $2x$ and $7x$.

QUESTION 6 Find an expression for 5 more than the sum of 3 and $2x$.

QUESTION 7 Find the area of a square with side length y metres.

QUESTION 8 If the first number is $4x$, write the next consecutive integer.

QUESTION 9 Increase $9x$ by 10 and then decrease this result by $2x$.

QUESTION 10 Find the volume of a cube with side length $2x$ cm.

QUESTION 11 Find the number $(x - 2)$ less than $4x + 5y$.

Basic algebra

TOPIC TEST — PART A

Instructions
- This part consists of 10 multiple-choice questions.
- Fill in only ONE CIRCLE for each question.
- Each question is worth 1 mark.
- Calculators are allowed.

Time allowed: 15 minutes — **Total marks: 10**

Marks

1 If $a = -4$ and $b = 3$ the value of a^2b is — 1

Ⓐ 48 Ⓑ 144 Ⓒ −24 Ⓓ −48

2 $a^3 + a^3 =$ — 1

Ⓐ a^6 Ⓑ a^3 Ⓒ $2a^6$ Ⓓ $2a^3$

3 $2a \times 3a \times 4a =$ — 1

Ⓐ $9a$ Ⓑ $9a^3$ Ⓒ $24a$ Ⓓ $24a^3$

4 $3x - (-x) =$ — 1

Ⓐ 3 Ⓑ $3x$ Ⓒ $3x^2$ Ⓓ $4x$

5 $3y^2 - 2y + 5y + 4y^2 =$ — 1

Ⓐ $7y^2 + 3y$ Ⓑ $7y^2 - 7y$ Ⓒ $7y^3 + 3y$ Ⓓ $7y^4 - 7y$

6 $x(x - 3) =$ — 1

Ⓐ $x^2 - 3$ Ⓑ $-3x$ Ⓒ $x^2 - 3x$ Ⓓ $3x^2$

7 $3(x + 4) - x =$ — 1

Ⓐ 12 Ⓑ 15 Ⓒ $2x + 4$ Ⓓ $2x + 12$

8 $3ab - b + ab - 2b =$ — 1

Ⓐ $2ab - b$ Ⓑ $2ab - 3b$ Ⓒ $4ab - b$ Ⓓ $4ab - 3b$

9 $5 + 3(m + 1) =$ — 1

Ⓐ $3m + 6$ Ⓑ $8m + 1$ Ⓒ $3m + 8$ Ⓓ $8m + 8$

10 The correct factorisation of $2ab - a$ is — 1

Ⓐ $2a(b - 1)$ Ⓑ $2a(b - a)$ Ⓒ $a(2b - 1)$ Ⓓ $a(2b - a)$

Total marks achieved for PART A

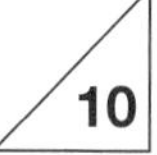

Basic algebra

TOPIC TEST — PART B

Instructions
- This part consists of 9 questions.
- Each question part is worth 1 mark.
- Write only the answer in the answer column.

Time allowed: 15 minutes **Total marks: 15**

Questions	Answers	Marks
1 Simplify $3y - 4y - y$		1
2 Simplify $6a^2 \times a$		1
3 Factorise $7x + 14y$		1
4 Simplify $12x^2 + x - 2x \times 5x + 6x$		1
5 Expand $3x(2x - 5y)$		1
6 If $a = 3$, $b = -2$ and $c = 0$, find the value of $3a + 2b - c$		1
7 Simplify the following.		
a $5x \div 5x$		1
b $-7a - 8a$		1
c $16a^2 \div 8a$		1
d $36m^3 - 12m^3$		1
e $(4abc)^2$		1
f $(-3a)^2 \times 2a$		1
8 Factorise $xy - 5x$		1
9 Expand and simplify.		
a $5(x + 2) - 3(x + 1)$		1
b $3(a - 4) + 4(a + 2b)$		1

Total marks achieved for PART B

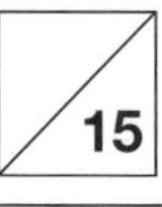

CHAPTER 6
Length, mass and time

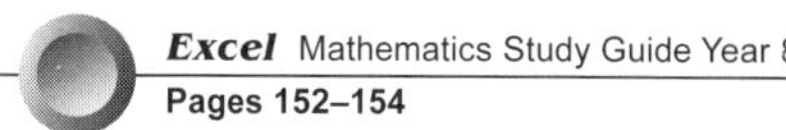

UNIT 1: Conversion of units of length

QUESTION 1 Convert the following to centimetres.

a 3 m = ______ **b** 15 m = ______ **c** 8 m = ______ **d** 18 m = ______
e 16 m = ______ **f** 31 m = ______ **g** 12 m = ______ **h** 64 m = ______
i 25 m = ______ **j** 5 m = ______ **k** 6 m = ______ **l** 20 m = ______
m 10 mm = ______ **n** 90 mm = ______ **o** 950 mm = ______ **p** 28 mm = ______

QUESTION 2 Convert the following to metres.

a 300 cm = ______ **b** 1400 cm = ______ **c** 5000 cm = ______ **d** 90 cm = ______
e 800 cm = ______ **f** 2100 cm = ______ **g** 3600 cm = ______ **h** 250 cm = ______
i 2500 cm = ______ **j** 840 cm = ______ **k** 1675 cm = ______ **l** 1260 cm = ______
m 3 km = ______ **n** 18 km = ______ **o** 5.65 km = ______ **p** 31.87 km = ______

QUESTION 3 Convert the following to millimetres.

a 2 m = ______ **b** 3.8 m = ______ **c** 5.9 m = ______ **d** 16.38 m = ______
e 10 m = ______ **f** 25 m = ______ **g** 63.8 m = ______ **h** 92.6 m = ______
i 37 m = ______ **j** 53 m = ______ **k** 85.9 m = ______ **l** 6.75 m = ______
m 60 cm = ______ **n** 92 cm = ______ **o** 35.6 cm = ______ **p** 38.9 cm = ______

QUESTION 4 Convert the following measurements to the units shown.

a 8 m = ______ cm **b** 6000 mm = ______ m **c** 9.82 km = ______ m
d 1236 mm = ______ cm **e** 12.5 m = ______ mm **f** 845 mm = ______ cm
g 9.36 cm = ______ mm **h** 3.5 km = ______ m **i** 13 m = ______ cm
j 900 cm = ______ mm **k** 80 cm = ______ mm **l** 92 m = ______ cm

QUESTION 5 Complete the following conversions.

a 8 km = ______ m **b** 8 dm = ______ m **c** 9 dm = ______ cm
d 10.3 m = ______ cm **e** 70 dm = ______ m **f** 7 Mm = ______ m
g 12 dm = ______ m **h** 33.58 m = ______ mm **i** 80 m = ______ dm
j 14 dm = ______ cm **k** 195 m = ______ cm **l** 63 cm = ______ mm

QUESTION 6 Complete the following.

a 58 dm = ______ cm **b** 236 m = ______ cm **c** 864 cm = ______ m
d 854 mm = ______ cm **e** 280 m = ______ dm **f** 56 hm = ______ m
g 4563 mm = ______ m **h** 935 cm = ______ mm **i** 15 Mm = ______ m
j 72.56 m = ______ mm **k** 8.35 km = ______ m **l** 8.35 m = ______ cm

UNIT 2: Pythagoras' theorem

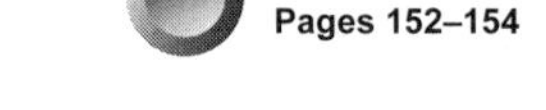

QUESTION 1 Find the value of the hypotenuse in each triangle by using Pythagoras' theorem.

a

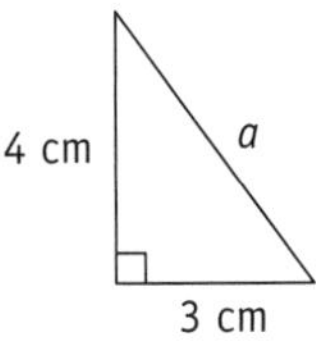

b

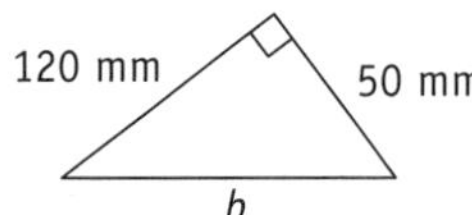

c

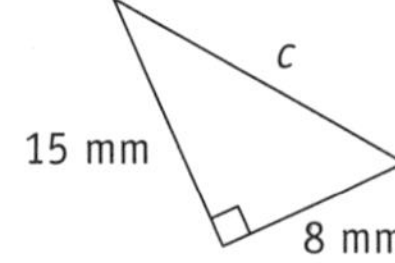

QUESTION 2 Find the value of the unknown side or hypotenuse correct to 1 decimal place.

a

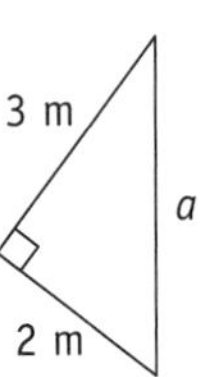

b

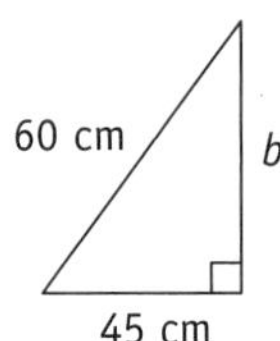

c

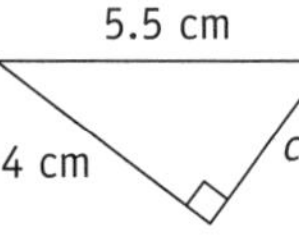

d

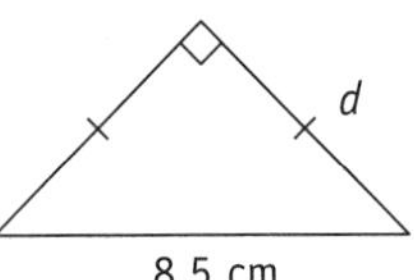

e

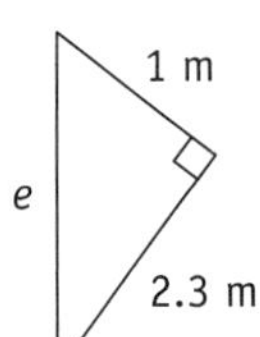

f

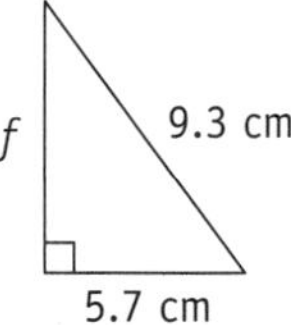

QUESTION 3 Find the value of the pronumeral correct to two significant figures.

a

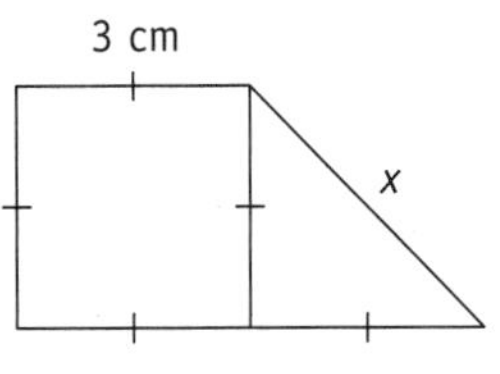

b

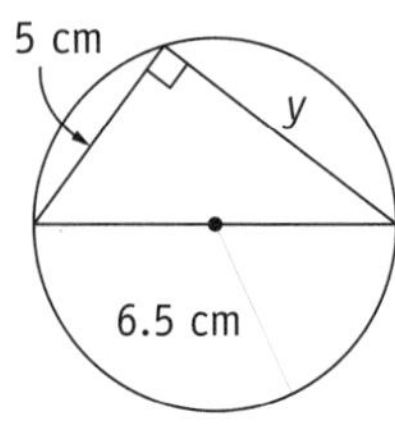

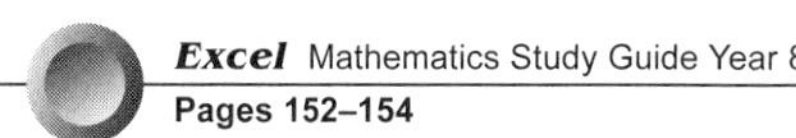

UNIT 3: Application of Pythagoras' theorem

QUESTION 1 A helium balloon tied to a weight is 1.2 m high and casts a shadow 0.5 m along the ground. What is the distance from the top of the balloon to the top of the shadow?

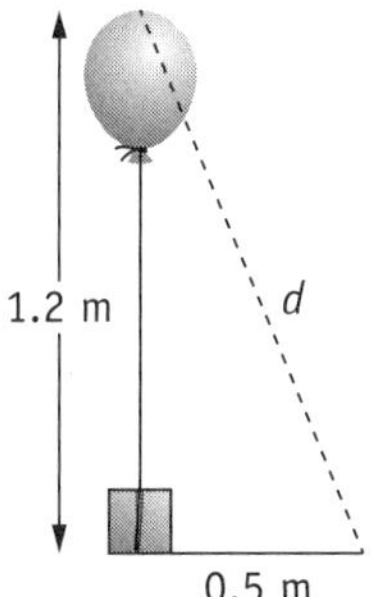

QUESTION 2 The base of a ladder leaning against a wall is 1.5 m away from the base of the wall. The top of the ladder reaches 2 m up the wall.

a Draw a diagram to represent this information.

b Find the length of the ladder.

QUESTION 3 A flagpole 8 m high is secured to the ground by a 10 m wire stretching from the top of the flagpole to an anchor point on the ground.

a Draw a diagram to represent this information.

b Calculate the distance from the base of the flagpole to the anchor point.

Excel Mathematics Study Guide Year 8
Pages 152–154

UNIT 4: Perimeters of regular shapes

QUESTION 1 Find the perimeter of each shape. All measurements are in centimetres.

a
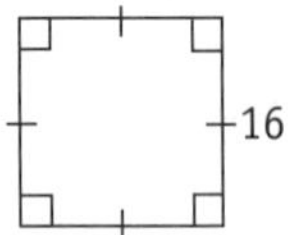

____

b
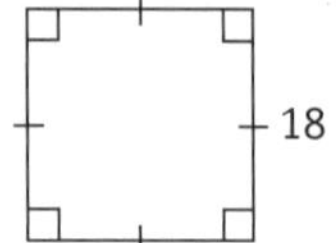

____

c
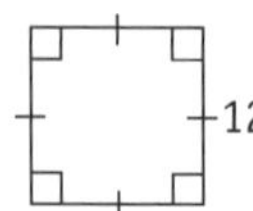

____

d
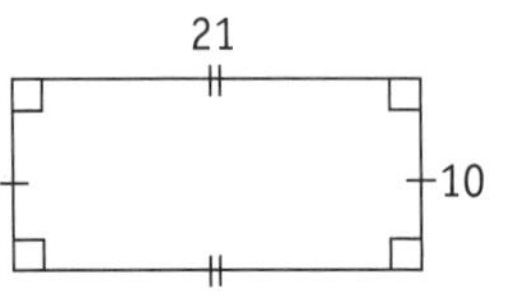

____

e
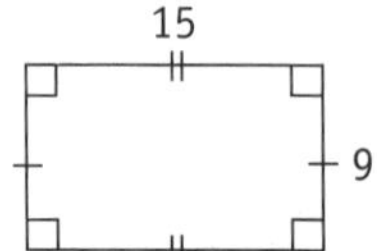

____

f
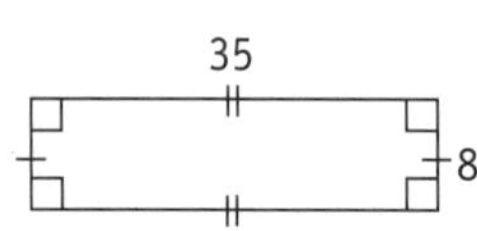

____

QUESTION 2 Find the perimeter of each triangle. All measurements are in centimetres.

a
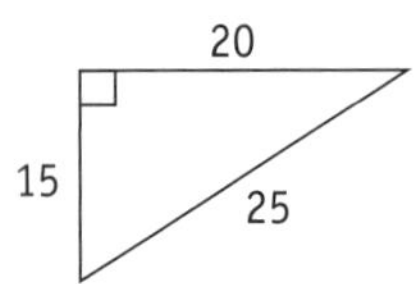

____

b
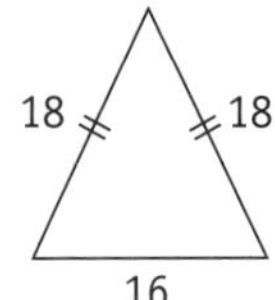

____

c
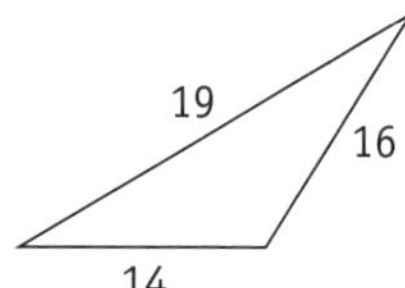

____

d
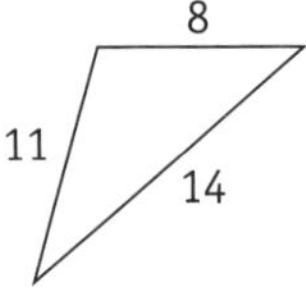

____

e
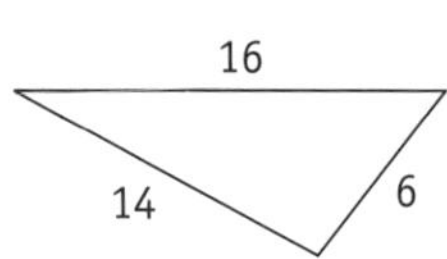

____

f
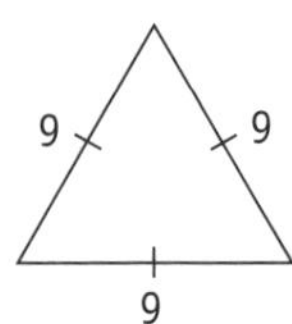

____

QUESTION 3 Find the perimeter of each shape. All measurements are in centimetres.

a
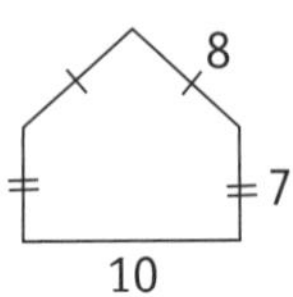

____

b
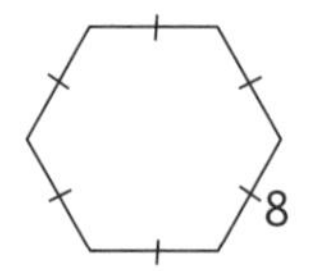

____

c
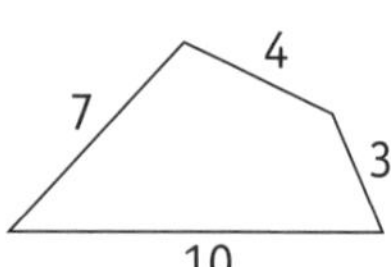

____

d
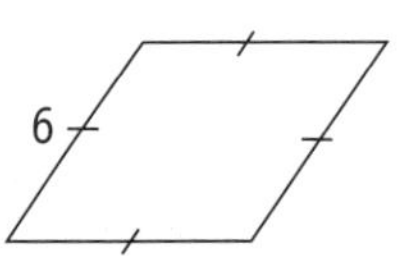

____

e
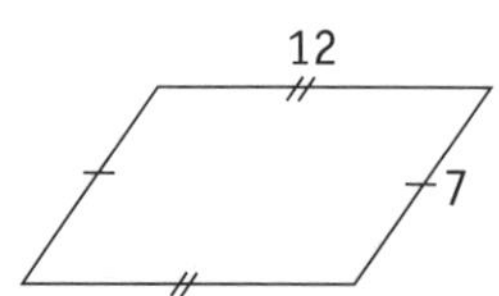

____

f
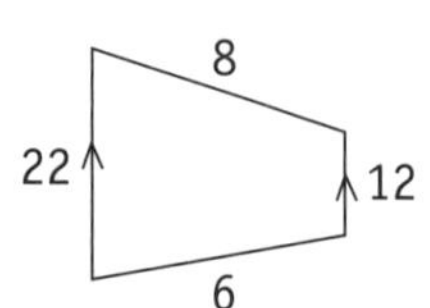

____

Length, mass and time

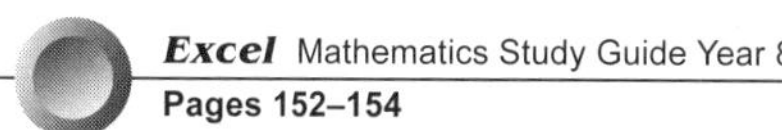

UNIT 5: Perimeters of irregular shapes

QUESTION 1 Find the perimeter of the following shapes. All measurements are in centimetres.

a

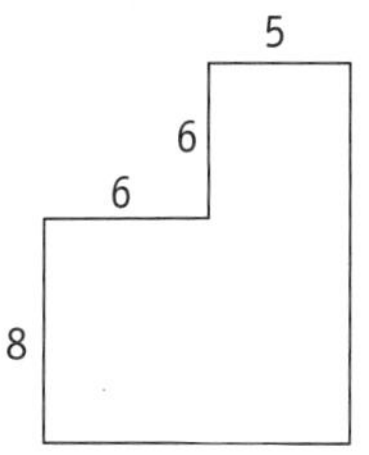

b

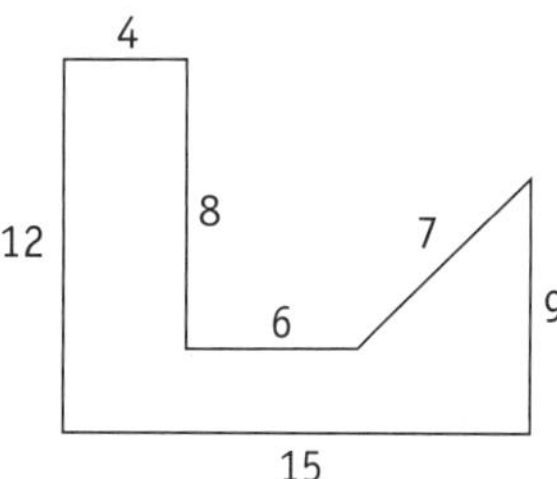

c

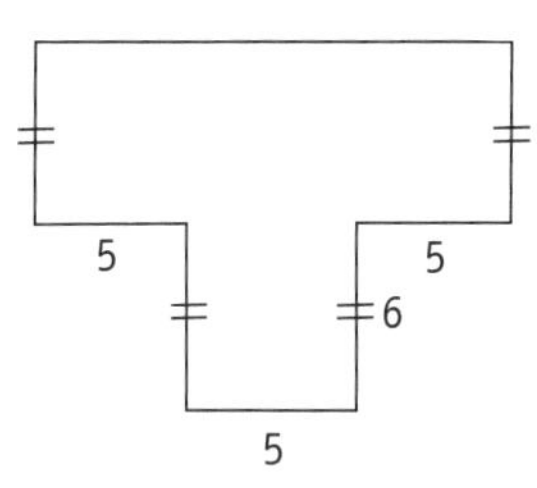

d

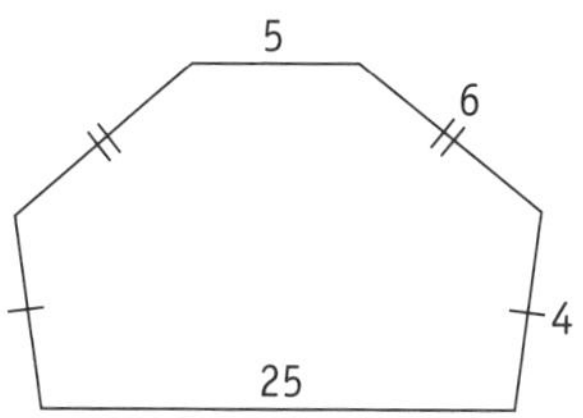

QUESTION 2 Find the perimeter.

a

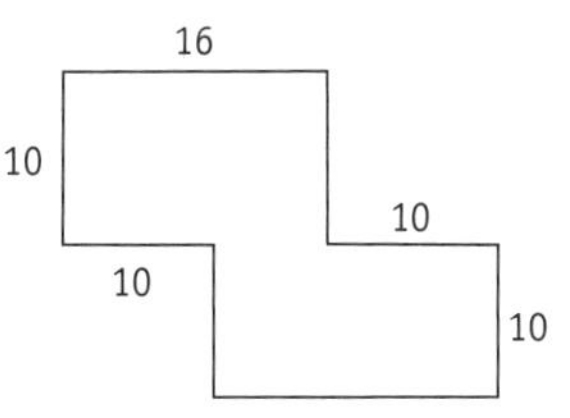

b

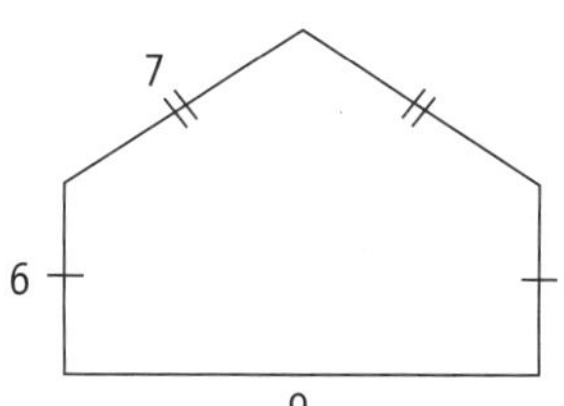

c

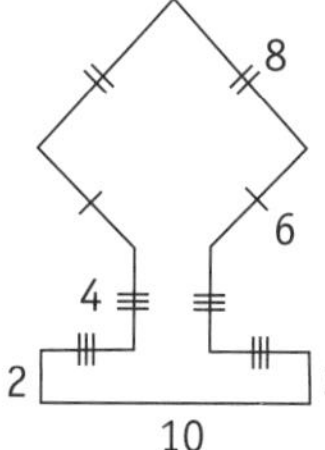

d

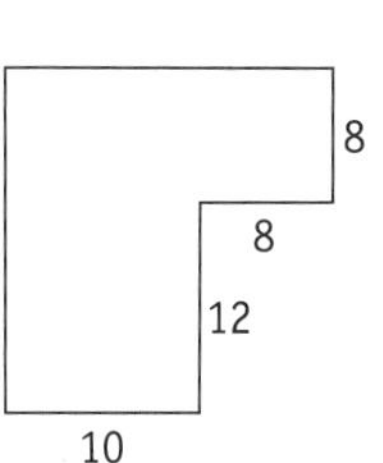

e

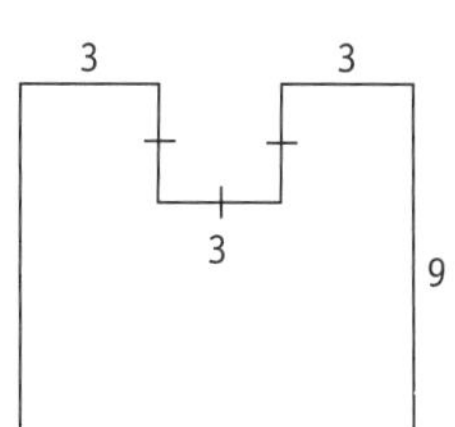

f

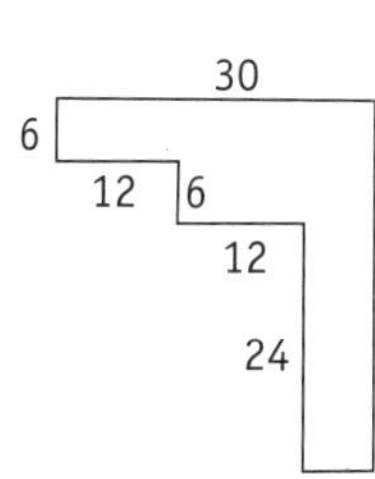

QUESTION 3

a If the perimeter of a square is 48 metres, find the length of each side.

b A room measures 9.86 metres long and 5.25 metres wide. Find the sum of the length and the width.

Length, mass and time

UNIT 6: Conversion of units of mass

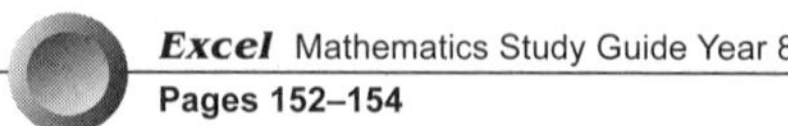

Excel Mathematics Study Guide Year 8
Pages 152–154

QUESTION **1** Convert the following to kilograms.

a 8 t = ____________ **b** 6438 g = ____________ **c** 6.35 t = ____________
d 12 670 g = ____________ **e** 8.16 t = ____________ **f** 9.135 t = ____________
g 728 g = ____________ **h** 9.25 t = ____________ **i** 7 200 000 mg = ____________
j 35 t = ____________ **k** 845 000 mg = ____________ **l** 0.43 t = ____________
m 24 890 000 mg = ____________ **n** 5 g = ____________ **o** 6.2 t = ____________

QUESTION **2** Change the following to grams.

a 9 kg = ____________ **b** 1 t = ____________ **c** 16.45 kg = ____________
d 1.2 kg = ____________ **e** 13.508 kg = ____________ **f** 4.5 t = ____________
g 0.7 kg = ____________ **h** 0.0875 t = ____________ **i** 3 t = ____________
j 12.69 kg = ____________ **k** 12 000 mg = ____________ **l** 52.125 kg = ____________
m 680 500 mg = ____________ **n** 0.93 t = ____________ **o** 4.5 kg = ____________

QUESTION **3** Convert the following to the units shown.

a 8000 g = ____________ kg **b** 63.5 t = ____________ kg **c** 9607 kg = ____________ t
d 30 000 g = ____________ kg **e** 4800 g = ____________ kg **f** 2.75 t = ____________ kg
g 42 000 kg = ____________ t **h** 965 g = ____________ kg **i** 67 500 mg = ____________ g
j 5000 g = ____________ t **k** 9.08 kg = ____________ mg **l** 70 500 kg = ____________ t
m 12 000 mcg = ____________ mg **n** 8563 g = ____________ kg **o** 3.25 t = ____________ kg

QUESTION **4** Complete the following.

a 3000 g = ____________ kg **b** 25 006 kg = ____________ t **c** 8.567 kg = ____________ g
d 2.306 t = ____________ g **e** 21 000 kg = ____________ t **f** 3.657 kg = ____________ g
g 9640 g = ____________ kg **h** 8.35 t = ____________ kg **i** 12.89 kg = ____________ g
j 25 630 kg = ____________ t **k** 125 000 g = ____________ t **l** 198 500 000 mcg = ____________ g
m 8005 kg = ____________ t **n** 0.037 t = ____________ g **o** 0.5 g = ____________ mg

QUESTION **5** Complete the following.

a 15 000 mcg = ____________ mg **b** 32.536 t = ____________ kg **c** 8360 kg = ____________ t
d 43 250 000 mcg = ____________ g **e** 0.67 t = ____________ g **f** 10 000 kg = ____________ t
g 52.6 g = ____________ mg **h** 53.69 t = ____________ kg **i** 560 kg = ____________ g
j 53 467 kg = ____________ t **k** 8.7 t = ____________ g **l** 0.004 t = ____________ g
m 8261.5 kg = ____________ t **n** 3.25 t = ____________ g **o** 5639 g = ____________ kg

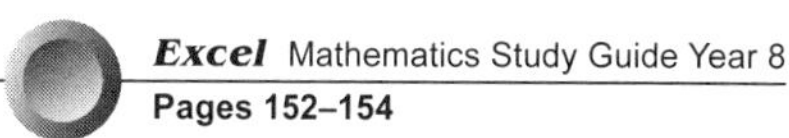

Length, mass and time

UNIT 7: Applications of mass

Question 1

a A packet of flour weighs 2.5 kg. How many of these packets would weigh 2 tonnes?

b There are six members in a family and the average mass is 45.6 kg. What is the total mass of the family?

c A truck and its load have a combined mass of 35 tonnes and 800 kilograms. The mass of the truck is two-fifths of this. What is the mass of the load?

d Find the mass of the container of a product if its gross mass is 5.8 kg and net mass (mass of contents) is 3650 g.

Question 2

a Find the net mass (mass of contents), if the gross mass of a package is 1980 g and the container's mass is 235 g.

b A ship has a mass of 45 600 000 kg. What is this in tonnes?

c Arrange the following masses in descending order (from largest to smallest).
460 g, 31.6 t, 26.8 kg, 4300 g, 3.6 kg, 8.65 t

Question 3

a A truck weighing 3.85 t is carrying a load of 1820 kg. What is the total weight of the truck and the load? Give your answer in tonnes.

b Clare buys 4.5 kg of flour and she uses 890 g for baking. How many grams of flour are left?

c Penny's cat weighed 2.9 kg in April and by September had put on 350 g. What was the cat's weight, in kilograms, in September?

Length, mass and time

UNIT 8: Conversion of units of time

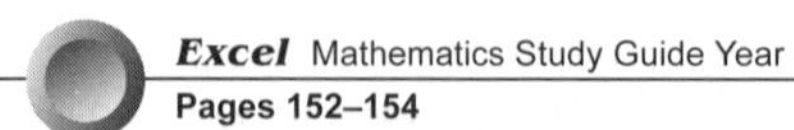
Excel Mathematics Study Guide Year 8
Pages 152–154

QUESTION 1 State which unit would be the best to use to measure the time for these activities.

a Walking 3 kilometres ____________

b Having a shower ____________

c Traveling from London to Sydney by plane ____________

d Playing a cricket match ____________

e Writing a two page essay ____________

QUESTION 2 Complete the following.

a 6 decades = ____________ years

b 8 minutes = ____________ seconds

c 5 centuries = ____________ years

d 15 hours = ____________ minutes

e 20 minutes = ____________ seconds

f 216 hours = ____________ days

g 9 weeks = ____________ days

h 30 minutes = ____________ seconds

i 96 months = ____________ years

j 6 fortnights = ____________ days

k 312 weeks = ____________ years

l 1 leap year = ____________ days

m 720 minutes = ____________ hours

n 7 years = ____________ months

o 3 years = ____________ days

p 2 weeks = ____________ hours

QUESTION 3 How many minutes in each of the following?

a 1 hour ____________

b 3 hours ____________

c 12 hours ____________

d 1 day ____________

e 1 week ____________

f 3 weeks ____________

g 30 days ____________

h 33 days ____________

QUESTION 4 How many hours in each of these?

a 2 days ____________

b 1 week ____________

c 1 fortnight ____________

d 1 month of 30 days ____________

e July ____________

f September ____________

QUESTION 5 How many days in each of the following?

a 1 week ____________

b 3 weeks ____________

c 10 weeks ____________

d 15 weeks ____________

e 1 fortnight ____________

f 2 fortnights ____________

g March ____________

h 1 year ____________

i 2 leap years ____________

j 1 decade (include 2 leap years) ____________

Length, mass and time

UNIT 9: Addition and subtraction of times

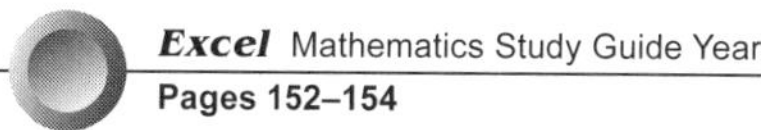

Question 1

a

h	min	
8	25	+
6	12	

b

h	min	
16	36	+
8	28	

c

h	min	
10	42	+
8	27	

d

h	min	
6	36	+
5	37	

e

h	min	
12	53	+
5	48	

f

h	min	
18	42	−
8	24	

g

h	min	
25	48	−
9	32	

h

h	min	
14	56	−
9	47	

Question 2

a Add 7 days, 5 hours and 28 minutes to 11 days, 16 hours and 48 minutes.

____

b From 25 h 37 min, take away 10 h 25 min.

____

c How many hours are there from Saturday noon to 8 pm (the same day)?

____

d How many weeks are there in half a year?

____

e How many hours are there from 8 am to 2 pm the same day?

____

Question 3 What will be the time:

a 10 minutes after 8:50 am? ______

b 20 minutes after 6:40 am? ______

c 40 minutes after 3:25 pm? ______

d 35 minutes after 8:25 am? ______

e 30 minutes after 6:47 pm? ______

f 21 minutes after 6:36 am? ______

g 25 minutes after 3:20 pm? ______

h 18 minutes after 5:38 am? ______

i 53 minutes after 10:00 am? ______

j 5 hours after 9:10 pm? ______

Question 4 What is the length of time between:

a 8:00 am and 11:00 am? ______

b 3:20 am and 7:35 am? ______

c 6 pm and 10 pm? ______

d 4:18 am and 7:46 am? ______

e 5 am and 9 am? ______

f 5:25 am and 2 pm? ______

g 3 am and 9 am? ______

h 6 am and 3 pm? ______

i 7 am and 1 pm? ______

j 8:15 am and 10:30 pm? ______

Length, mass and time

UNIT 10: Working with time

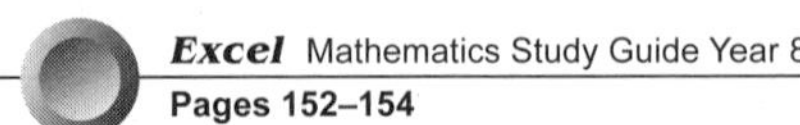

Question 1 Write the following times using am or pm.

a 10 in the morning ______________

b 11:25 at night ______________

c 8:20 in the evening ______________

d 2 hours after 6 am ______________

e 9:30 in the morning ______________

f 4 hours before noon ______________

Question 2 Change the following 24-hour times to 12-hour times (to their am or pm equivalents).

a 0900 ______________

b 2219 ______________

c 1932 ______________

d 1833 ______________

e 0631 ______________

f 0800 ______________

Question 3 Change the following 12-hour times to 24-hour times.

a 8 am ______________

b 2:40 pm ______________

c 5:20 pm ______________

d 8:20 am ______________

e 4:10 am ______________

f 9:35 am ______________

g 6:40 pm ______________

h 6:50 pm ______________

Question 4 Write the following times in words.

a 10:10 ______________

b 8:55 ______________

c 9:20 ______________

d 3:50 ______________

Question 5 Change the following to digital times.

a five minutes past three ______________

b twenty-five past eight ______________

c six o'clock ______________

d midday ______________

e a quarter past five ______________

f half past six ______________

Question 6 What would be the following times?

a 3 hours after 10 am ______________

b 2 hours before 3 pm ______________

c 2 hours before 1 am ______________

d $1\frac{1}{2}$ hours after 11:30 am ______________

UNIT 11: Problem solving using 12- and 24-hour time

QUESTION 1 Write these plane arrival times in 12-hour times. State whether it is am or pm.

a 0835 hours ______ **b** 0943 hours ______ **c** 1025 hours ______

d 1520 hours ______ **e** 2036 hours ______ **f** 2340 hours ______

QUESTION 2 Write the following times in 24-hour notation.

a Morning ______

b Afternoon ______

c Morning ______

d Night ______

QUESTION 3 Write the times in 24-hour notation for the following digital clocks.

a AM 5:30 ______

b PM 2:45 ______

c PM 11:20 ______

d AM 2:15 ______

QUESTION 4 Work out the time in hours and minutes between:

a 1130 hours and 1750 hours ______ **b** 0820 hours and 1320 hours ______

QUESTION 5 Write the time in 24-hour notation.

a 15 minutes after 1630 hours ______ **b** 20 minutes before 2300 hours ______

QUESTION 6 Write 2120 h in 12-hour time. ______

QUESTION 7 How long is it between 1800 h and 1955 h? ______

QUESTION 8 Convert the following times into 12-hour times indicating whether the time is am or pm.

a 1610 ______ **b** 0935 ______ **c** 2330 ______

QUESTION 9 Convert the following 12-hour times into 24-hour times.

a 5:30 am ______ **b** 3:20 pm ______ **c** 10:40 pm ______

QUESTION 10 Liam is to fly from Sydney to Melbourne on a flight that leaves Sydney at 1830. The flight time is 1 hour and 50 min. What is the time in Sydney when Liam's flight arrives in Melbourne? Give the time in 24-hour time.

______

Length, mass and time

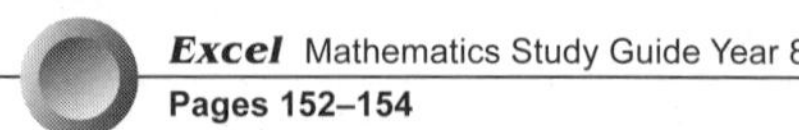

UNIT 12: Problem solving with length, mass and time

QUESTION 1 A room measures 9.38 m long and 7.63 m wide. Calculate the difference between the length and the width in metres.

____

QUESTION 2 An equilateral triangle has a side of 12 cm. Find its perimeter.

____

QUESTION 3 I walked 6.358 km on Monday and 5821 metres on Tuesday. How many kilometres did I walk altogether?

____

QUESTION 4 Find the perimeter of a square whose side length is 9.35 cm.

____

QUESTION 5 Find the perimeter of a rectangular block of land 300 m long and 160 m wide.

____

QUESTION 6 A ship has a mass of 42 838 000 kg. What is this in tonnes?

____

QUESTION 7 Find the mass of the load, if the mass of a loaded truck is 21.58 tonnes and the mass of the empty truck is 13.52 tonnes.

____

QUESTION 8 A packet of salt weighs 800 g. How many of these packets would weigh 5 tonnes?

____

QUESTION 9 There are 11 members in a team and the average mass is 52.4 kg. What is the total mass of the team?

____

QUESTION 10 Find the mass of the container of a product, if the product's gross mass is 3.85 kg and the net mass is 2738 g.

____

QUESTION 11 If a pulse beats 75 times per minute, how many beats would there be in 45 minutes?

____

QUESTION 12 How many hours are there in 5 weeks?

____

Length, mass and time

TOPIC TEST

PART A

Instructions
- This part consists of 10 multiple-choice questions.
- Fill in only ONE CIRCLE for each question.
- Each question is worth 1 mark.
- Calculators are NOT allowed.

Time allowed: 15 minutes **Total marks: 10**

Marks

1 What is the perimeter of this figure?

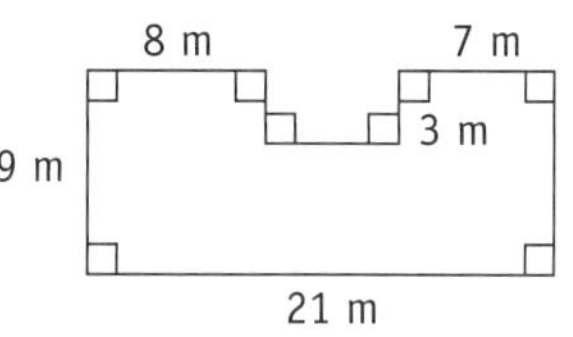

Ⓐ 48 m Ⓑ 57 m Ⓒ 63 m Ⓓ 66 m 1

2 How many centimetres are there in 3.4 metres?

Ⓐ 34 Ⓑ 3400 Ⓒ 430 Ⓓ 340 1

3 The number of minutes from 11:30 am to 1:12 pm on the same day is

Ⓐ 152 Ⓑ 84 Ⓒ 102 Ⓓ 94 1

4 Change 850 g to kg.

Ⓐ 8.5 kg Ⓑ 0.85 kg Ⓒ 0.085 kg Ⓓ 85 kg 1

5 Convert 485 mm into metres.

Ⓐ 48.5 m Ⓑ 4.85 m Ⓒ 0.485 m Ⓓ 485 000 m 1

6 Convert 1.05 kilograms into grams.

Ⓐ 0.001 05 g Ⓑ 105 g Ⓒ 1050 g Ⓓ 10.5 g 1

7 What is the time 9 hours and 35 minutes after a quarter to six in the morning?

Ⓐ 3:40 pm Ⓑ 3:20 pm Ⓒ 4:40 pm Ⓓ 4:20 pm 1

8 A $1\frac{1}{2}$ hour test starts at 9:35 am. It should finish at

Ⓐ 10.05 am Ⓑ 11:35 am Ⓒ 11:30 am Ⓓ 11:05 am 1

9 A tap drips every $1\frac{1}{2}$ seconds. How many times will it drip in 2 minutes?

Ⓐ 90 Ⓑ 80 Ⓒ 70 Ⓓ 60 1

10 $5\frac{1}{4}$ kilograms can be written as

Ⓐ 5.005 kg Ⓑ 5250 g Ⓒ $\frac{23}{4}$ kg Ⓓ 5125 g 1

Total marks achieved for PART A

Length, mass and time

TOPIC TEST — PART B

Instructions
- This part consists of 15 questions.
- Each question is worth 1 mark.
- Write only the answer in the answer column.

Time allowed: 15 minutes **Total marks: 15**

	Questions	Answers	Marks
1	Find the perimeter of a square with side length 9 metres.		1
2	Find the perimeter of a rectangle with length 28 cm and width 15 cm.		1
3	The perimeter of a rectangular block of land is 200 m. If the land is 40 m wide, how long is it?		1
4	How many metres in $5\frac{1}{2}$ km?		1
5	What fraction of 8 m is 75 cm?		1
6	A packet of sugar weighs 2000 grams. How many of these packets would weigh 2 tonnes?		1
7	Convert 15 600 kg to tonnes.		1
8	A cruise ship has a mass of 38 500 000 kg. What is this in tonnes?		1
9	How many minutes are there from 3:00 pm to 6:00 pm?		1
10	Express 8 days in hours.		1
11	Write 1850 hours in standard 12-hour time.		1
12	Change 1580 seconds into minutes and seconds.		1
13	Add 8 days, 9 hours and 48 minutes to 6 days, 14 hours and 27 minutes.		1
14	A carton has a mass of 15 kg. If it contains 20 cans of beans, what is the mass of each can (ignoring the weight of the carton)?		1
15	The weight of a dog is $4\frac{1}{2}$ kg. It puts on 840 g in 8 months. What is its weight now?		1

Total marks achieved for PART B

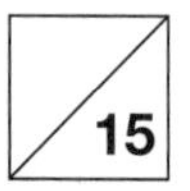

CHAPTER 7
Area, volume and capacity

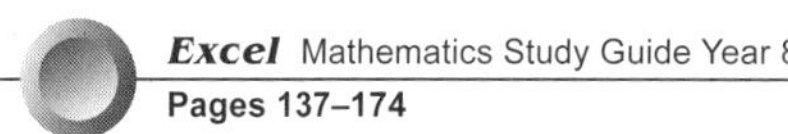

UNIT 1: Review of units of measurement for area and its conversion

QUESTION **1** Complete the following.

a 1 cm = __________ mm **b** 1 m = __________ cm **c** 1 km = __________ m

d $1\ \text{cm}^2 = 1\ \text{cm} \times 1\ \text{cm}$
= _____ mm × _____ mm
= __________ mm^2

e $1\ \text{m}^2 = 1\ \text{m} \times 1\ \text{m}$
= _____ cm × _____ cm
= __________ cm^2

f $1\ \text{km}^2 = 1\ \text{km} \times 1\ \text{km}$
= _____ m × _____ m
= __________ m^2

g 1 ha = __________ m^2 **h** $1\ \text{km}^2$ = __________ ha

QUESTION **2** Complete each of the following.

a $20\ \text{cm}^2$ = __________ mm^2 **b** $215\ \text{cm}^2$ = __________ mm^2 **c** $800\ \text{mm}^2$ = __________ cm^2

d $9\ \text{cm}^2$ = __________ mm^2 **e** $60\ \text{mm}^2$ = __________ cm^2 **f** $20\ \text{m}^2$ = __________ cm^2

g $240\ \text{m}^2$ = __________ cm^2 **h** $72\,000\ \text{cm}^2$ = __________ m^2 **i** $4000\ \text{cm}^2$ = __________ m^2

j $3.8\ \text{m}^2$ = __________ cm^2 **k** 500 ha = __________ m^2 **l** 8 ha = __________ m^2

m $20\ \text{km}^2$ = __________ ha **n** 350 ha = __________ m^2 **o** $12.8\ \text{km}^2$ = __________ ha

p $8\,000\,000\ \text{m}^2$ = __________ km^2 **q** $120\ \text{cm}^2$ = __________ mm^2 **r** $10\,000\ \text{mm}^2$ = __________ cm^2

s $3600\ \text{m}^2$ = __________ cm^2 **t** $9\ \text{km}^2$ = __________ ha

QUESTION **3** The area of a farm is 50 000 ha. Calculate:

a its area in m^2 ____

____

b its area in km^2 ____

____

UNIT 2: Area of a trapezium

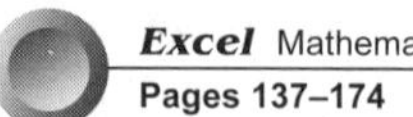

QUESTION 1 Find the area of the following trapeziums.

a

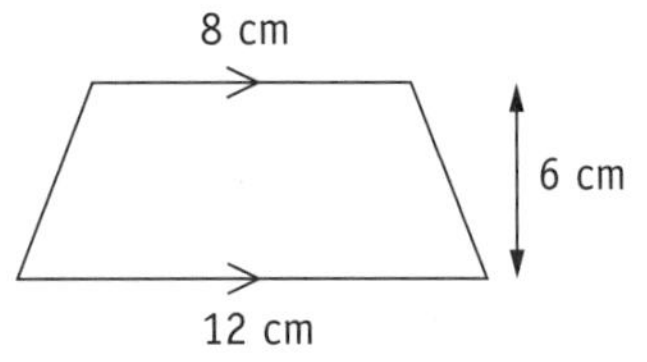

b

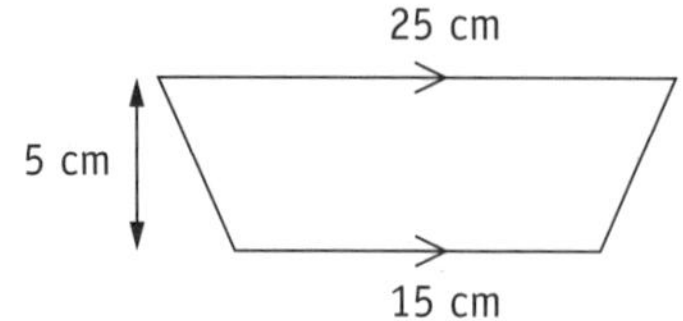

c

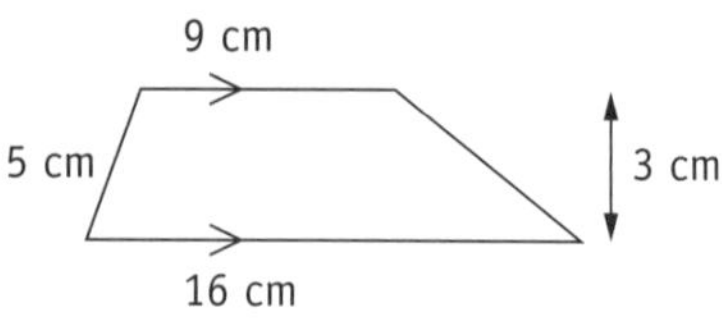

QUESTION 2 Calculate the area of these shapes.

a

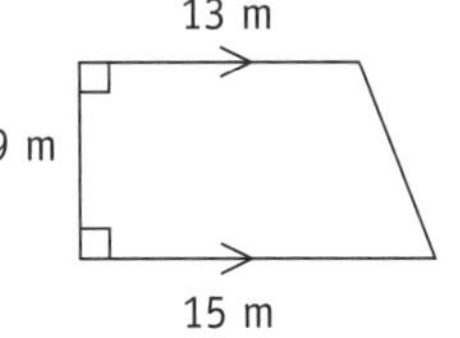

b

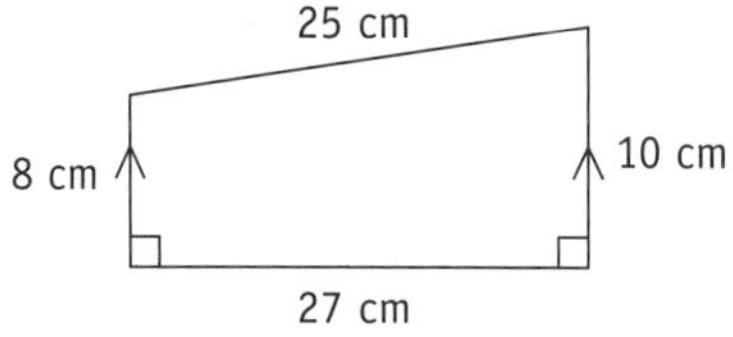

c

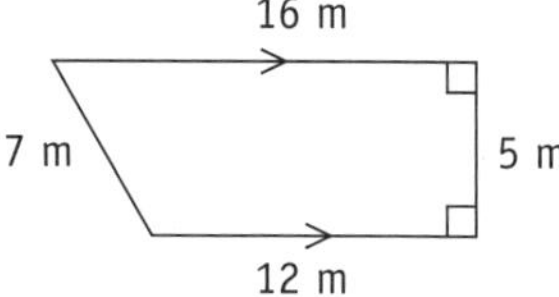

QUESTION 3 Work out the area of these trapeziums.

a

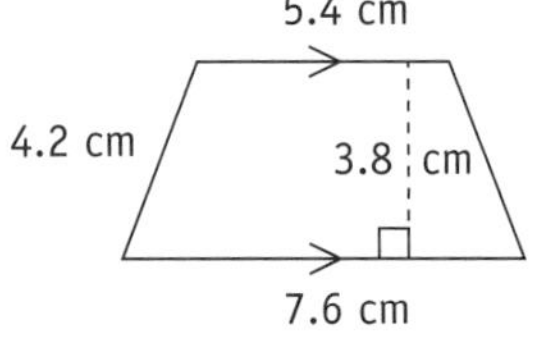

b

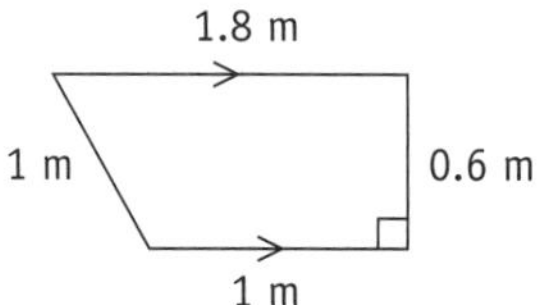

c

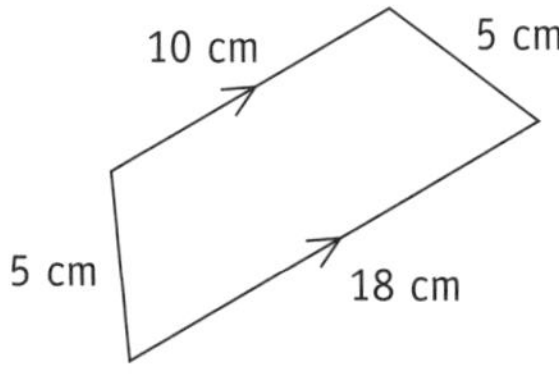

d

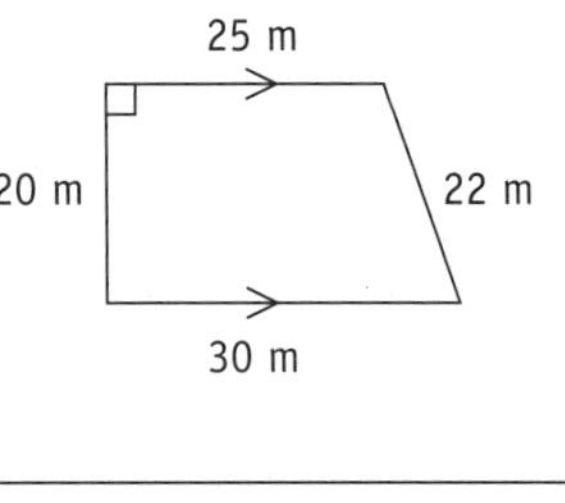

e

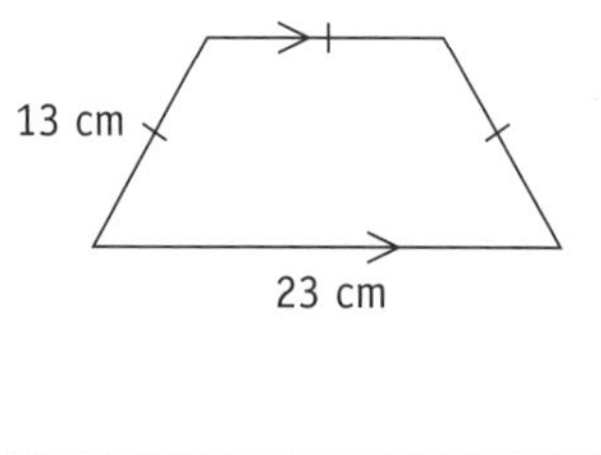

f

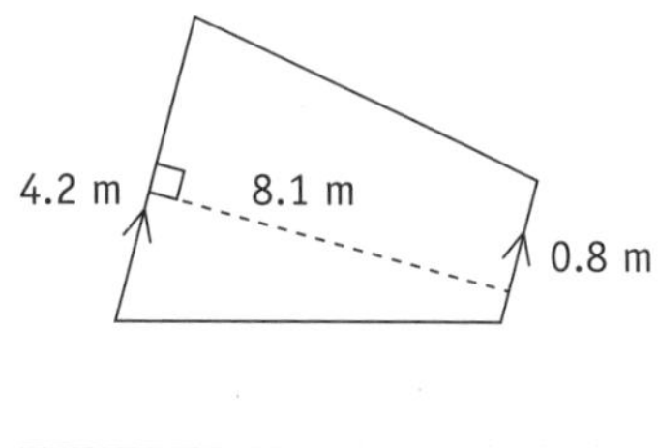

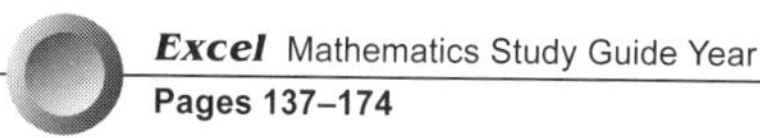

UNIT 3: Area of a parallelogram

QUESTION 1 Find the area of each parallelogram.

a

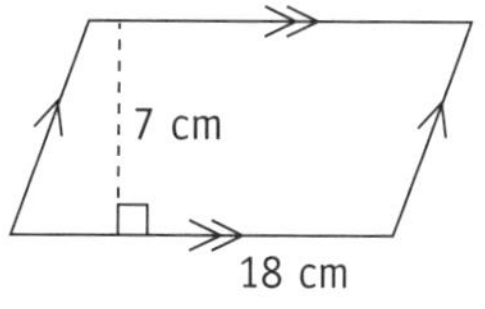

b

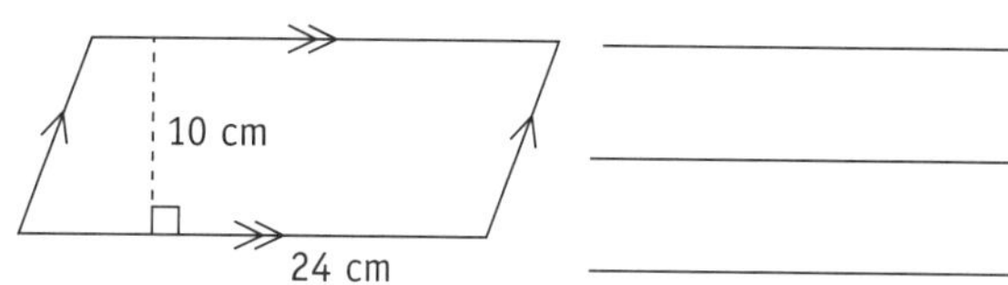

c

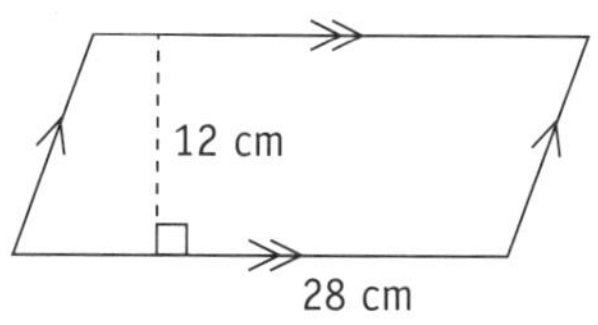

d

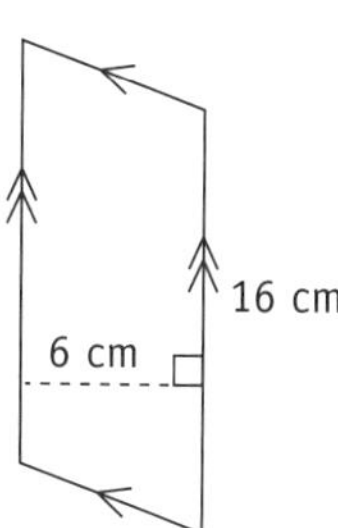

QUESTION 2 Use the formula $A = bh$ to find the area of each parallelogram given below.

a

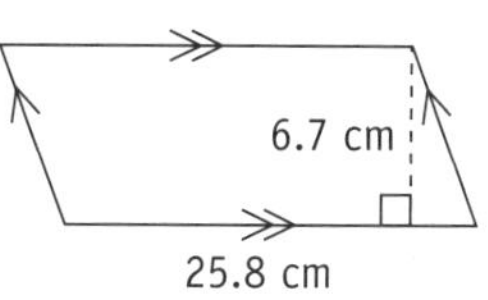

b

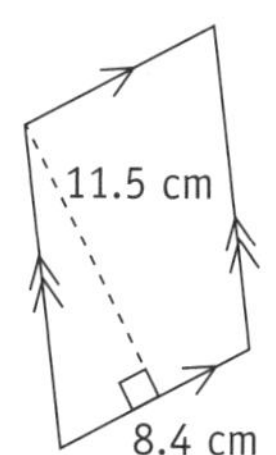

c

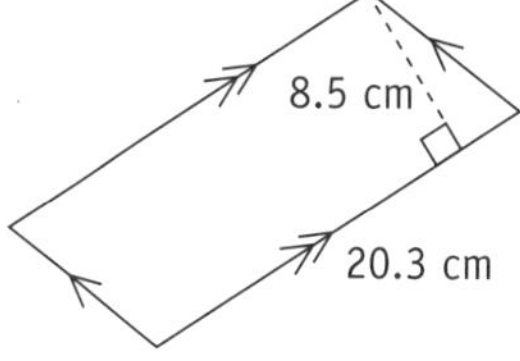

QUESTION 3 Complete the following table for the base, height or area of each parallelogram.

	Base	Height	Area
a	10 m		190 m^2
b		30 cm	150 cm^2
c	12 mm	18 mm	
d	20 cm	24 cm	
e	18 m		216 m^2
f	25 mm	18 mm	
g		15 cm	120 cm^2
h	24 m		432 m^2
i		23 cm	345 cm^2
j	28 cm	15 cm	

Area, volume and capacity

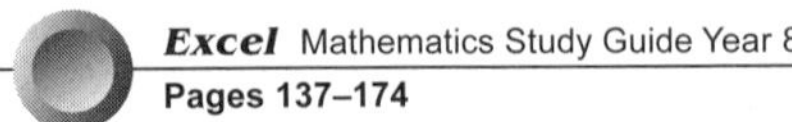

UNIT 4: Area of a rhombus

Question 1 Find the area of each rhombus. All measurements are in cm.

a

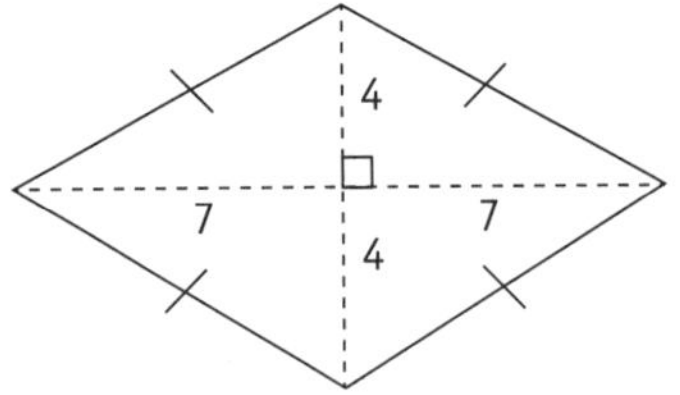

b

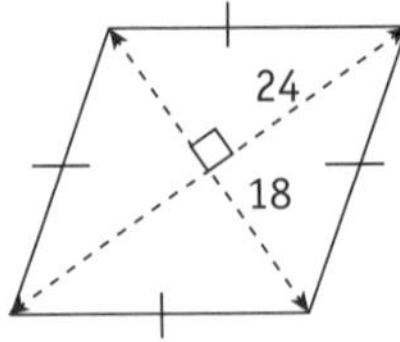

c

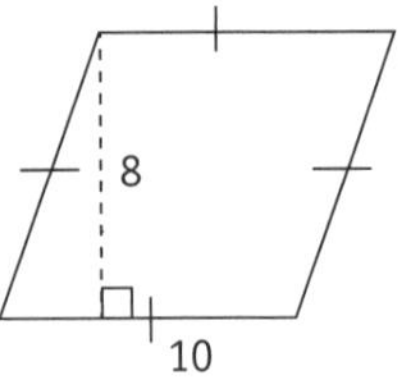

Question 2 Calculate the area of each rhombus given below.

a

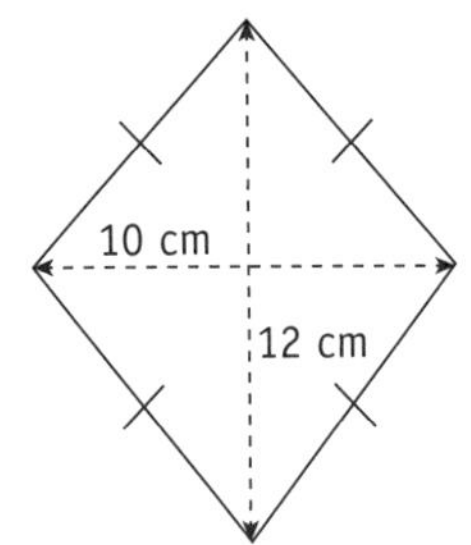

b

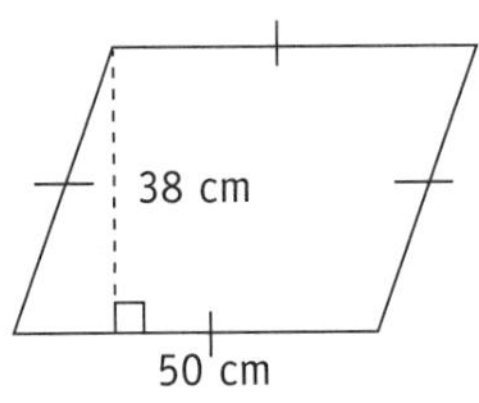

c

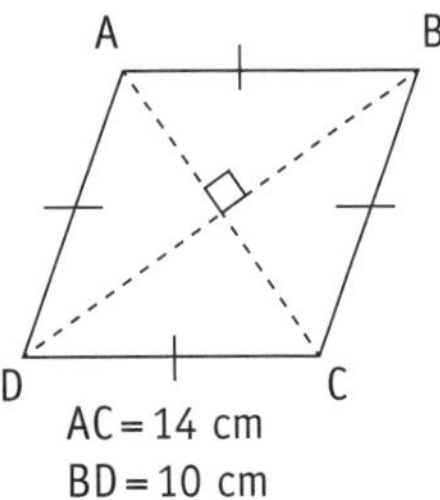

AC = 14 cm
BD = 10 cm

Question 3 Complete the following tables for the area, height or diagonal lengths of each rhombus.

	Base	Height	Area
a	16 m	8m	
b	6 cm		72 cm^2
c	20 cm	6 cm	
d	24 cm		384 cm^2
e	18 m		198 m^2

	Diagonal 1 (d_1)	Diagonal 2 (d_2)	Area
f	17m		238 m^2
g	26 cm	18 cm	
h		14 cm	188 cm^2
i	19 mm		285 mm^2
j		18 cm	378 cm^2

UNIT 5: Area of a kite

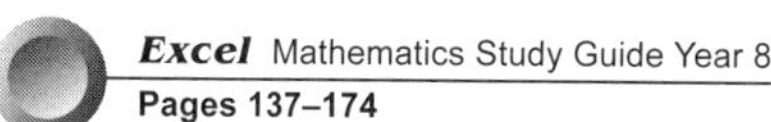

QUESTION 1 Find the area of the following kites. All measurements are in cm.

a

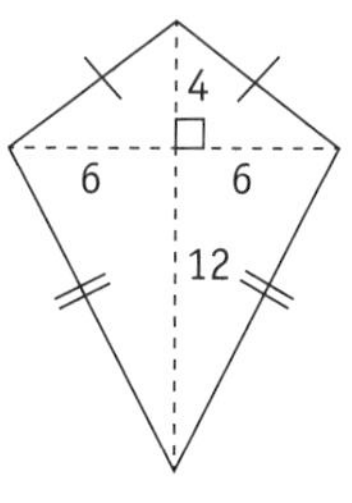

b

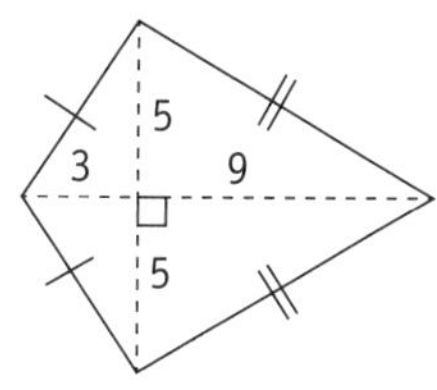

c

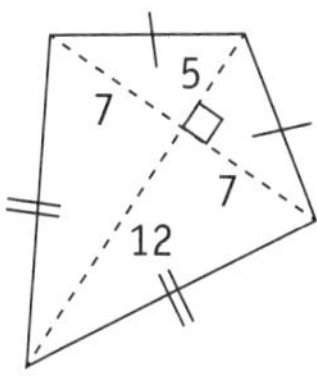

d

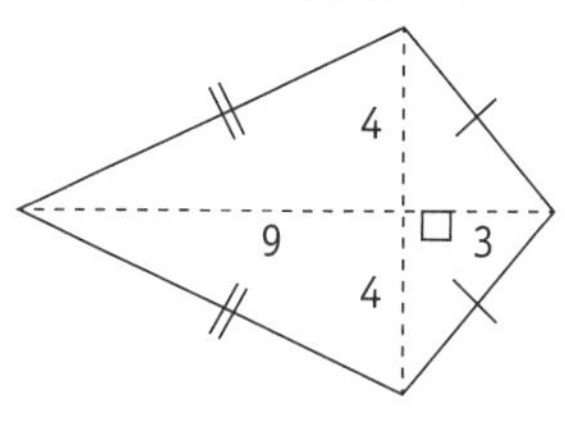

e

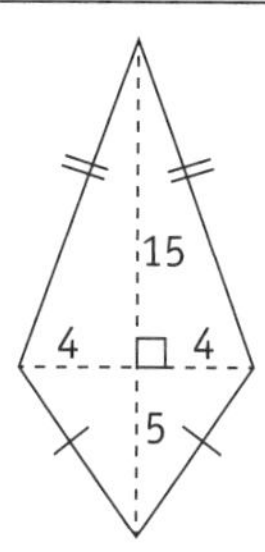

f

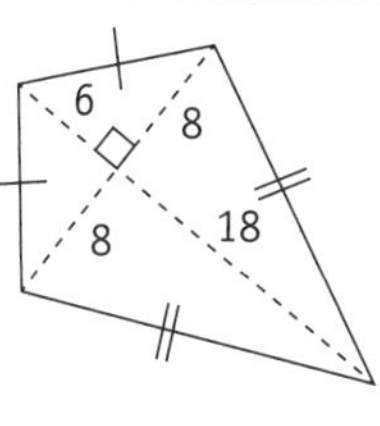

QUESTION 2 Find the area of each of the following kites.

a

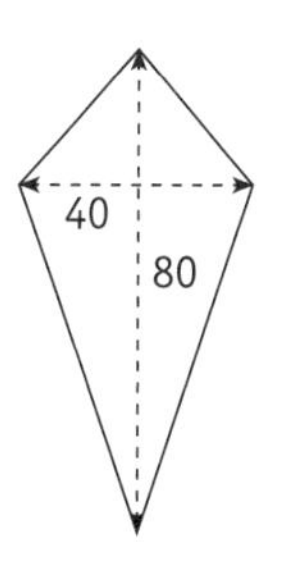

b

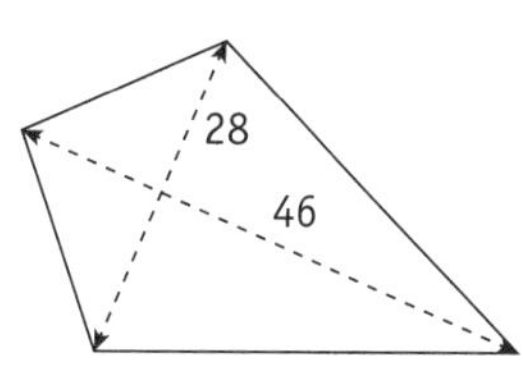

c

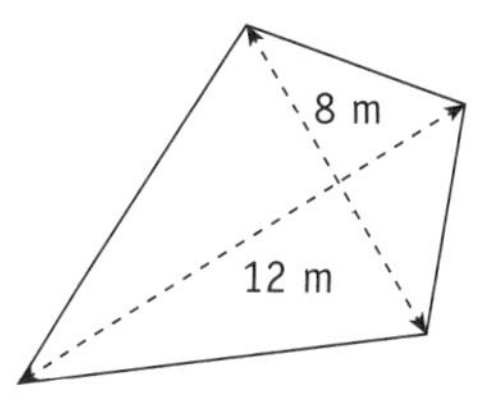

QUESTION 3 Complete the following tables for the diagonal lengths or area of each kite.

	Diagonal 1 (d_1)	Diagonal 2 (d_2)	Area
a	10 m	8 m	
b	15 cm		180 cm^2
c	20 cm	11 cm	
d	22 cm		374 cm^2
e	25 m		250 m^2

	Diagonal 1 (d_1)	Diagonal 2 (d_2)	Area
f	15 m		135 m^2
g	17 cm	10 cm	
h		12 cm	180 cm^2
i	22 mm		220 mm^2
j		18 cm	180 cm^2

Area, volume and capacity

UNIT 6: Areas of plane shapes

QUESTION 1 Write the area formula next to the shapes given below.

a

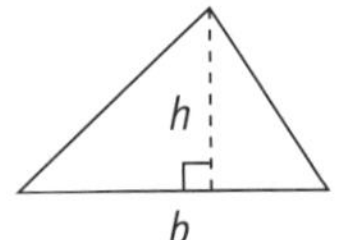

b

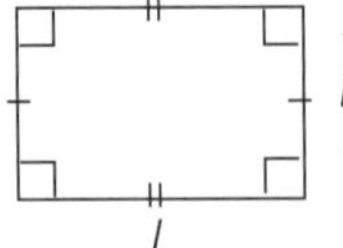

c

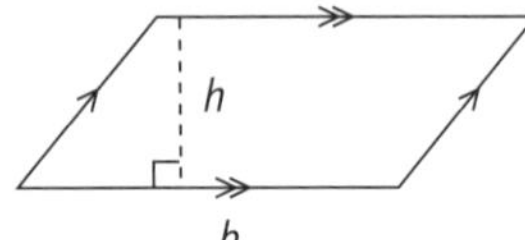

d

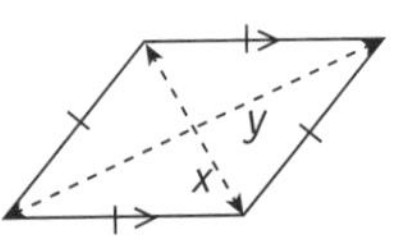

e

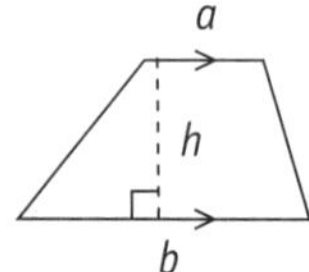

f

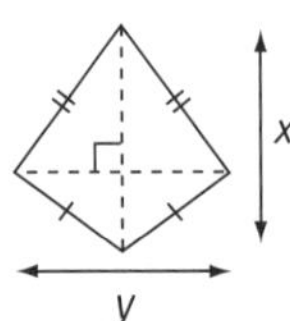

QUESTION 2 Find the area of the following shapes. All measurements are in centimetres.

a

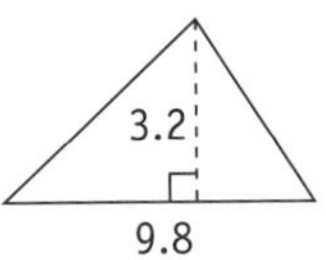

b

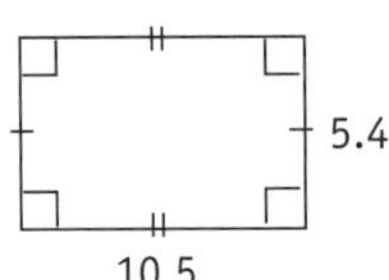

c

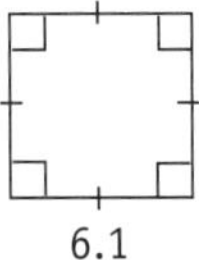

d

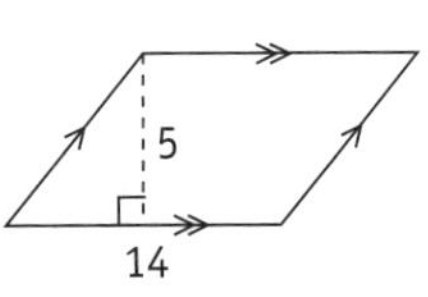

e

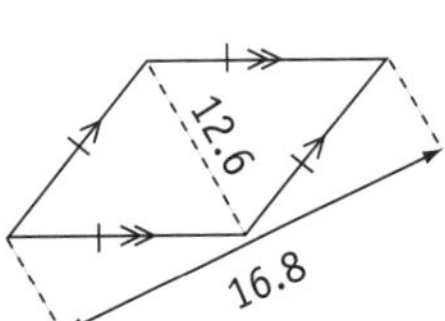

f

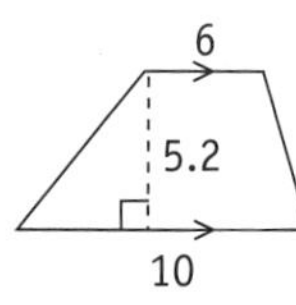

g

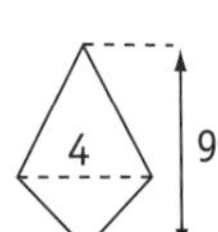

h

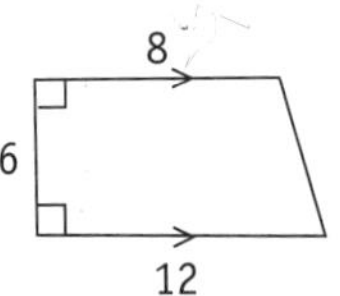

i

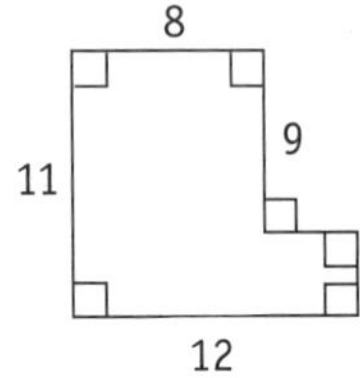

Area, volume and capacity

UNIT 7: Problem solving with area

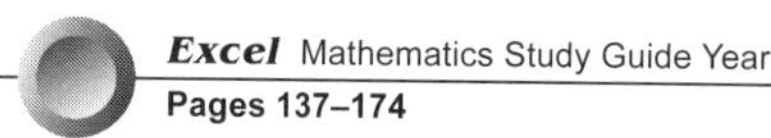
Excel Mathematics Study Guide Year 8
Pages 137–174

QUESTION 1 Calculate the number of square centimetres (cm^2) in 3 square metres. ______

______

QUESTION 2 Calculate the number of square metres (m^2) in 5 square kilometres (km^2). ______

______

QUESTION 3 Calculate the number of square metres (m^2) in 6 hectares. ______

______

QUESTION 4 A rectangular playground measures 450 m by 350 m. Find its area in hectares (ha). ______

______

QUESTION 5 A garden is in the shape of a parallelogram with base 50 m and height 30 m. What is the cost of top dressing the garden at $35 per square metre? ______

______

QUESTION 6 In a hotel, a small reception area is in the form of a rhombus with its base length 60 m and height as 35 m.

a Calculate its area in square metres. ______

______

b Find the cost of tiling the floor if it costs $180 per square metre. ______

______

QUESTION 7 The area of a parallelogram is the same as the area of a square with side length 40 cm. Calculate the height of the parallelogram if its base is 50 cm.

______

______

QUESTION 8 A farm has an area of 230 km^2. Calculate this area in hectares. ______

______

QUESTION 9 Another farm has an area of 450 ha. Calculate its area in:

a m^2 ______

______

b km^2 (square kilometres) ______

______

QUESTION 10 A rectangular mirror measures 50 cm by 40 cm and is surrounded by a frame of width 8 cm. Calculate:

a the area of the mirror ______

b the area of the frame ______

c the total area ______

Area, volume and capacity

UNIT 8: Volume of a cube

QUESTION 1 Calculate the volume of each of the following cubes whose side length is given.

a 8 cm

b 9 cm

c 21 cm

d 10 m

e 13 m

f 25 m

g 15 cm

h 18 m

i 17 m

QUESTION 2 Find the volume of each cube. All measurements are in metres.

a

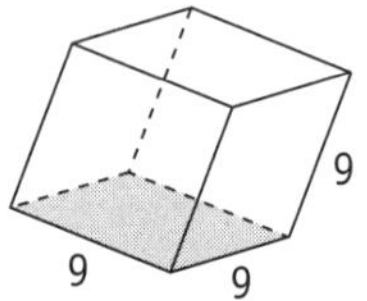

b

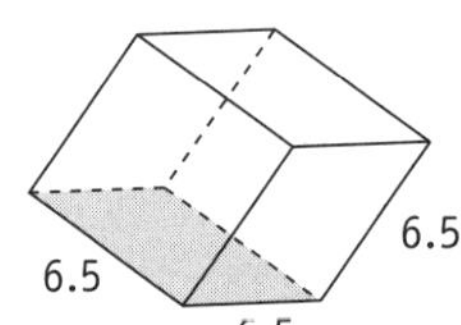

c

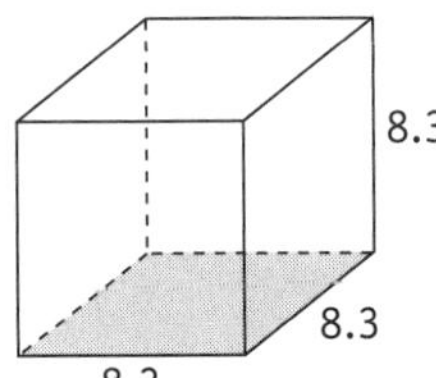

d

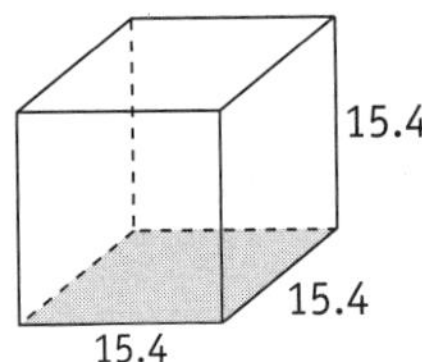

e

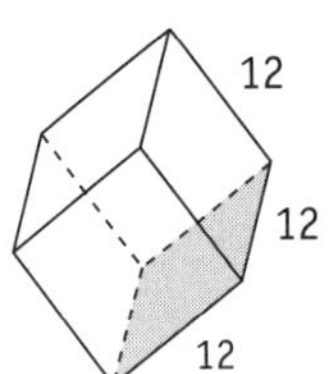

f

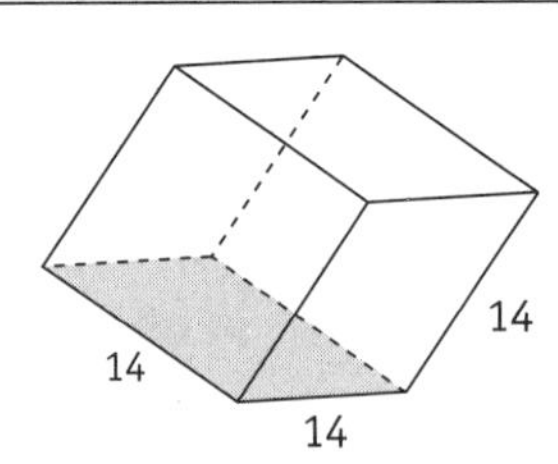

QUESTION 3

a The volume of a cube is 9261 cm^3. What is the length of each side?

b The volume of a cube is 4096 cm^3. Find its side length.

Area, volume and capacity

UNIT 9: Volume of a rectangular prism

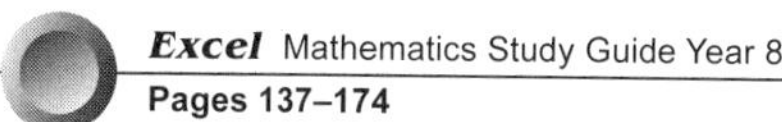

QUESTION 1 Find the volume of each rectangular prism whose dimensions are given below. All given measurements are in cm.

	Length	Breadth	Height	Volume
a	8	12	7	
b	10	9	5	
c	14	11	6	
d	9	5	6	
e	7	10	8	
f	6	7	9	

QUESTION 2 Complete the following. Measurements for length, breadth and height are in cm, for volume in cm^3

	Length	Breadth	Height	Volume
a	11	8	7	
b	9	12	14	
c	8	7		336
d		12	14	1344
e	24		16	4224
f	12	8	9	

QUESTION 3 Find the volume of each rectangular prism. All measurements are in cm.

a

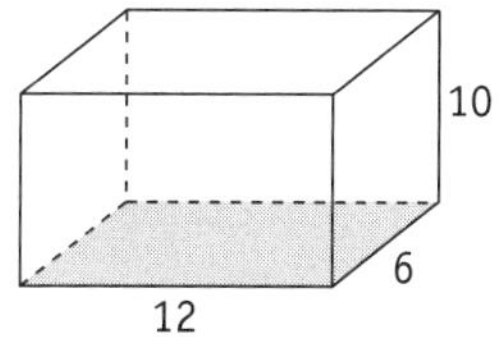

b

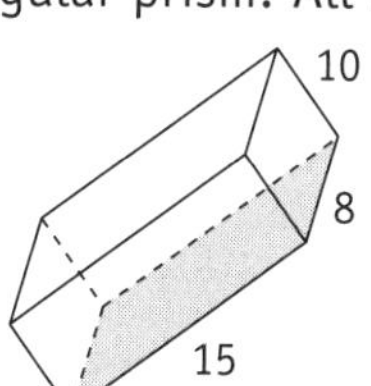

c

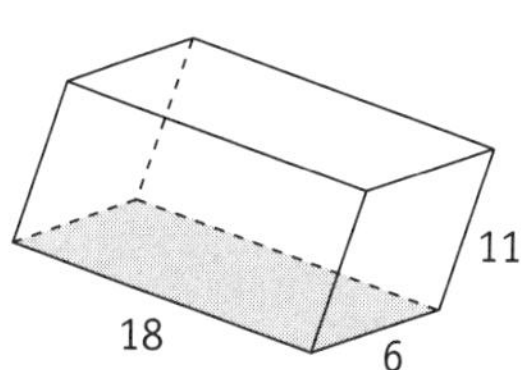

d

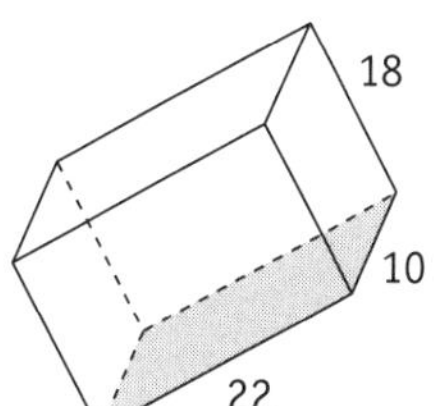

e

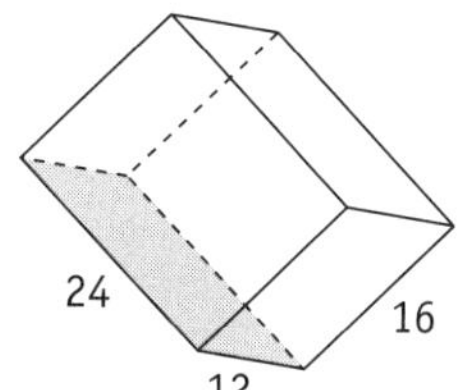

f

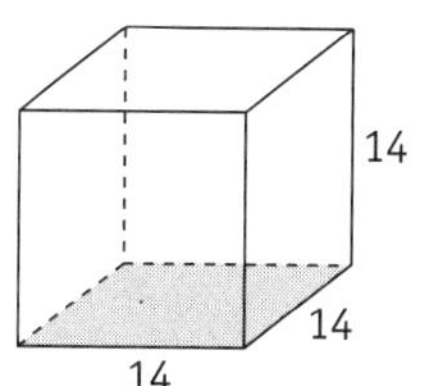

Area, volume and capacity

UNIT 10: Volume of a triangular prism

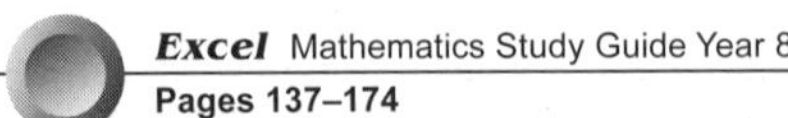

QUESTION **1** Find the volume of the following triangular prisms. All lengths are in centimetres, areas in square centimetres.

a

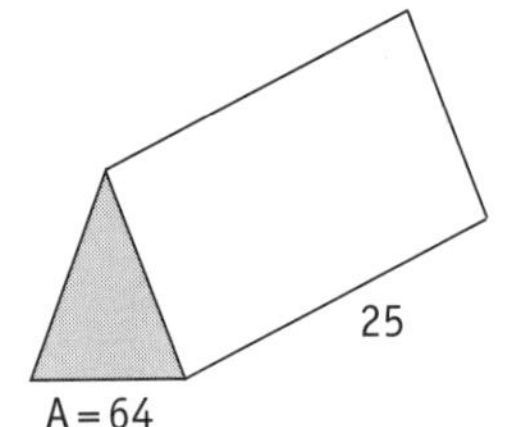

b

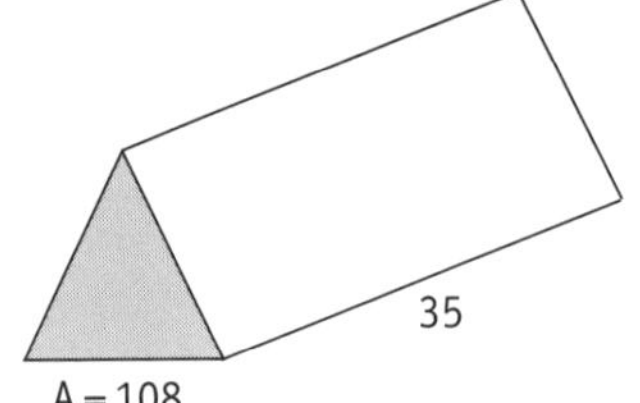

c

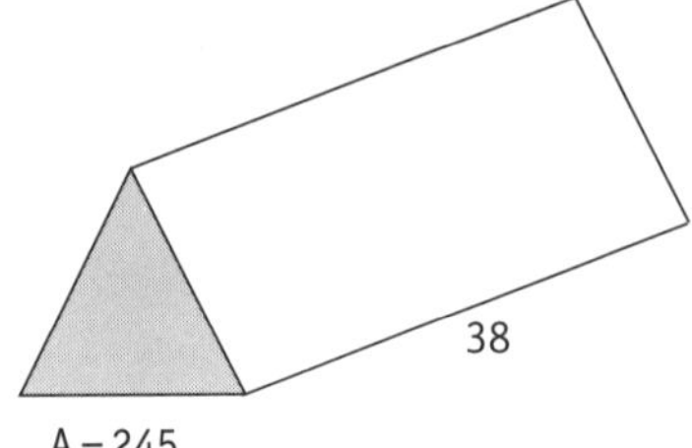

d

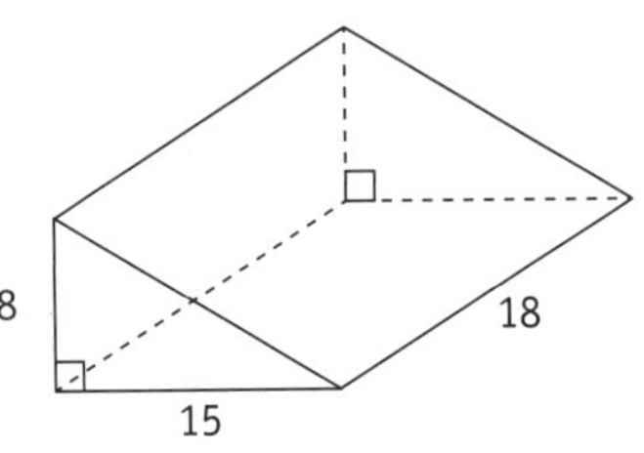

e

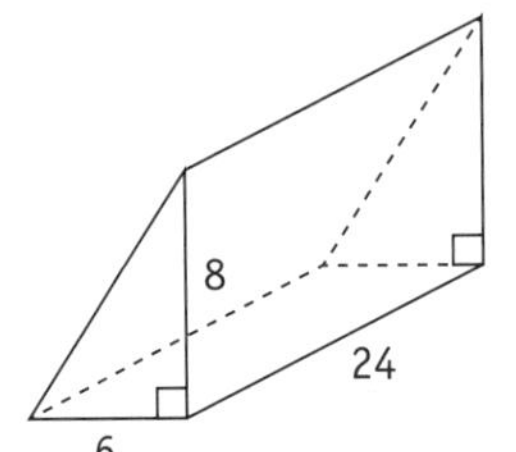

f

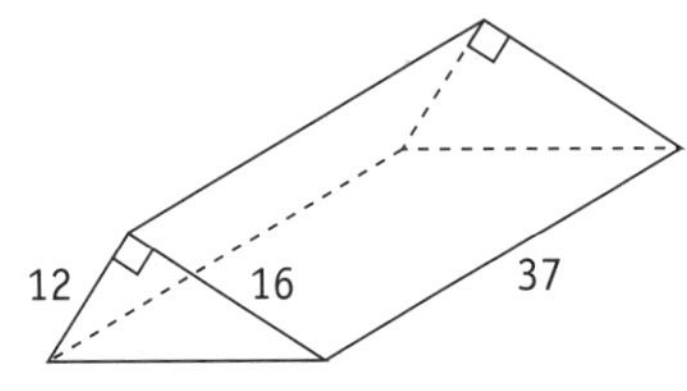

QUESTION **2** Find the volume of each of the following triangular prisms whose dimensions are given below.

	Cross-sectional area	Height	Volume
a	68 cm^2	12 cm	
b	79 m^2	11 m	
c	120 cm^2	14 cm	
d	98 cm^2	16 cm	
e	156 m^2	16 m	
f	96 m^2	12 m	

QUESTION **3** Complete the following table for each triangular prism.

	Cross-sectional area	Height	Volume
a	230 m^2	14 m	
b		12 cm	816 cm^3
c	42 m^2	12 m	
d	88 m^2		2464 m^3
e		19 cm	608 cm^3
f	74 cm^2		1554 cm^3
g	68 m^2	22 m	
h	96 m^2	18 m	

Area, volume and capacity

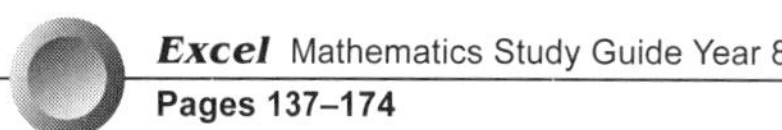

UNIT 11: Volumes of prisms—miscellaneous

QUESTION **1** Find the volume of the following prisms. Lengths are in cm, areas in cm^2.

a

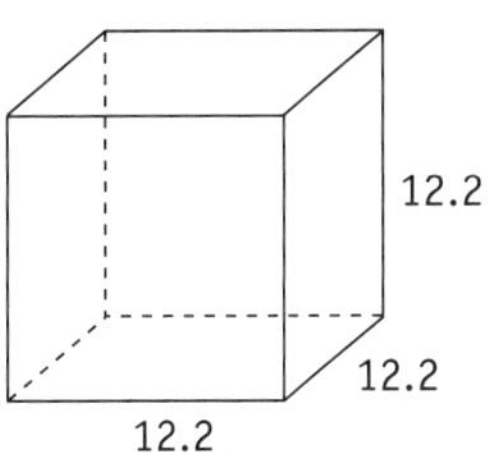

b

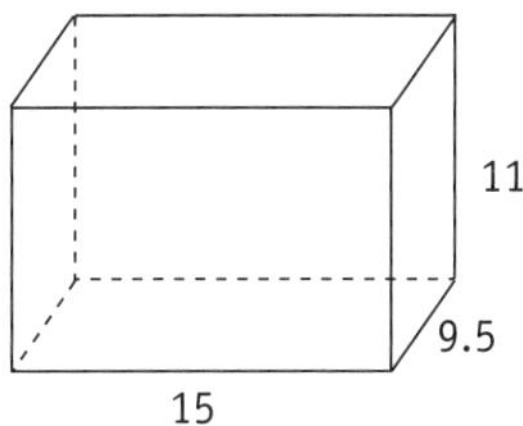

c

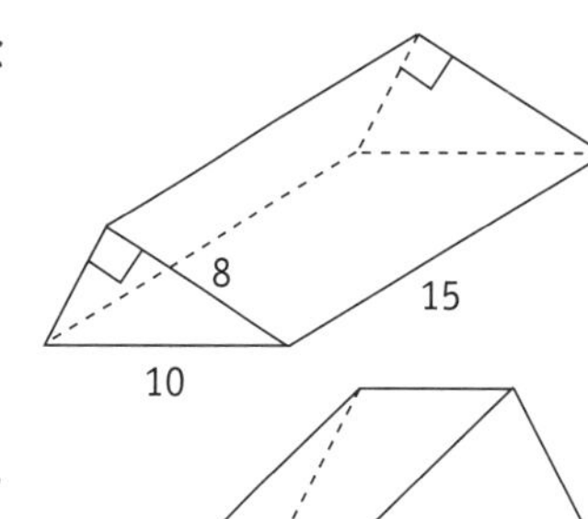

d

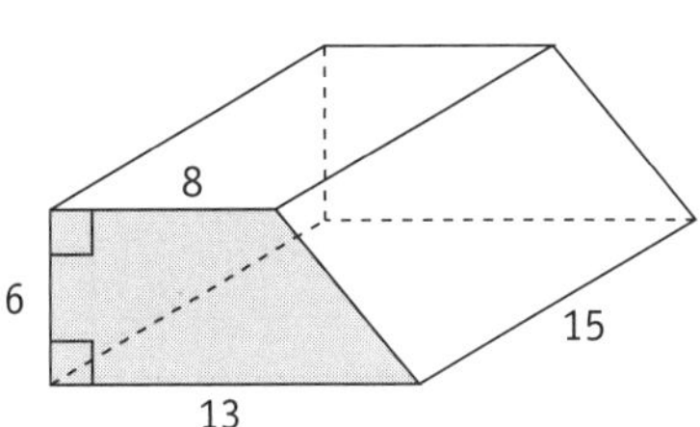

e

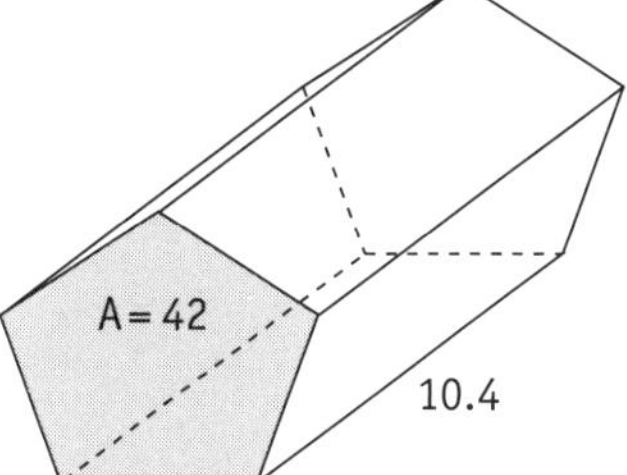

f

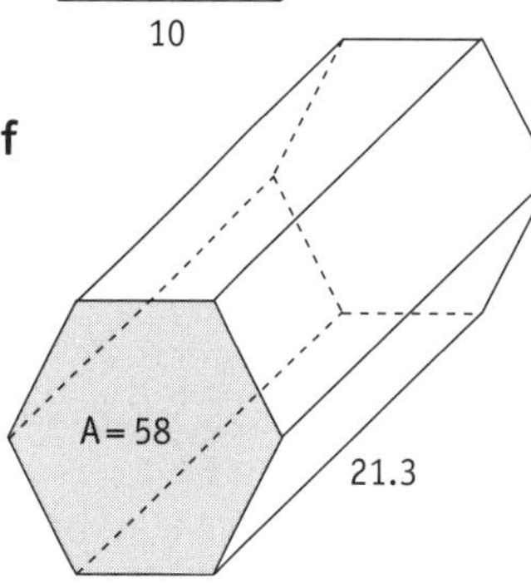

QUESTION **2** For the given solid:

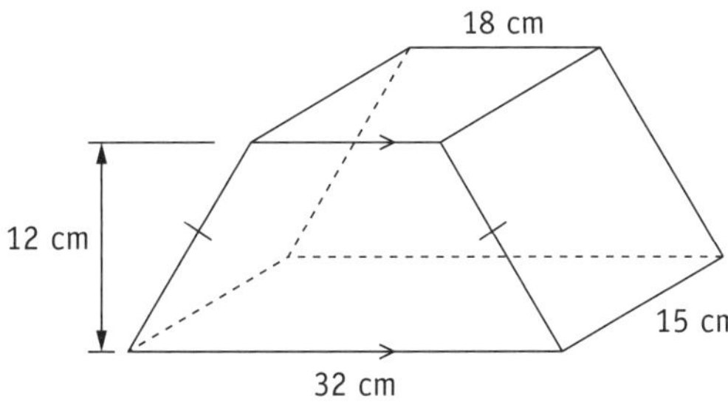

a draw it as a rectangular prism and two triangular prisms.

b find the volume of the rectangular prism. ____

c find the volume of each triangular prism. ____

d find the total volume of the solid. ____

QUESTION **3** For the given solid:

a find the shaded area. ____________________________

____

b find the volume of the solid. ____________________________

____

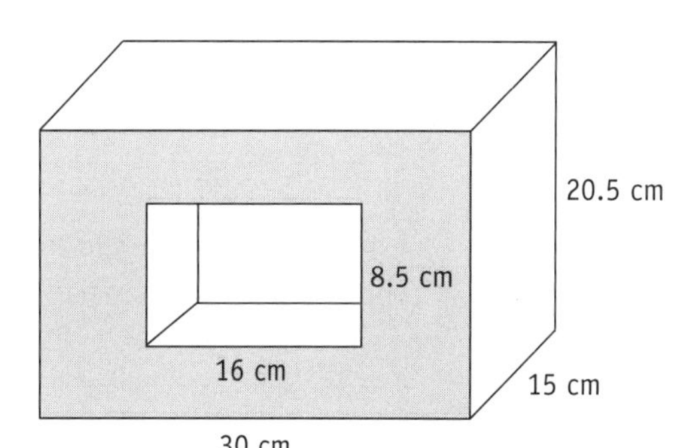

Area, volume and capacity

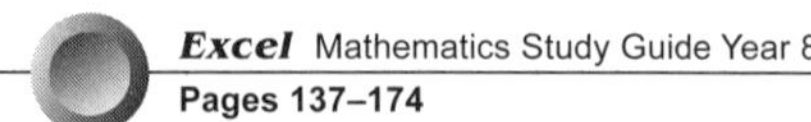

Excel Mathematics Study Guide Year 8
Pages 137–174

UNIT 12: Problem solving with volume

QUESTION 1 Calculate the volume of a rectangular prism with a base area of 384 cm^2 and a height of 18 cm.

QUESTION 2 Find the side length of a cube whose volume is 729 cm^3.

QUESTION 3 A rectangular prism has dimensions 80 mm × 110 mm × 160 mm. Calculate the volume of the prism in:

a cubic millimetres

b cubic centimetres

QUESTION 4 A small cube has dimensions 2 cm × 2 cm × 2 cm. Calculate the volume of:

a 1 small cube.

b 50 small cubes.

c Calculate the number of small cubes of this size that would fit into a rectangular box with dimensions 28 cm × 40 cm × 60 cm.

QUESTION 5 If the side length of a cube is 14 cm, find its volume.

QUESTION 6 If the cross-sectional area of a triangular prism is 360 cm^2 and its height is 50 cm. Calculate its volume.

QUESTION 7 A triangular prism has a volume of 4320 cm^3 and height of 12 cm. Find its cross-sectional area.

QUESTION 8 A prism has its side face in the form of a parallelogram with its base 30 cm and a height of 14 cm. If the depth of the prism is 42 cm, find its volume.

Area, volume and capacity

UNIT 13: Capacity and volume

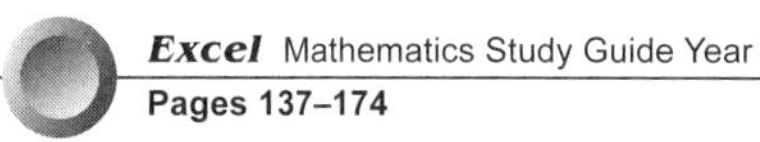

Question 1 Convert the following to the units given.

a 60 000 mL = ____________ L **b** 18.76 L = ____________ mL **c** 19.8 kL = ____________ L

d 25.3 L = ____________ mL **e** 12.5 kL = ____________ L **f** 16.54 L = ____________ mL

g 7 kL = ____________ L **h** 12 865 mL = ____________ L **i** 3486 mL = ____________ L

j 10 580 mL = ____________ L **k** 18.7 L = ____________ mL **l** 10.7 kL = ____________ L

Question 2 Convert the following to the units given.

a 10 000 mm^3 = ____________ cm^3 **b** 8 mL = ____________ cm^3 **c** 3580 cm^3 = ____________ L

d 30 000 L = ____________ kL **e** 7000 mm^3 = ____________ cm^3 **f** 38 000 cm^3 = ____________ L

g 4000 cm^3 = ____________ L **h** 26 000 cm^3 = ____________ L **i** 9635 mm^3 = ____________ cm^3

j 80 000 cm^3 = ____________ m^3 **k** 9653 mL = ____________ cm^3 **l** 8786 L = ____________ cm^3

Question 3

a A jug has a volume of 27 000 cm^3. How many litres of water can it hold?

____

b A swimming pool holds 43 000 L of water. How many kilolitres is this?

____

c A bottle contains $\frac{7}{8}$ of a litre of drink. How many millilitres is this?

____

Question 4 A fish tank measures 90 cm × 56 cm × 35 cm.

a Find the volume of the tank in cm^3. ______________________________

b How many litres of water will the tank hold? ______________________________

Question 5 Find the volume of the following prisms and then find how many millilitres of liquid each would hold.

a

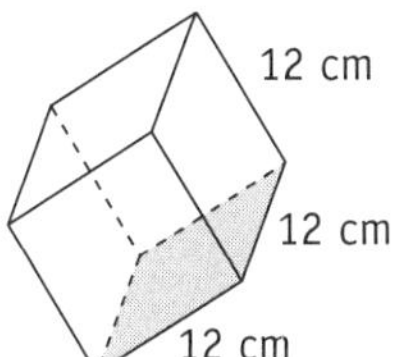

____

____

b

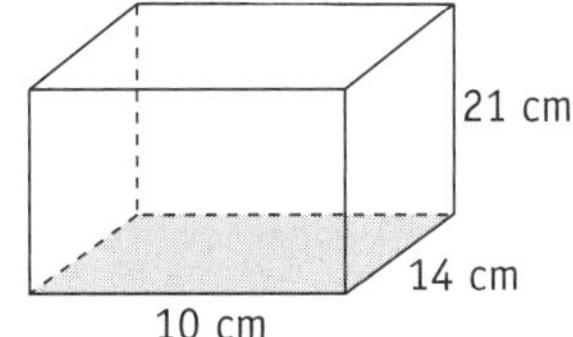

____

____

c

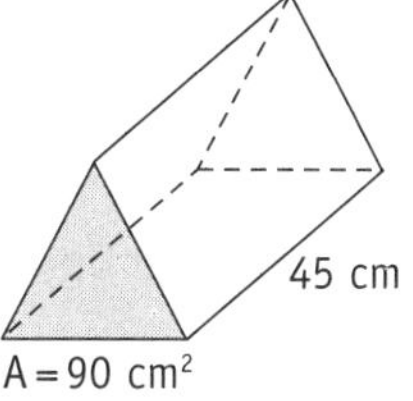

____

____

Area, volume and capacity

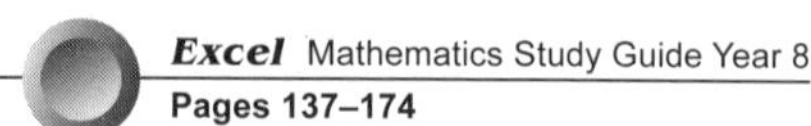

UNIT 14: Problem solving with area, volume and capacity

QUESTION **1** A rectangular tank with dimensions 80 cm × 70 cm × 58 cm is filled with water. How much water will it hold?

____

QUESTION **2** A dripping tap loses 8 mL of water every 30 seconds. How much water will be lost in 5 hours?

____

QUESTION **3** Find the side length of a cube that has a volume of 9261 cm^3.

____

QUESTION **4** Madeleine drank $\frac{3}{4}$ of a litre of milk. How many millilitres is this?

____

QUESTION **5** Eighty children went on a picnic. Each child drank a can of drink containing 350 mL. How many litres was this?

____

QUESTION **6** A rectangular prism has dimensions 14 cm x 12 cm x 25 cm.

a What is its volume? ______________________________

b What would be its volume if all the dimensions were doubled. ______________________________

QUESTION **7** Find the area of a photograph that is 80 cm long and 50 cm wide.

____

QUESTION **8** The perimeter of a square is 70 cm. Find its area.

____

QUESTION **9** A triangle has an area of 2960 cm^2. If the base of the triangle is 80 cm long, find its height.

____

QUESTION **10** Jasbir drank 2.5 litres of milk. How many millilitres is this?

____

QUESTION **11** One of the faces of a cube has an area of 324 cm^2. What is the volume of the cube?

____

QUESTION **12** Find the total area of the four walls of a room 12 m long, 7 m wide and 4 m high.

____

Area, volume and capacity

TOPIC TEST — PART A

Instructions
- This part consists of 10 multiple-choice questions.
- Fill in only ONE CIRCLE for each question.
- Each question is worth 1 mark.

Time allowed: 15 minutes — **Total marks: 10**

Question	Marks
1 A rectangular field measures 650 m by 760 m. Its area in hectares is Ⓐ 494 ha Ⓑ 49.4 ha Ⓒ 4.94 ha Ⓓ 0.494 ha	1
2 When half full, a petrol tank holds 32 litres. How much more petrol does it hold when it is three-quarters full? Ⓐ 16 L Ⓑ 48 L Ⓒ 50 L Ⓓ 64 L	1
3 The area of a parallelogram with base 12 mm, side 11 mm and height 10 mm is Ⓐ 110 mm^2 Ⓑ 132 mm^2 Ⓒ 120 mm^2 Ⓓ 1320 mm^2	1
4 How many kL can be held in a container with a volume of 6.5 m^3? Ⓐ 6.5 kL Ⓑ 65 kL Ⓒ 650 kL Ⓓ 6500 kL	1
5 Find the difference in area between the square and kite shown. Ⓐ 9 cm^2 Ⓑ 29 cm^2 Ⓒ 69 cm^2 Ⓓ 89 cm^2	1
6 What is double the volume of the prism shown? Ⓐ 66 m^3 Ⓑ 132 m^3 Ⓒ 264 m^3 Ⓓ 1056 m^3	1
7 How many tiles 15 cm × 15 cm are needed to cover a floor measuring 6 m × 9 m? Ⓐ 240 000 Ⓑ 24 000 Ⓒ 2400 Ⓓ 240	1
8 How many mL in 0.04 L ? Ⓐ 4 mL Ⓑ 40 mL Ⓒ 400 mL Ⓓ 4000 mL	1
9 What is the side length of a cube that holds 1L? Ⓐ 1 cm Ⓑ 10 cm Ⓒ 100 cm Ⓓ 1000 cm	1
10 What is the height of a rhombus if its area is 216 cm^2 and its base is 180 mm? Ⓐ 1.2 cm Ⓑ 1.2 mm Ⓒ 120 cm Ⓓ 120 mm	1

Figures for question 5:

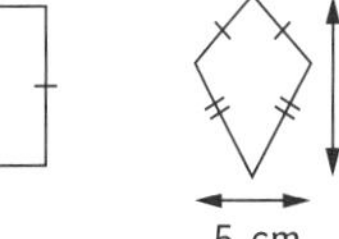

Figure for question 6:

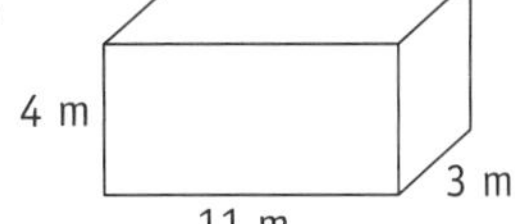

Total marks achieved for PART A

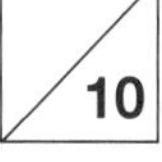

Area, volume and capacity

TOPIC TEST — PART B

Instructions
- This part consists of 13 questions.
- Each question is worth 1 mark unless indicated otherwise.
- Write only the answer in the answer column.

Time allowed: 20 minutes | **Total marks: 15**

	Questions	Answers	Marks
1	Find the volume of a triangular prism with base 5 cm, perpendicular height 8 cm and depth 9 cm.		1
2	Covert 375 mL to litres.		1
3	If the area of a triangle is 240 mm^2 and its base is 60 mm, find its height.		1
4	Find the volume of a cube if the area of one face is 49 cm^2.		1
5	How many millilitres are in 7.2 kilolitres?		1
6	A tub has a volume of 25 000 cm^3. How many litres of water can it hold?		1
7	Find the number of cubic millimetres (mm^3) in one cubic metre.		2
8	What is the area of this trapezium? (Trapezium with parallel sides 6.2 m and 7.4 m, base 8 m, slant side 8.3 m)		1
9	Find the area of a triangle with base 12 mm and height 4 cm. Answer in cm^2.		1
10	There are 27 students in a Food Technology class and each one needs 450 mL of milk to make custard. How many 2L bottles of milk will need to be purchased so that each student has sufficient milk?		1
11	How many complete 5 mL doses can be given from a medicine bottle with a volume of 58 cm^3?		1
12	Find the area of a kite with diagonals 18 cm and 26 cm.		1
13	What volume of sand will be needed to fill a sandpit in the shape of a square to a depth of 16 cm, if the side length of the sandpit is 2.2 m? Answer in cm^3.		2

Total marks achieved for PART B

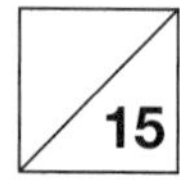

CHAPTER 8

Circles

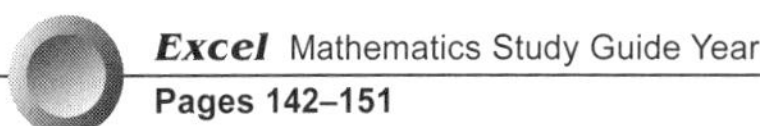

Excel Mathematics Study Guide Year 8
Pages 142–151

UNIT 1: Parts of a circle

QUESTION **1** Name each part of the circle indicated.

a

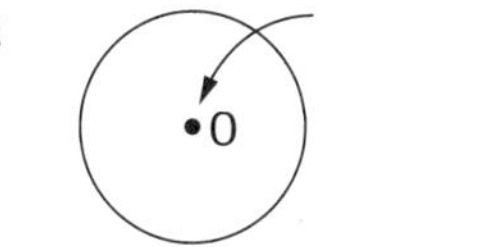

____

b

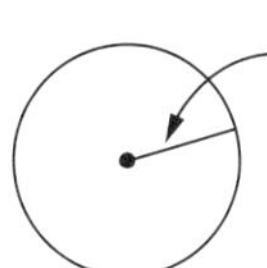

____

c

____

d

____

e

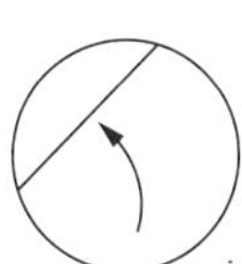

____

QUESTION **2** Name the part of the circle indicated below.

a

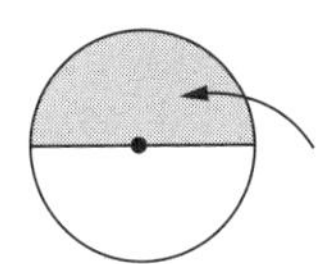

____

b

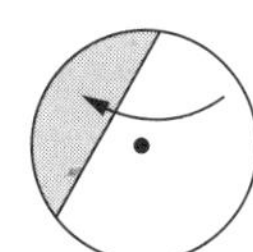

____

c

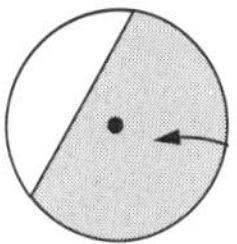

____

d

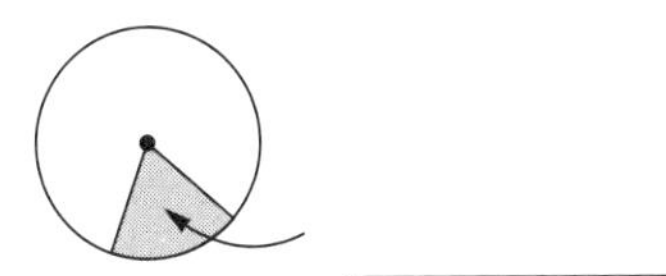

____

e

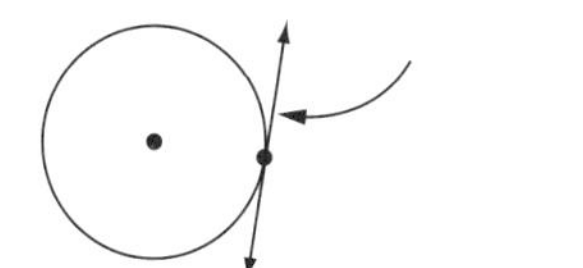

____

f

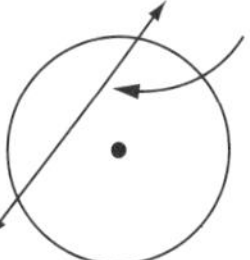

____

g

____

h

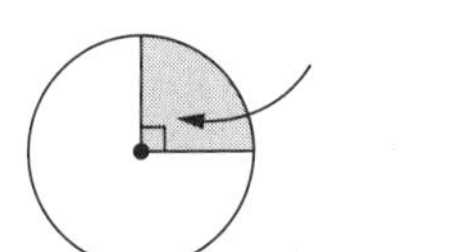

____

QUESTION **3** Find the fraction of the circle given below.

a

____

b

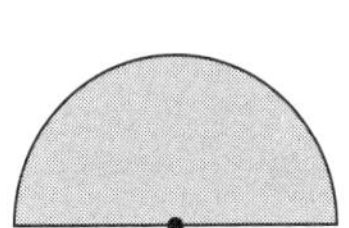

____

c

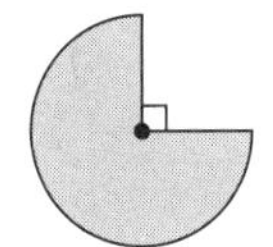

____

d

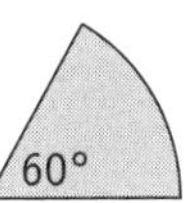

____

QUESTION **4**

a If the diameter of a circle is 16 cm, find its radius. ______________

b If the radius of a circle is 7.5 cm, find its diameter. ______________

c The formula for the circumference of a circle is ______________ .

d The formula for the area of a circle is ______________ .

e A circle can be divided into ______________ quadrants.

Circles

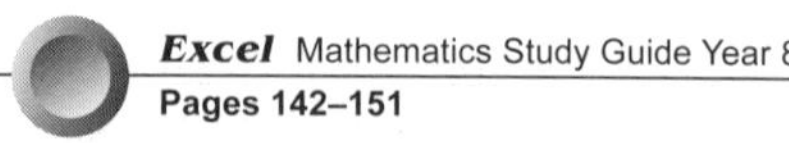

UNIT 2: Circumference of a circle when radius is given

QUESTION 1 By using as the value of π, find the approximate length of the circumference for these circles. All measurements are in cm.

a 

b 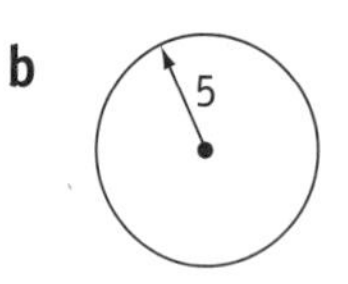

c 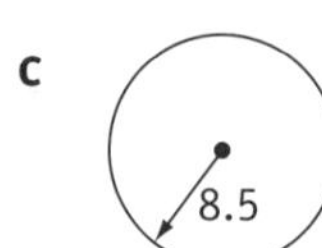

d 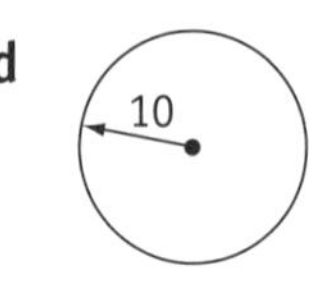

QUESTION 2 Calculate the circumference of the following circles correct to 2 decimal places. All measurements are in cm.

a 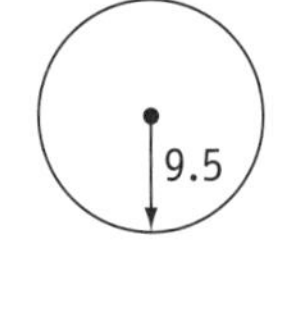

b 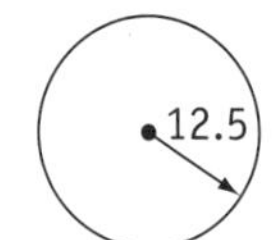

c 

d 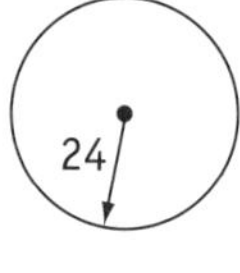

QUESTION 3 Calculate the circumference of each circle, giving answers correct to 3 significant figures. All measurements are in cm.

a 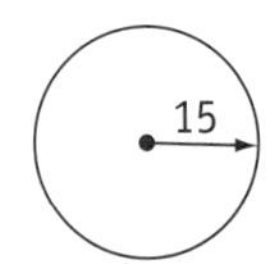

b 

c 

d 

QUESTION 4 Calculate the perimeter of each figure correct to 2 decimal places. All measurements are in cm.

a 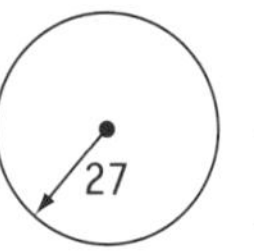

b 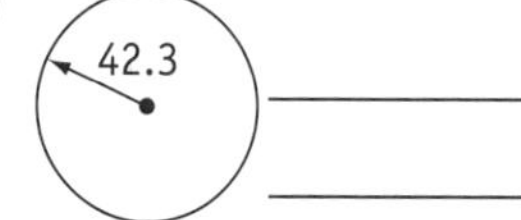

c 

Circles

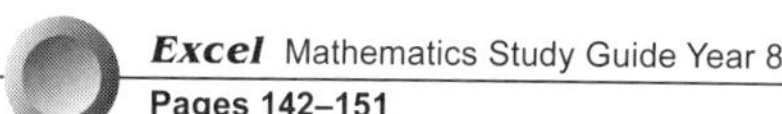

UNIT 3: Circumference of a circle when diameter is given

QUESTION 1 By using as the value of π, find the approximate length of the circumference for these circles.

a 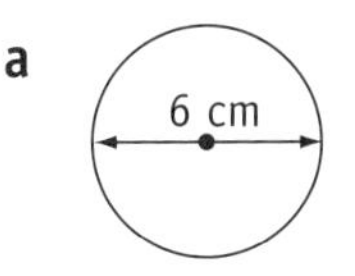

b 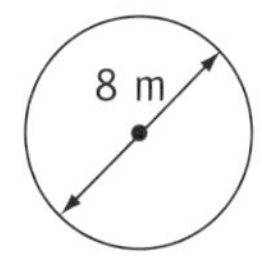

c 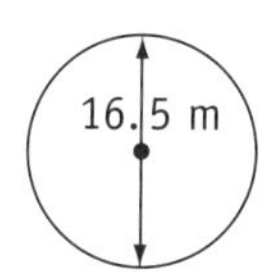

d 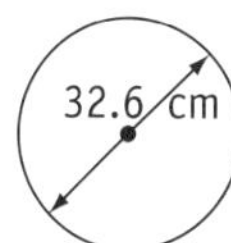

QUESTION 2 Calculate the circumference of the following circles correct to 2 decimal places.

a 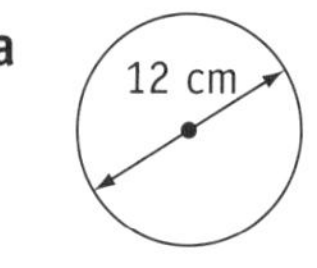

b 

c 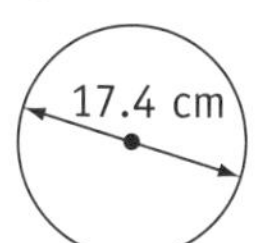

d 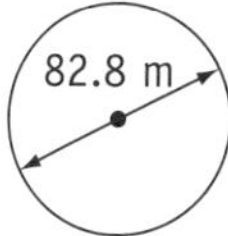

QUESTION 3 Calculate the circumference of these circles. Give your answers correct to 3 significant figures.

a 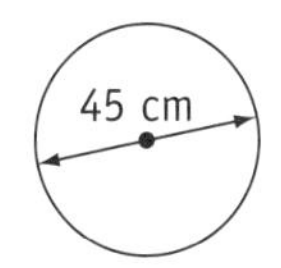

b 

c 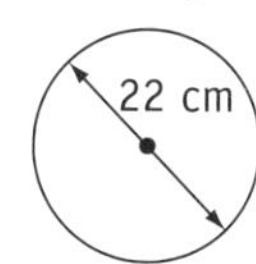

d 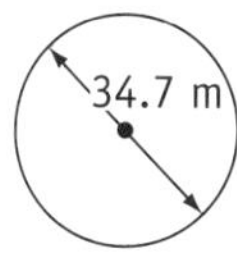

QUESTION 4 Calculate the perimeter of these figures correct to 1 decimal place.

a 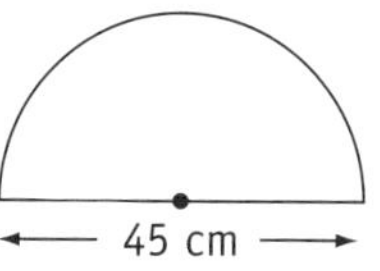

b 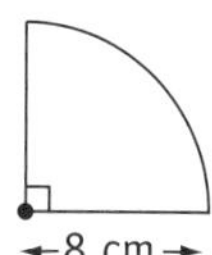

c 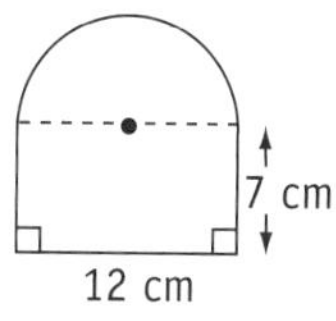

UNIT 4: Applications of circumference

Excel Mathematics Study Guide Year 8
Pages 142–151

QUESTION 1 Calculate the total perimeter of the following semi-circles correct to one decimal place.

a
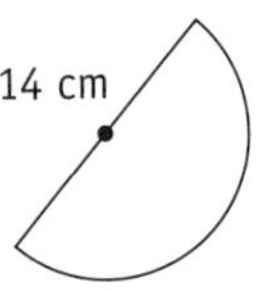

b
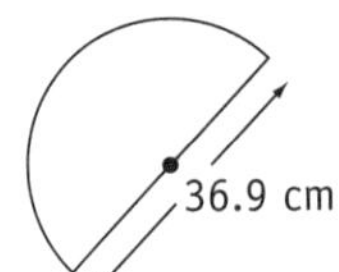

c
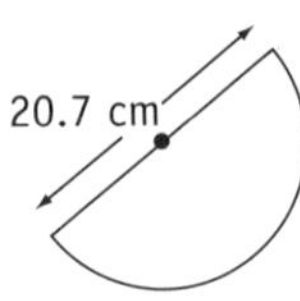

QUESTION 2 Find the total perimeter of these quadrants correct to two decimal places.

a
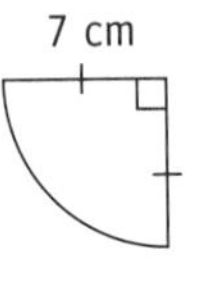

b
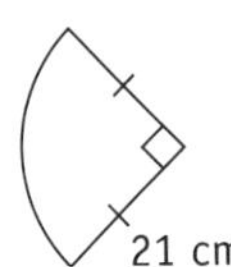

c
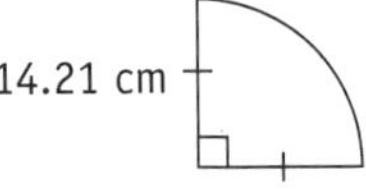

QUESTION 3 Calculate the perimeter of each of these figures correct to three significant figures.

a
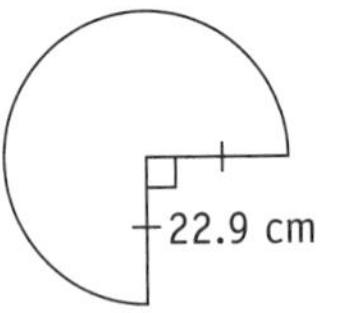

b
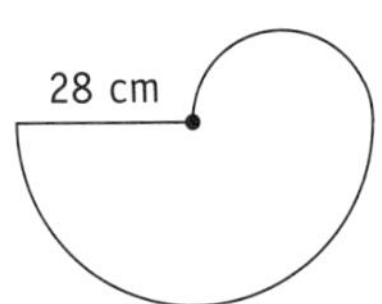

c
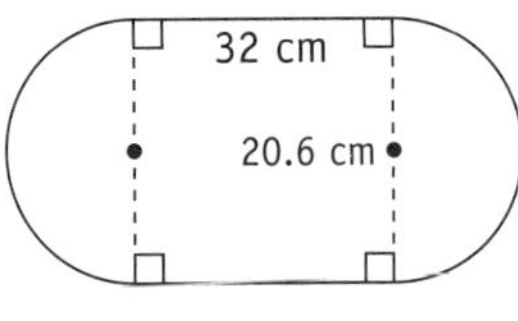

d
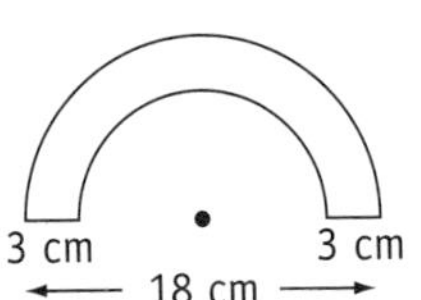

e
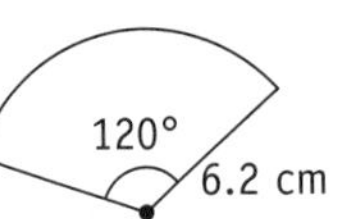

f
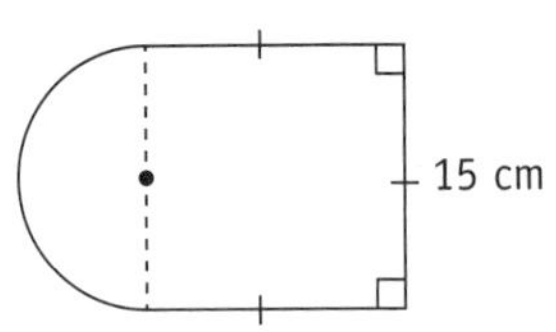

g
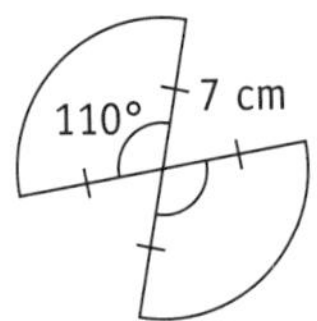

h
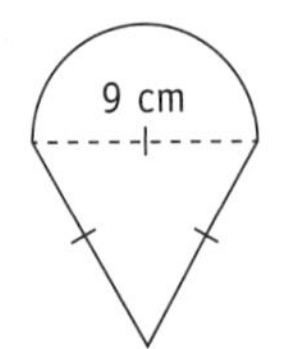

i
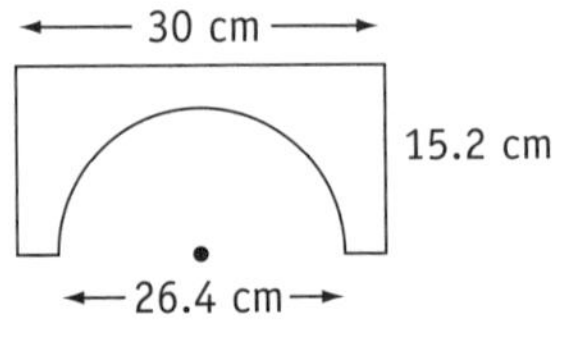

Circles

UNIT 5: Area of a circle

Excel Mathematics Study Guide Year 8
Pages 142–151

Question 1 By using as the value of π, find the approximate area of these circles.

a

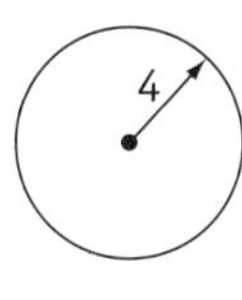

b

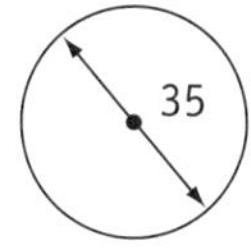

c

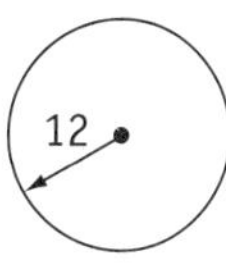

Question 2 Calculate the area of each circle correct to 2 decimal places. All measurements are in cm.

a

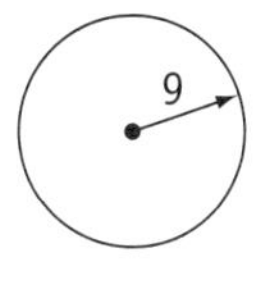

b

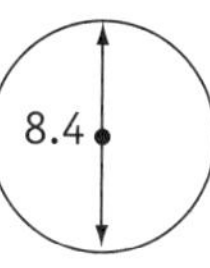

c

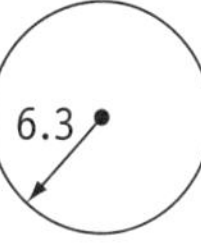

d

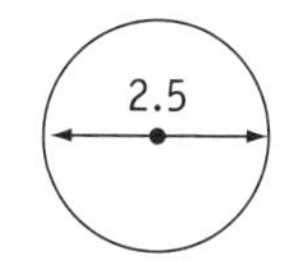

e

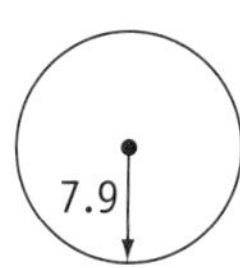

f

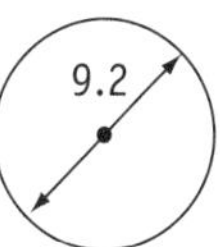

Question 3 Calculate the area of these circles correct to 3 significant figures. All measurements are in cm.

a

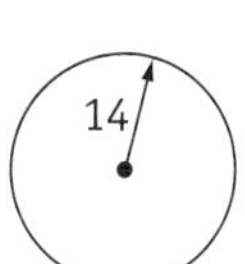

b

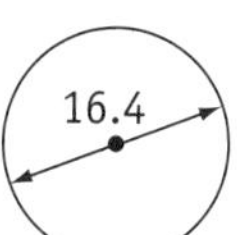

c

Question 4 Calculate the shaded area correct to 1 decimal place.

a

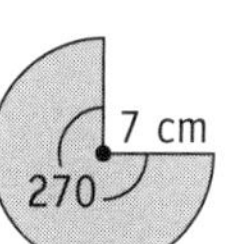

b

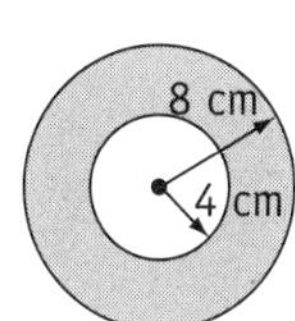

Circles

UNIT 6: Applications of area

QUESTION **1** Calculate the area of the following semi-circles correct to one decimal place.

a

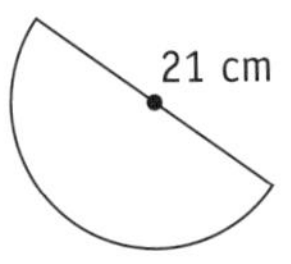

b

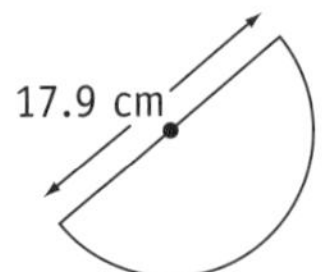

c

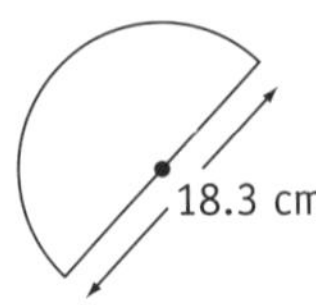

QUESTION **2** Calculate the area of these quadrants correct to two decimal places.

a

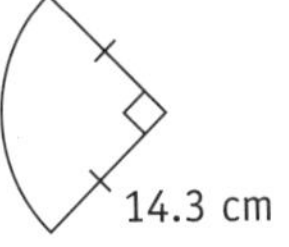

b

c

QUESTION **3** Calculate the shaded area of these shapes correct to 2 decimal places.

a

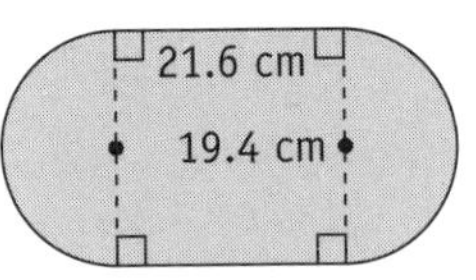

b

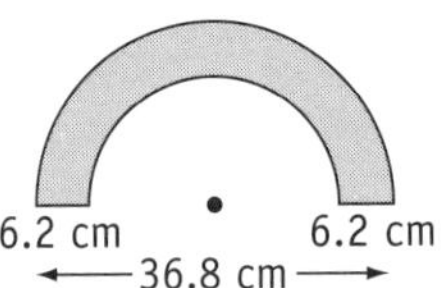

c

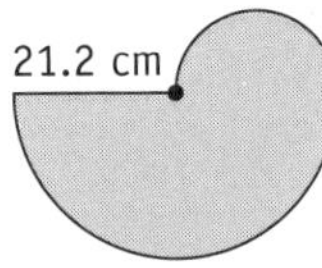

d

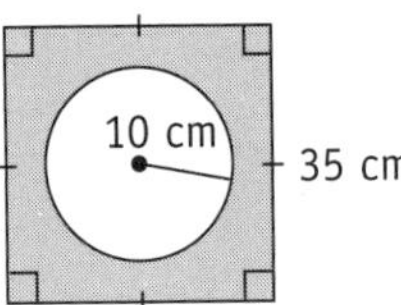

e

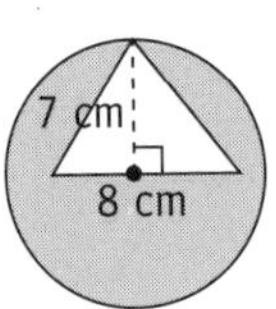

f

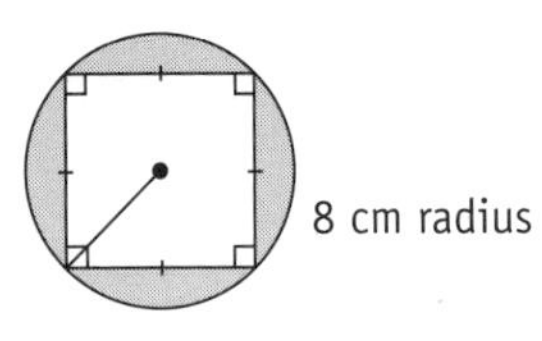

g

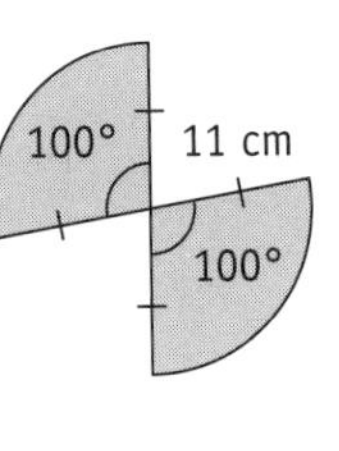

h

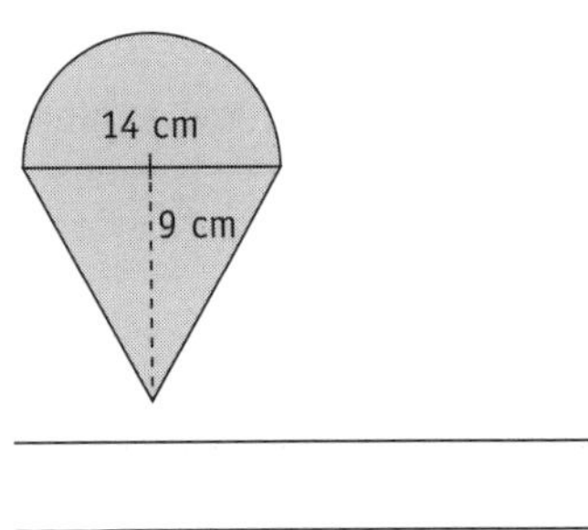

i

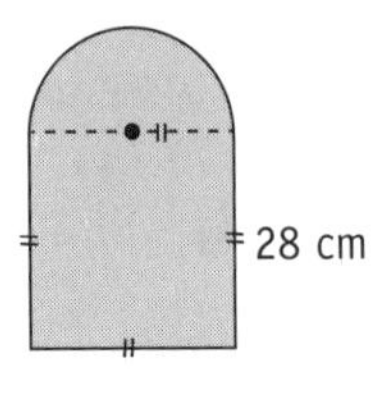

UNIT 7: Problem solving with circles

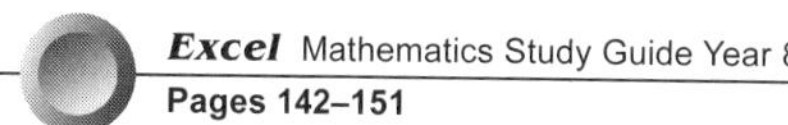

Answer the following questions to one decimal place where necessary.

Question 1 Calculate the circumference of a 20 cent coin if its diameter is 2.85 cm.

____

Question 2 Calculate the diameter of a circle with a circumference 572 cm.

____

Question 3

a A circular track has radius 42 m. Caclulate the distance around the track.

____

b How many laps must an athlete run to complete the 1500 m event?

____

Question 4 Find the perimeter of a circular cake tin with diameter 27 cm.

____

Question 5 How many complete circles with radius 4 cm can be cut from a piece of fabric measuring 5 m by 1.3 m?

____

Question 6 Find the area of a round tablecloth if its diameter is 5 metres.

____

Question 7 The radius of a car wheel is 24 cm. How far does the car travel with 100 complete turns?

____

Question 8 What is the cooking area of a circular camping barbeque grill having radius 14 cm?

____

Question 9 What distance do you ride in one turn of a merry-go-round when you sit 8.5 metres from the centre?

____

Question 10 Find the circumference of the blade of a circular saw with diameter 22 cm.

____

Question 11 What is the area of the largest circle that can be cut from a piece of cardboard 15 cm by 18 cm?

____

Circles

TOPIC TEST

PART A

Instructions
- This part consists of 10 multiple-choice questions.
- Fill in only ONE CIRCLE for each question.
- Each question is worth 1 mark.
- Calculators are allowed.

Time allowed: 10 minutes **Total marks: 10**

Marks

1 The diameter of a circle is 24 cm. Its radius is 1

Ⓐ 6 cm Ⓑ 8 cm Ⓒ 12 cm Ⓓ 16 cm

2 The shaded area in the figure is called a 1

Ⓐ semi-circle Ⓑ segment Ⓒ chord Ⓓ sector

3 The area of a circle of radius 13 cm is closest to 1

Ⓐ 530 cm^2 Ⓑ 130 cm^2 Ⓒ 80 cm^2 Ⓓ 2120 cm^2

4 The circumference of a circle with diameter 39 m is closest to 1

Ⓐ 240 m Ⓑ 60 m Ⓒ 120 m Ⓓ 1200 m

5 A circle has a circumference 132 km in length. Its radius is closest to 1

Ⓐ 13 km Ⓑ 21 km Ⓒ 42 km Ⓓ 207 km

6 A circle has an area of 23 m^2. Its radius is closest to 1

Ⓐ 6 m Ⓑ 2.4 m Ⓒ 7.3 m Ⓓ 2.7 m

7 Which is closest to the area of a semi-circle of diameter 20 cm? 1

Ⓐ 157 cm^2 Ⓑ 314 cm^2 Ⓒ 1257 cm^2 Ⓓ 628 cm^2

8 Which is closest to the area of a quadrant of radius 82 mm? 1

Ⓐ 528 mm^2 Ⓑ 53 cm^2 Ⓒ 2 m^2 Ⓓ 10 562 mm^2

9 A large circle has its diameter 4 times the length of the diameter of a smaller circle. How many times larger is the circumference of the larger circle than that of the smaller circle? 1

Ⓐ 2 Ⓑ 4 Ⓒ 8 Ⓓ 16

10 A large circle has its diameter 4 times the length of the diameter of a smaller circle. How many times larger is the area of the larger circle than that of the smaller circle? 1

Ⓐ 2 Ⓑ 4 Ⓒ 8 Ⓓ 16

Total marks achieved for PART A /10

Circles

TOPIC TEST

PART B

Instructions
- This part consists of 4 questions.
- Each question is worth the marks indicated.
- Show all working.
- Answer to one decimal place.

Time allowed: 20 minutes **Total marks: 20**

Marks

1 a

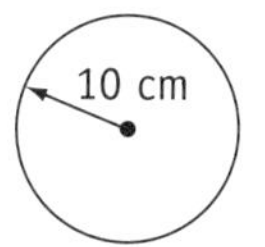

circumference = ______

area = ______

b

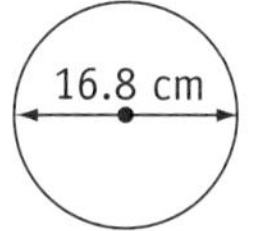

circumference = ______

area = ______

c

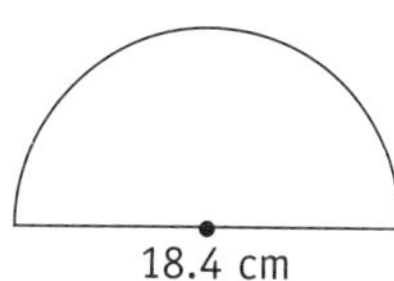

perimeter = ______

5

2 Calculate the perimeter of each shape.

a

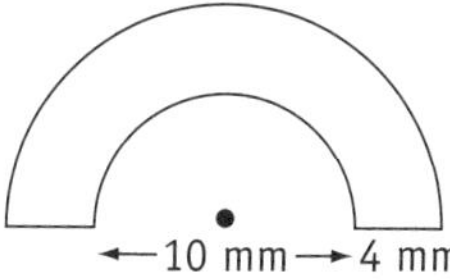

______

b

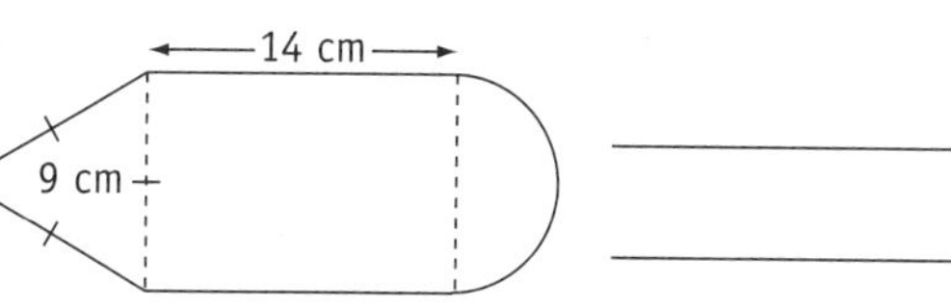

______

4

3 Find the (shaded) area of each figure.

a

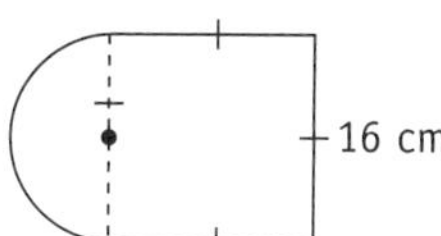

______

b

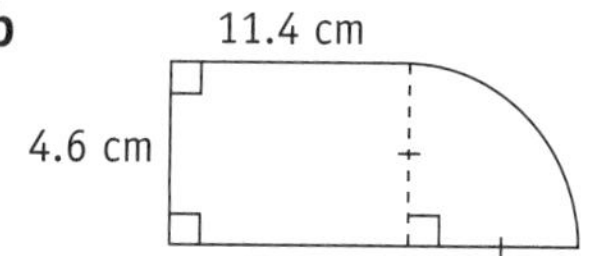

______

4

c

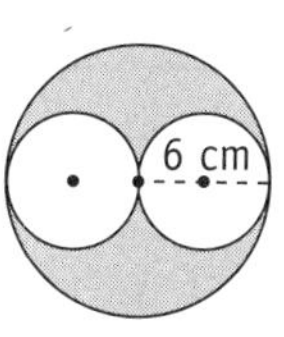

______

d

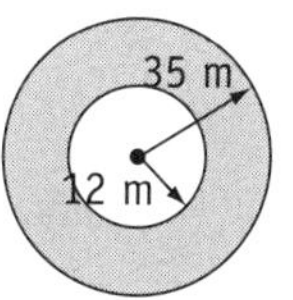

______

4

4 A circle with diameter 1.2 m is cut from a square piece of wood with sides 1.5 m long. What is the area of the leftover wood once the circle is cut out?

______

3

Total marks achieved for PART B

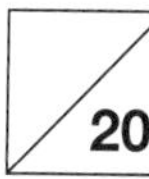

Chapter 9

Linear relationships

Excel Mathematics Study Guide Year 8
Pages 66–86

UNIT 1: The number plane

Question 1 Write the letter used to name the following points.

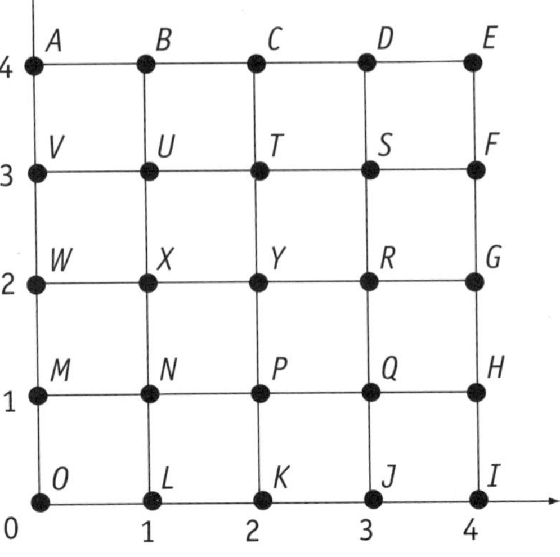

a (0, 0) ______ **b** (1, 1) ______
c (2, 2) ______ **d** (3, 3) ______
e (4, 4) ______ **f** (0, 1) ______
g (0, 2) ______ **h** (0, 3) ______
i (0, 4) ______ **j** (1, 0) ______
k (2, 0) ______ **l** (3, 0) ______
m (4, 0) ______ **n** (2, 1) ______
o (2, 3) ______ **p** (2, 4) ______
q (1, 2) ______ **r** (1, 3) ______
s (1, 4) ______ **t** (3, 1) ______
u (3, 2) ______ **v** (3, 4) ______ **w** (4, 1) ______ **x** (4, 2) ______
y (4, 3) ______

Question 2 Write down the coordinates of the following points shown in the number plane.

a *A* ______
b *B* ______
c *C* ______
d *D* ______
e *E* ______
f *F* ______
g *G* ______
h *H* ______
i *I* ______
j *J* ______
k *K* ______
l *L* ______
m *M* ______
n *N* ______
o *O* ______
p *P* ______
q *Q* ______ **r** *R* ______ **s** *S* ______ **t** *T* ______
u *U* ______ **v** *V* ______ **w** *W* ______ **x** *X* ______
y *Y* ______ **z** *Z* ______

Linear relationships

UNIT 2: Tables of values

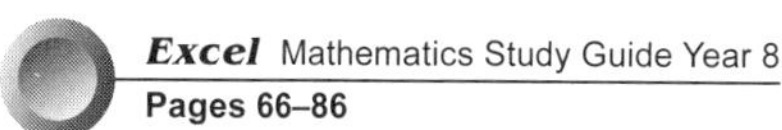

QUESTION 1 Complete each table of values.

a $y = x + 3$

x	y
0	
1	
2	
3	

b $y = 2x + 1$

x	y
–1	
0	
1	
2	

c $y = 3x - 2$

x	y
–2	
–1	
0	
1	

d $y = x + 2$

x	y
–1	
0	
1	
2	

e $y = x - 2$

x	y
–2	
–1	
0	
1	

f $y = 2x - 3$

x	y
–1	
0	
1	
2	

QUESTION 2 Complete each table of values.

a $y = 2x + 2$

x	y
–1	
0	
1	
2	

b $p = 2q$

q	p
–1	
0	
1	
2	

c $c = 3d - 4$

d	c
–1	
0	
1	
2	

d $m = 2n - 1$

n	m
–1	
0	
1	
2	

e $s = t + 4$

t	s
–1	
0	
1	
2	

f $a = b - 3$

b	a
–1	
0	
1	
2	

Excel Mathematics Study Guide Year 8
Pages 66–86

UNIT 3: Graphing ordered pairs and number patterns

Question 1

a Plot the following ordered pairs on the number plane:
$A(-2, -1)$ $B(-1, 0)$ $C(0, 1)$ $D(2, 3)$ $E(3, 4)$

b Are the points are collinear?

____

c Join the collinear points on the number plane.

d What is the rule? ______________

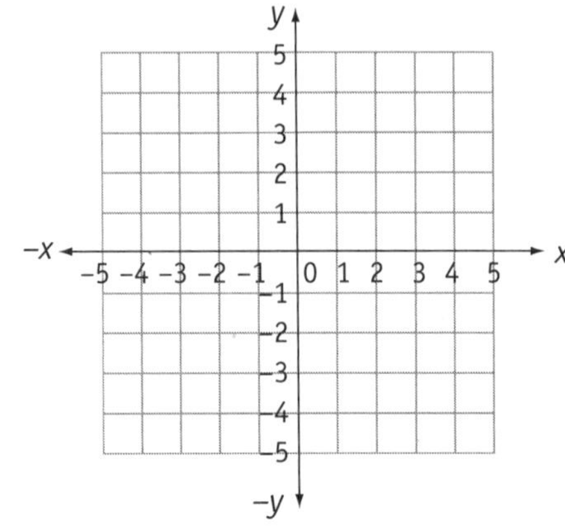

Question 2

a Complete the table for the rule $y = 2x + 1$.

b Write the set of ordered pairs formed.

____

c Plot the set of ordered pairs on the number plane.

d Are the points collinear?

____

x	y
−1	
0	
1	
2	

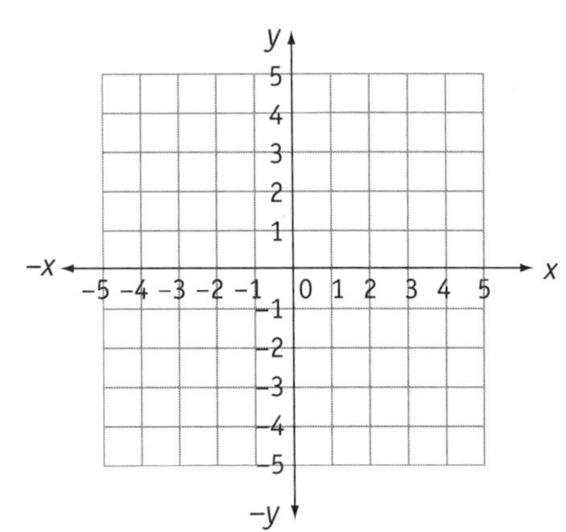

Question 3

a Plot the set of ordered pairs:
$A(-1, 0)$ $B(2, 4)$ $C(2, 0)$

b What type of triangle is $\triangle ABC$? ______________

c What is the length of side AC? ______________

d What is the length of side BC? ______________

e Use Pythagoras' theorem to calculate the length of side AB.

____

f What is the perimeter of $\triangle ABC$? ______________

g What is the area of $\triangle ABC$? ______________

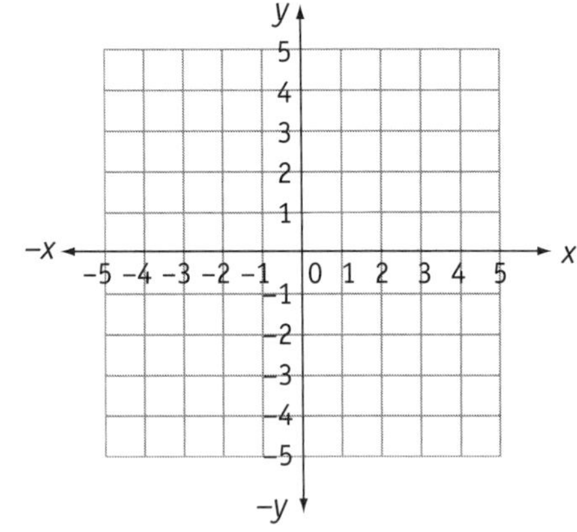

Question 4 Complete each table and write the set of ordered pairs formed alongside.

a $y = x + 5$

x	y	(x, y)
0		
1		
2		
3		

b $y = 2x + 3$

x	y	(x, y)
−1		
0		
1		
2		

c $y = 3x - 2$

x	y	(x, y)
−1		
0		
1		
2		

UNIT 4: Graphing lines on the number plane

QUESTION 1 Complete the tables of values and then graph the equation on the number plane.

a $y = x + 2$

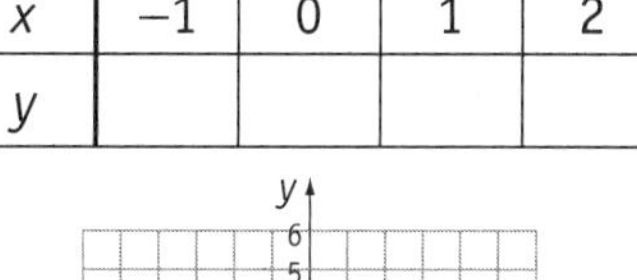

x	−1	0	1	2
y				

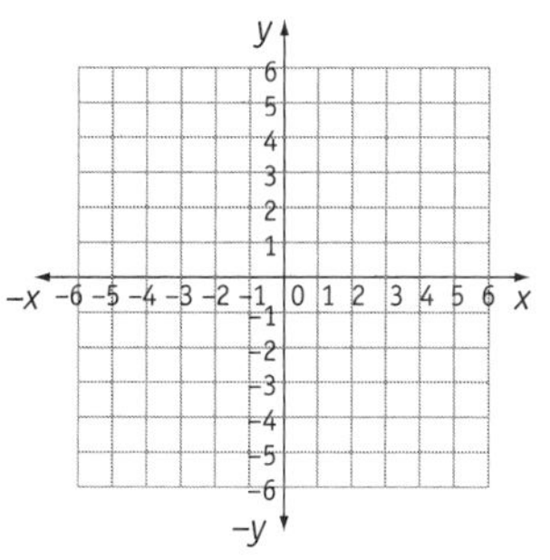

b $y = 2x - 1$

x	−1	0	1	2
y				

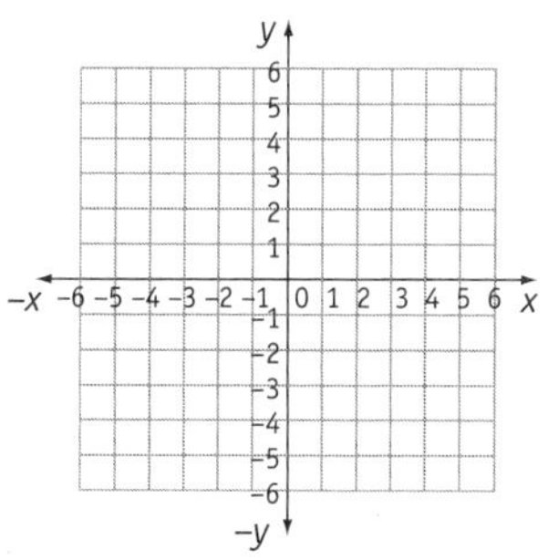

c $y = -2x + 1$

x	−1	0	1	2
y				

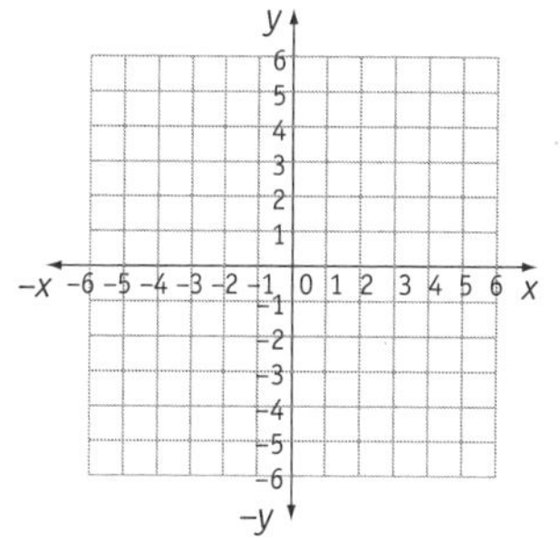

QUESTION 2 Complete the tables of values and then graph the equation on the number plane.

a $x = -1$

x				
y	−1	0	1	2

b  $y = 2$

x	−1	0	1	2
y				

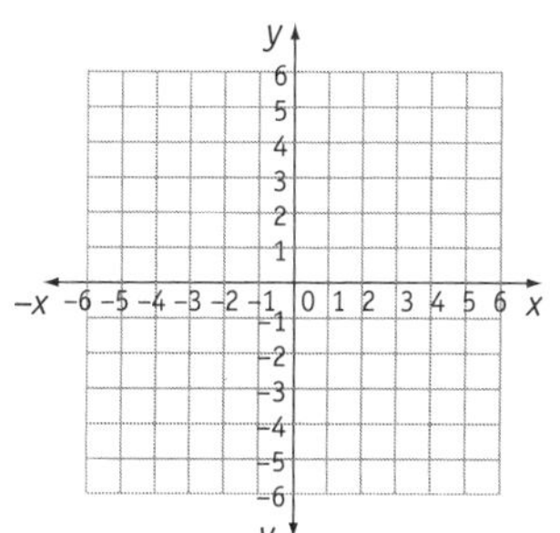

c $x = -4$

x				
y	−1	0	1	2

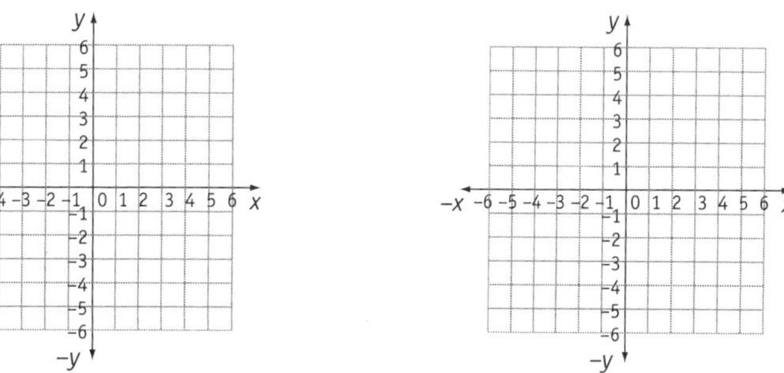

QUESTION 3 On the same number plane, graph the following equations by first completing the tables of values. Find their point of intersection.

a $y = 2x$

x	−1	0	1	2
y				

$y = -2x$

x	−1	0	1	2
y				

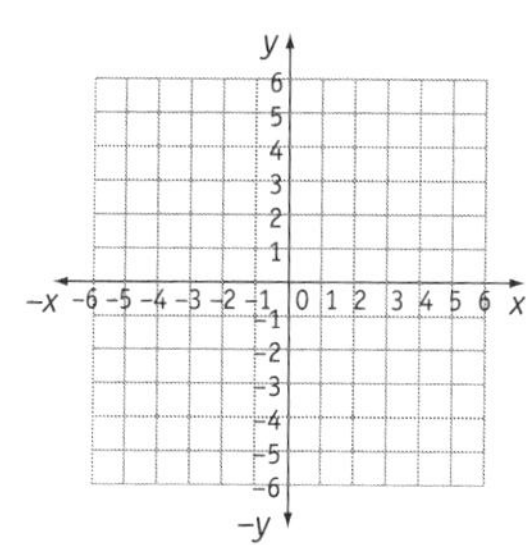

____

b $y = 2x + 2$

x	−1	0	1	2
y				

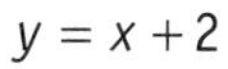 $y = x + 2$

x	−1	0	1	2
y				

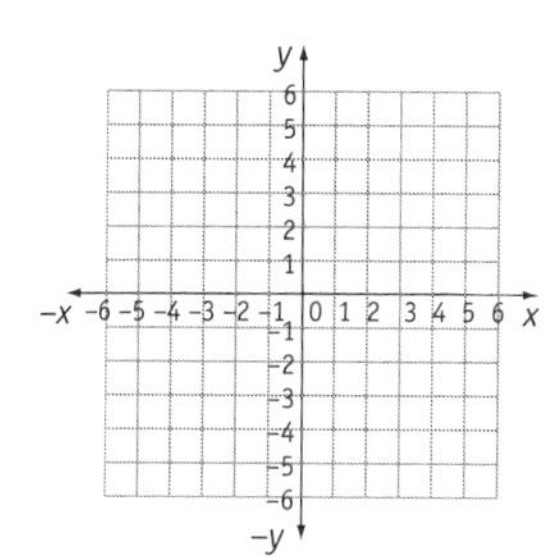

____

Linear relationships

Excel Mathematics Study Guide Year 8
Pages 66–86

UNIT 5: Using the intercept method to graph lines

QUESTION 1 Find the x-intercept for each equation.

a $2x + y = 2$ ______________________
b $2x - y = 4$ ______________________
c $x + y = 6$ ______________________
d $x - y = 8$ ______________________
e $4x - 3y = 6$ ______________________
f $2x - 3y = 12$ ______________________

QUESTION 2 Find the y-intercept for each equation.

a $2x - 2y = 4$ ______________________
b $3x - y = 8$ ______________________
c $2x + y = 6$ ______________________
d $4x - y = 3$ ______________________
e $3x + 4y = 12$ ______________________
f $3x + y = 3$ ______________________

QUESTION 3 Draw the graph of each equation, given the x-intercept and the y-intercept.

a x-intercept = 2, y-intercept = 4

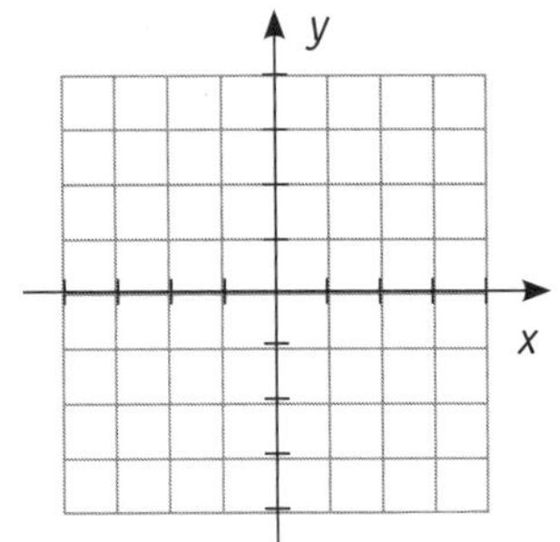

b x-intercept = −2, y-intercept = 3

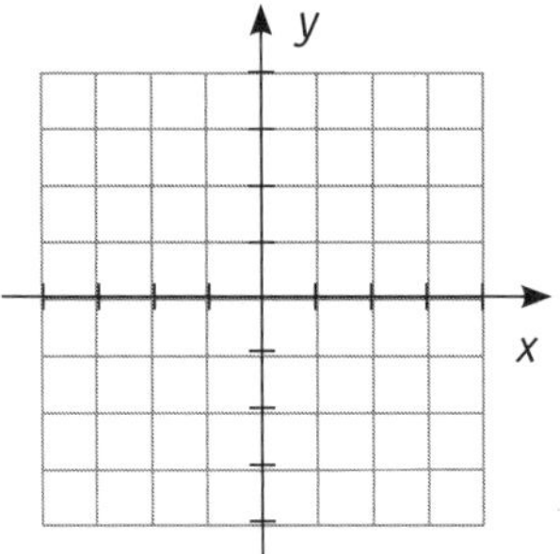

QUESTION 4 For each equation, find the x-intercept and the y-intercept and then draw its graph.

a $y = -2x + 1$

x	0	
y		0

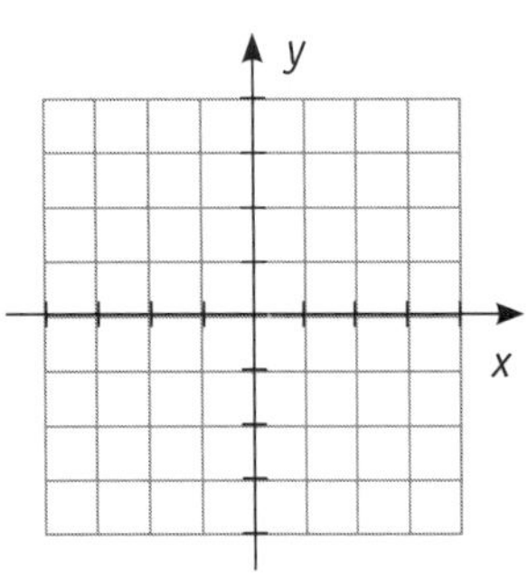

b $x + y - 3 = 0$

x	0	
y		0

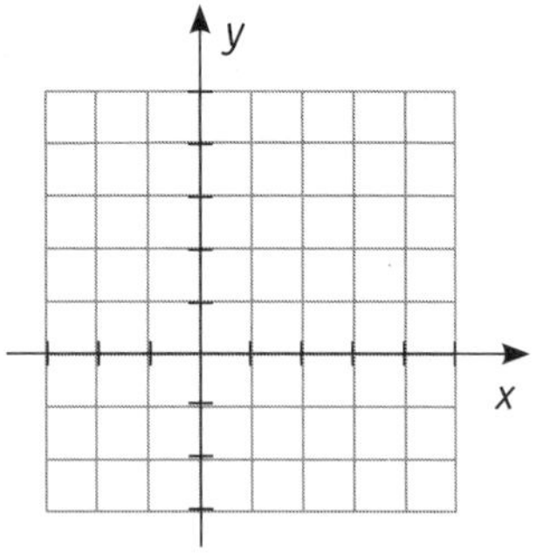

c $y = 3x + 1$

x	0	
y		0

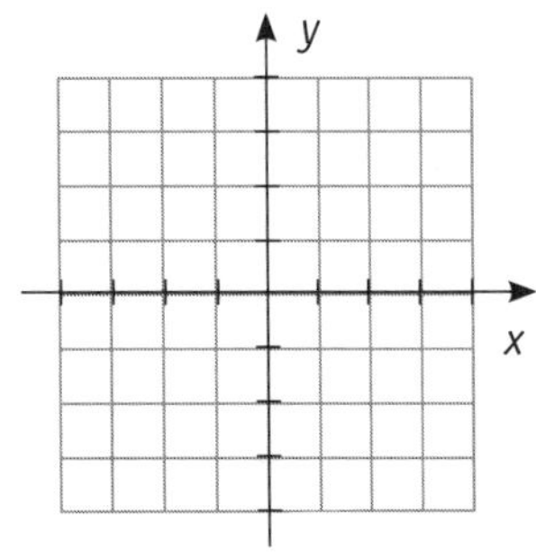

d $y = \frac{2}{3}x - 1$

x	0	
y		0

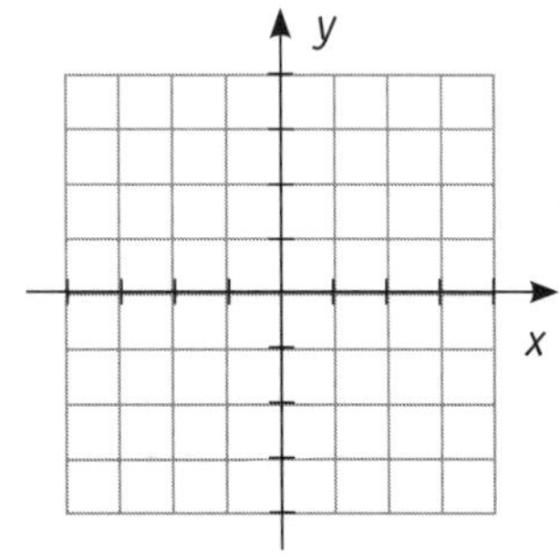

Linear relationships

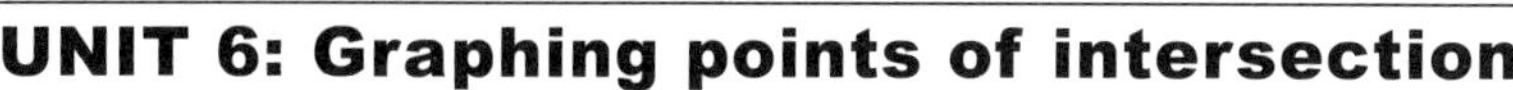

UNIT 6: Graphing points of intersection

Excel Mathematics Study Guide Year 8
Pages 66–86

QUESTION 1 Graph each pair of lines on the same number plane and find their point of intersection.

a $x = -1$

x				
y	−1	0	1	2

$y = 2$

x	−1	0	1	2
y				

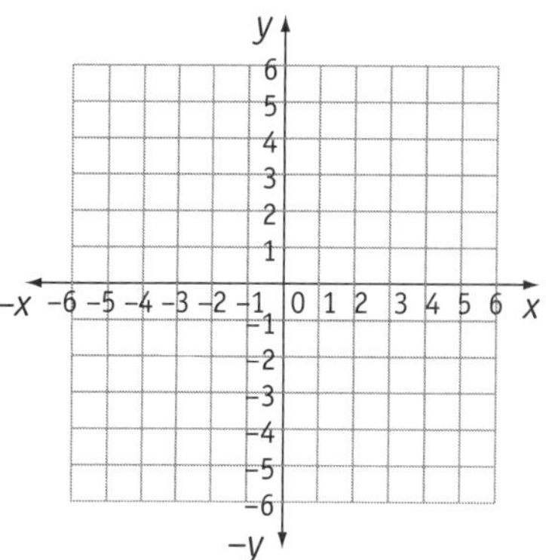

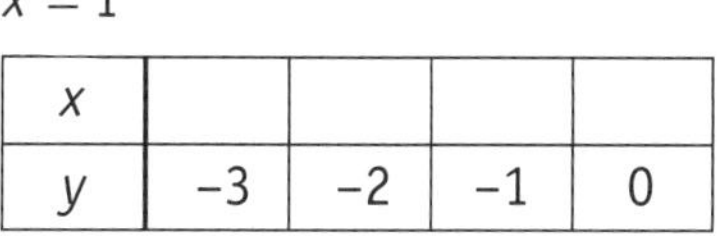

____

b $x = 1$

x				
y	−3	−2	−1	0

$y = -3$

x	−1	0	1	2
y				

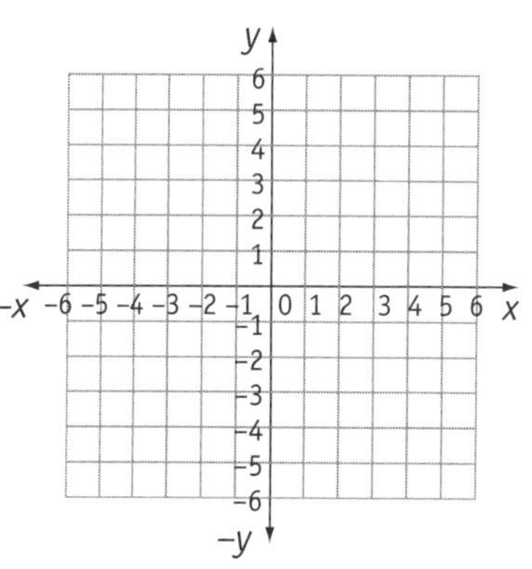

____

QUESTION 2 Graph each pair of lines on the same number plane and find their point of intersection.

a $y = 3x - 2$

x	0	1	2	3
y				

$y = 2x + 1$

x	0	1	2	3
y				

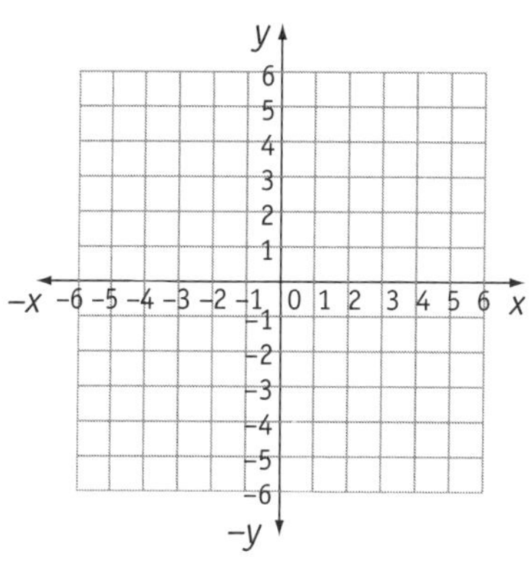

____

b $y = 2x - 1$

x	−1	0	1	2
y				

$y = -2x + 3$

x	−1	0	1	2
y				

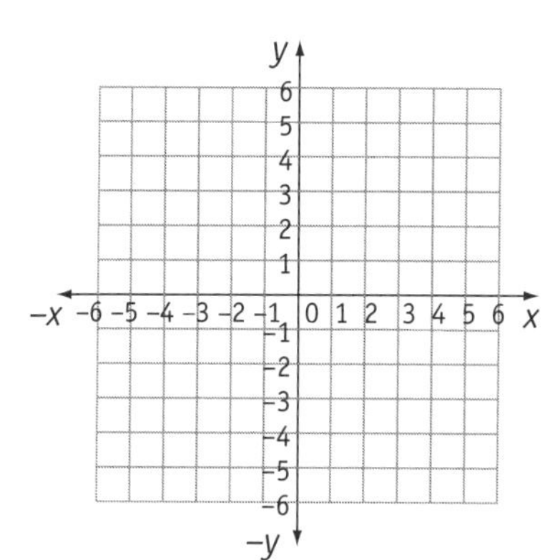

____

Excel Mathematics Study Guide Year 8
Pages 66–86

UNIT 7: Determining whether or not a point lies on a line

QUESTION 1 Which of the following points lie on the line $2x + 3y = 6$?

a $(0, 2)$ ______ **b** $(0, 0)$ ______ **c** $(-3, 4)$ ______

d $(3, 0)$ ______ **e** $(4, 0)$ ______ **f** $(6, -2)$ ______

QUESTION 2 Which of the following lines pass through the origin, $(0, 0)$?

a $2x - y = 0$ ______ **b** $y = 3x$ ______ **c** $x - 5y = 0$ ______

d $2x + 3y = 10$ ______ **e** $3y = -2x$ ______ **f** $3x + 4y = 12$ ______

QUESTION 3 Does the given point lie on the given line?

a $x + 3y = 7$ $(7, 0)$ ______ **b** $x + y = 8$ $(3, 5)$ ______

c $2x + y = 6$ $(0, 6)$ ______ **d** $y = 2x - 5$ $(1, -3)$ ______

e $y = -x + 3$ $(4, 3)$ ______ **f** $4x + y = 5$ $(2, -1)$ ______

QUESTION 4 A straight line $y = mx + 3$ passes through the point $(2, 5)$. Find the value of m.

______

______

QUESTION 5 If the point $(-2, -8)$ is on the line $ax - 2y - 4 = 0$, what is the value of a?

______

______

QUESTION 6 Find the missing coordinates to make each of the following points satisfy the equation $y = 2x - 6$.

a $(0, \ )$ ______ **b** $(\ , 4)$ ______ **c** $(1, \ )$ ______

d $(5, \ )$ ______ **e** $(\ , -5)$ ______ **f** $(\ , -8)$ ______

Linear relationships

UNIT 8: Using graphs to solve linear equations

QUESTION 1 The graph of $y = x + 1$ is drawn below.
Use the graph to write the x value, when:

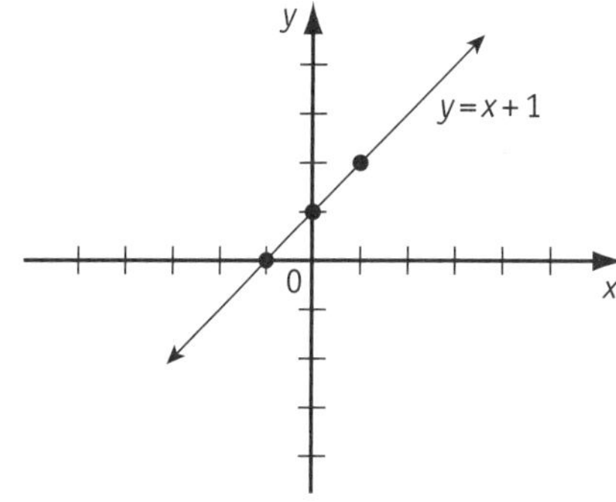

a $y = 3$ $x =$ ______________

b $y = -2$ $x =$ ______________

c $y = 5$ $x =$ ______________

QUESTION 2 Use your answers from Question 1 to solve the following equations.

a $x + 1 = 3$ ______________

b $x + 1 = -2$ ______________

c $x + 1 = 5$ ______________

QUESTION 3 Use the graph of $y = 3x + 2$ to solve the following linear equations.

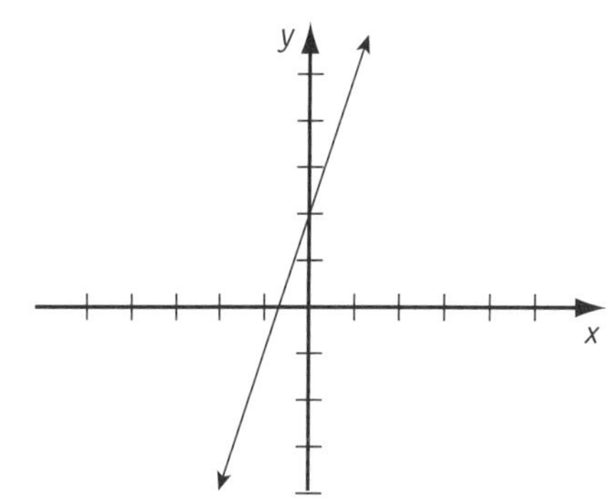

a $3x + 2 = -1$

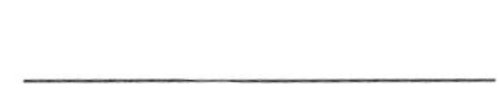

b $3x + 2 = 5$

____

c $3x + 2 = -4$

d $3x + 2 = 2$

QUESTION 4 Draw the graph of $y = 2x - 1$ and use the graph to solve each of the following equations.

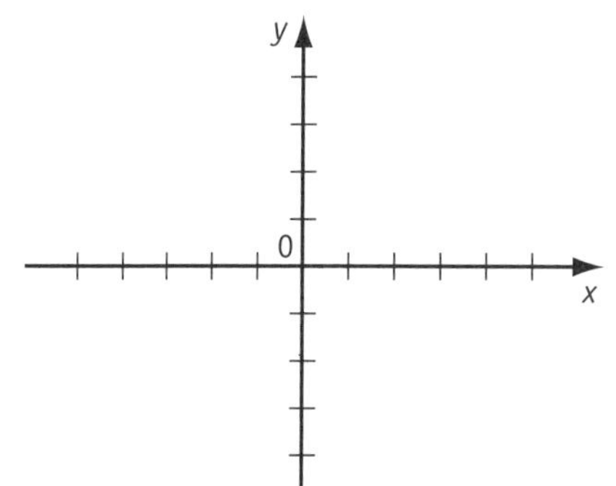

a $2x - 1 = 3$

b $2x - 1 = -4$

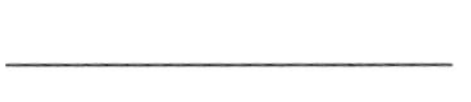

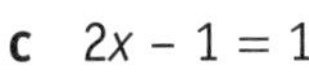

c $2x - 1 = 1$

d $2x - 1 = 5$

Linear relationships

TOPIC TEST PART A

Instructions
- This part consists of 10 multiple-choice questions.
- Fill in only ONE CIRCLE for each question.
- Each question is worth 1 mark.
- Calculators are NOT allowed.

Time allowed: 15 minutes **Total marks: 10**

Marks

1 Which graph best represents $y = -x$ 1

Ⓐ 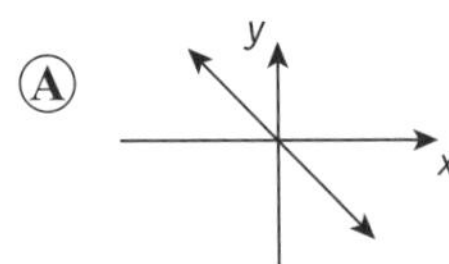
Ⓑ 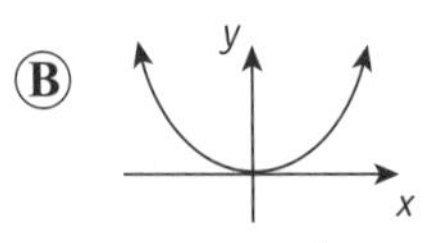
Ⓒ 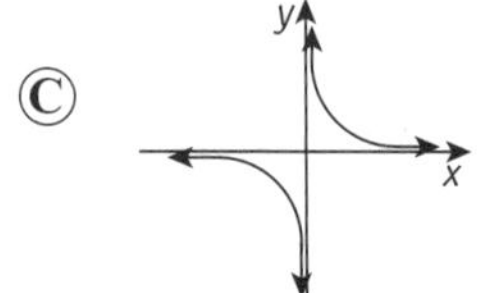
Ⓓ 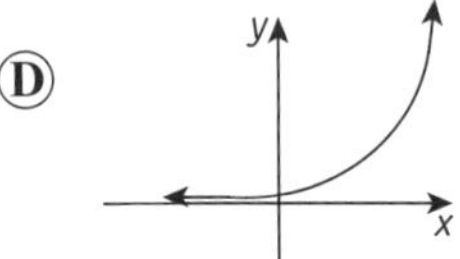

2 The equation of the line l is 1

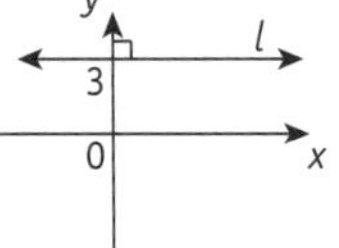

Ⓐ $x = -3$ Ⓑ $x = 3$
Ⓒ $y = -3$ Ⓓ $y = 3$

3 If the x-coordinate of a point on the line $y = 3x - 2$ is 4, what is the y-coordinate? 1

Ⓐ 5 Ⓑ 9 Ⓒ 10 Ⓓ 12

4 For the equation $y = -3x + 8$, find the y-intercept. 1

Ⓐ -3 Ⓑ 8 Ⓒ $\frac{3}{8}$ Ⓓ $\frac{8}{3}$

5 Which one of the following points lie on the x-axis? 1

Ⓐ (2, 5) Ⓑ (2, –5) Ⓒ (0, 7) Ⓓ (7, 0)

6 The equation of the line l is 1

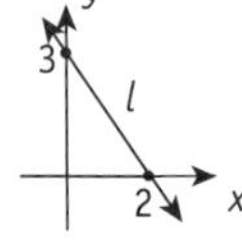

Ⓐ $3x - 2y + 6 = 0$ Ⓑ $3x + 2y - 6 = 0$
Ⓒ $2x - 3y + 6 = 0$ Ⓓ $2x + 3y - 6 = 0$

7 The point (0, –3) lies on which line? 1

Ⓐ $4x + 5y = 15$ Ⓑ $5x - 4y = 15$ Ⓒ $4x - 5y = 15$ Ⓓ $5x + 4y = 15$

8 For the equation $2x + y = 6$ find the x-intercept. 1

Ⓐ (3, 0) Ⓑ (0, 3) Ⓒ (1, 0) Ⓓ (0, 1)

9 What is the equation of a line with a y-intercept of 3 and an x-intercept of –1? 1

Ⓐ $y = -x + 3$ Ⓑ $y = 3x - 1$ Ⓒ $y = 3x + 3$ Ⓓ $y = -x - 1$

10 The coordinates of the point of intersection of the lines $x = 2$ and $y = -7$ are 1

Ⓐ (7, 2) Ⓑ (–7, 2) Ⓒ (2, 7) Ⓓ (2, –7)

Total marks achieved for PART A

Linear relationships

TOPIC TEST — PART B

Instructions
- This part consists of 3 questions.
- Each question part is worth 1 mark.
- Write only the answer in the answer column.

Time allowed: 15 minutes — **Total marks: 15**

Questions	Answers	Marks

1 Match each table of values with an equation.

a

x	−2	0	2	4
y	−1	0	1	2

b

x	−1	0	1	2
y	−7	−2	3	8

c

x	−1	0	1	2
y	7	4	1	−2

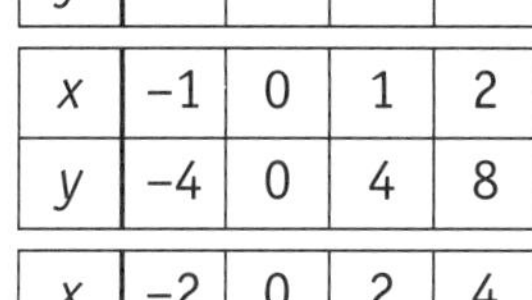

d

x	−1	0	1	2
y	−4	0	4	8

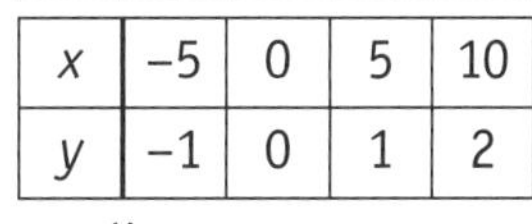

e

x	−5	0	5	10
y	−1	0	1	2

f

x	−2	0	2	4
y	4	0	−4	−8

i $y = \frac{x}{5}$ **ii** $y = 5x - 2$ **iii** $y = -2x$

iv $y = \frac{x}{2}$ **v** $y = 4x$ **vi** $y = 4 - 3x$

Answers: a ______ b ______ c ______ d ______ e ______ f ______

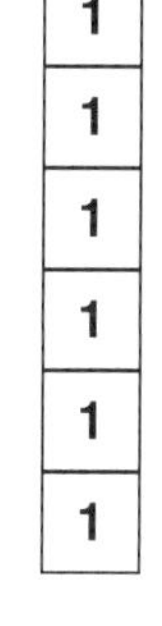

2 **a** On the same number plane, draw the graphs of the following lines.

i $x = 2$ **ii** $y = -3$

iii $x = -3$ **iv** $y = 2$

b Write the coordinates of the point of intersection of the lines $x = -3$ and $y = 2$. ______ 1

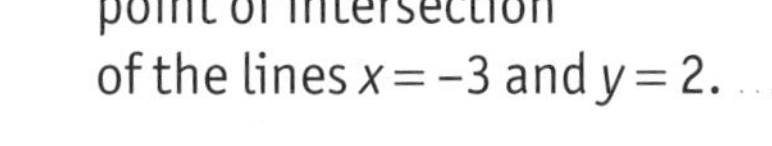

c Which lines are parallel to the x-axis? ______ 1

d Which lines are parallel to the y-axis? ______ 1

3 **a** Complete the table of values below for the equation $y = 2x + 1$. 1

x	−2	−1	0	1	2
y					

b Draw the graph of $y = 2x + 1$ 1

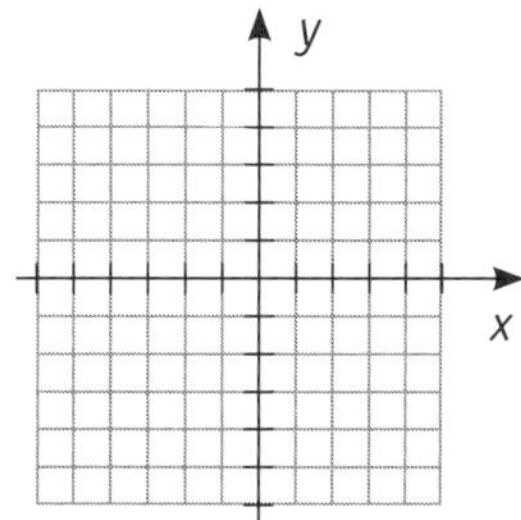

c Where does the graph cut the x-axis? ______ 1

d Where does the graph cut the y-axis? ______ 1

e Using the graph, solve $2x + 1 = -5$. ______ 1

Total marks achieved for PART B /15

CHAPTER 10

Equations

UNIT 1: Equations by inspection

Excel Mathematics Study Guide Year 8
Pages 88–108

QUESTION **1** Solve the following one-step equations by inspection and then check by substitution.

a $x + 5 = 12$

b $a + 3 = 15$

c $m - 2 = 11$

d $b + 8 = 24$

e $a + 4 = 16$

f $n + 3 = 7$

g $x - 4 = 10$

h $x + 1 = 8$

i $y + 3 = 13$

j $p - 7 = 27$

k $t - 2 = 24$

l $a - 8 = 16$

m $y - 9 = -10$

n $x - 3 = -5$

o $8 + x = 16$

QUESTION **2** Solve the following equations by inspection and then check by substitution.

a $m + 3 = 9$

b $x + 2 = 12$

c $8 + y = 32$

d $7 + n = -8$

e $x - 6 = 18$

f $x - 3 = 21$

g $m + 4 = 15$

h $y - 7 = 23$

i $a + 6 = 19$

j $m - 5 = 15$

k $n - 4 = 14$

l $a - 1 = 8$

m $y - 6 = -9$

n $t - 3 = -7$

o $19 + a = 9$

Equations

UNIT 2: One-step equations

Excel Mathematics Study Guide Year 8
Pages 88–108

QUESTION **1** Solve the following one-step equations.

a $x + 6 = 16$

b $a + 3 = 24$

c $m - 2 = 12$

d $\frac{x}{5} = 10$

e $\frac{y}{3} = 9$

f $4x = -28$

g $x - 5 = 11$

h $x + 3 = 8$

i $y + 7 = 31$

j $3p = -12$

k $\frac{y}{4} = -8$

l $9t = 36$

m $y - 6 = -26$

n $5a = 20$

o $2 + x = 28$

QUESTION **2** Solve the following equations. Check your solution by substitution.

a $m + 4 = 18$

b $\frac{x}{5} = -8$

c $9 + y = 19$

d $\frac{a}{4} = -20$

e $x - 8 = 19$

f $\frac{b}{4} = 9$

g $m + 12 = 29$

h $\frac{x}{3} = 7$

i $a + 5 = 7$

j $5a = 45$

k $n - 5 = 33$

l $-2x = -20$

m $y - 6 = -8$

n $2x = 18$

o $10 + a = 27$

Equations

UNIT 3: Two-step equations

Excel Mathematics Study Guide Year 8
Pages 88–108

QUESTION 1 Solve the following two-step equations.

a $2x + 7 = 17$

b $5 = 3x - 7$

c $24 = 5y - 6$

d $\frac{5m}{3} = 15$

e $\frac{x-2}{8} = 7$

f $2a + 4 = 18$

g $\frac{x}{3} - 5 = 9$

h $2x + 3 = 17$

i $\frac{a-3}{6} = 4$

j $8x - 9 = 23$

k $2x + 5 = 19$

l $6t - 4 = 20$

m $7y - 3 = 18$

n $\frac{m}{3} - 5 = 6$

o $3k + 7 = 22$

QUESTION 2 Solve the following equations. Verify your solution by substitution.

a $5x - 4 = 16$

b $\frac{m}{3} + 7 = 11$

c $\frac{x-5}{6} = 3$

d $\frac{x-2}{7} = 3$

e $2x - 7 = 13$

f $\frac{m}{4} = 9$

g $27 - 3m = 0$

h $5y - 4 = 11$

i $2y + 5 = -30$

j $3x + 7 = 16$

k $5x - 3 = 17$

l $3x - 6 = 24$

m $2a - 16 = 4$

n $5a - 2\frac{1}{2} = 7\frac{1}{2}$

o $b + 0.6 = 2.8$

Equations

UNIT 4: Three-step equations

QUESTION 1 Solve the following three-step equations.

a $4x + 12 = 3x - 19$

b $2x - 5 = x - 9$

c $6t - 4 = 4t + 20$

d $11m - 8 = 7m + 16$

e $3m - 5 = 2m + 6$

f $4a - 3 = 3a + 10$

g $8y - 4 = 7y + 16$

h $5x - 6 = 4x + 9$

i $7y - 16 = 4y + 5$

j $3x - 7 = 2x - 4$

k $5a + 6 = 48 - a$

l $4p - 9 = 3p + 16$

m $2m - 7 = m + 8$

n $6y - 12 = y + 23$

o $8 + m = 12 - 2m$

QUESTION 2 Solve the following equations. Check your solution by substitution.

a $3x - 7 = 2x + 15$

b $2x - 8 = 13 - x$

c $6x - 8 = 3x - 23$

d $3y - 14 = 2y + 36$

e $2x - 5 = x - 18$

f $5x + 15 = 60 - 4x$

g $6m - 7 = 4m + 9$

h $8x - 24 = 2x - 12$

i $3y + 15 = 2y + 19$

j $2m + 7 = 4m + 23$

k $3x + 8 = x - 56$

l $4y - 10 = 2y + 28$

Equations

UNIT 5: Equations with grouping symbols

Excel Mathematics Study Guide Year 8
Pages 88–108

Question 1 Solve the following equations.

a $3(x + 2) = 21$

b $2(m + 7) = 30$

c $6(m - 3) = 12$

d $2(x + 4) = 16$

e $3(a - 7) = 18$

f $4(n - 3) = 36$

g $5(2n - 1) = 25$

h $2(3p - 4) = 16$

i $5(2x + 1) = -35$

j $4(a - 2) = 24$

k $3(m + 5) = m + 17$

l $2(3x + 2) = 34$

Question 2 Solve the following equations.

a $5(a + 2) = 4(a - 9)$

b $3(x - 2) = 2(x + 5)$

c $4(y - 6) = 3(y + 2)$

d $7(x - 8) = 6(x + 5)$

e $8(6 - x) = 7(x - 6)$

f $2(a + 6) + a + 3 = 0$

g $3(x + 1) + x + 3 = 18$

h $5(a + 9) = 4(a + 7)$

i $3(3a - 5) = 4(10 - a)$

j $2(a + 3) - a + 7 = 4$

k $3(a - 7) - 2a + 9 = 15$

l $2(5a - 3) - 9a + 12 = 0$

m $6(a + 4) = 5(a - 3)$

n $8(2a + 7) = 5(3a - 3)$

o $9(3a - 8) = 13(2a - 3)$

Equations

UNIT 6: Simple equations with one fraction

QUESTION 1 Solve the following equations.

a $\frac{x}{3} = 9$

b $\frac{a}{5} = 8$

c $\frac{y}{9} = 7$

d $\frac{m}{4} = 10$

e $\frac{m}{6} = 7$

f $\frac{x}{8} = 9$

g $\frac{a+5}{3} = 2$

h $\frac{3x}{2} = 9$

i $\frac{5x-4}{3} = 7$

j $\frac{x}{7} = 2$

k $\frac{x+2}{9} = 6$

l $\frac{m+5}{3} = 9$

QUESTION 2 Solve the following equations. Check your solution by substitution.

a $\frac{2x}{3} = 8$

b $\frac{3y-1}{4} = 10$

c $\frac{t-5}{6} = 5$

d $\frac{m-7}{3} = 4$

e $\frac{2a+8}{7} = 6$

f $\frac{3p-5}{3} = 8$

g $\frac{3x}{5} + 1 = 7$

h $\frac{4x}{3} - 2 = 3$

i $\frac{3x+4}{2} = 2$

j $\frac{3m-1}{4} = 8$

k $\frac{x}{6} = 5$

l $\frac{m}{5} = 2$

m $\frac{3x}{9} = -2$

n $\frac{6p}{5} + 3 = 9$

o $\frac{3a+4}{2} = 14$

Equations

UNIT 7: Formulae

Question 1 Given the formula $V = Ah$, find V if:

a $A = 12,\ h = 3.6$

b $A = 29,\ h = 4.2$

c $A = 4.7,\ h = 5$

Question 2 Given the formula $P = 2L + 2B$, find P if:

a $L = 20,\ B = 10$

b $L = 4.8,\ B = 2.2$

c $L = 6.5,\ B = 3.3$

Question 3 If $y = mx + b$, find y for the given values of m, x and b.

a $m = -2,\ x = 8,\ b = 3$

b $m = 4,\ x = 3,\ b = 5$

c $m = -4,\ x = 10,\ b = 6$

Question 4 If $A = \frac{1}{2}bh$, find A for the given values of b and h.

a $b = 12,\ h = 5.6$

b $b = 10,\ h = 4.7$

c $b = 11,\ h = 8.2$

Question 5 If $\Delta = b^2 - 4ac$, find Δ if:

a $a = 2$, $b = 10$ and $c = 12$

b $a = 3$, $b = -26$ and $c = 16$

c $a = 1$, $b = -1$ and $c = -6$

Question 6 If $S = ut + \frac{1}{2}at^2$, find S if:

a $u = 5$, $a = 12$ and $t = 6$

b $u = 7.2$, $a = 0.8$ and $t = 4$

c $u = 22$, $a = 15$ and $t = 0.5$

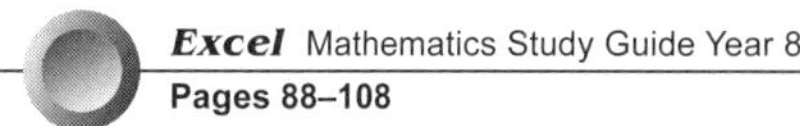

UNIT 8: Problem solving with equations

In the following questions, suppose the number is represented by x. Write the statement as an equation and find the value of x.

Question 1 The sum of a number and 15 is 37. ____________________

Question 2 5 subtracted from 4 times a number is 23. ____________________

Question 3 I think of a number and add 21. The result is 47. ____________________

Question 4 A number increased by 7 is 48. ____________________

Question 5 The difference between 4 times a number and 9 is 19. ____________________

Question 6 I think of a number, double it, and the result is 64. ____________________

Question 7 I think of a number, double it and add 3. The result is 29. ____________________

Question 8 I think of a number, divide it by 5, subtract 7 and the result is 8. ____________________

Question 9 The difference between 3 times a number and 7 is 23. ____________________

Question 10 A rectangle is 10 cm longer than it is wide. If its perimeter is 72 cm, find the width and the length of the rectangle.

____

Question 11 If the perimeter of an equilateral triangle is 42 cm, what is the length of each side?

____

Question 12 The length of a rectangle is 12 cm and its perimeter is 40 cm.

Find its width.

____

Question 13 If I subtract 10 from a certain number the result is 18.

What is the number?

____

Question 14 Find x if the sum of 8, 12, 16, x is 50.

____

Question 15 I think of a number, add 6 to it, multiply this sum by 3 and then subtract 7. The result is 26.

What is the number?

____

Equations

TOPIC TEST — PART A

Instructions
- This part consists of 10 multiple-choice questions.
- Fill in only ONE CIRCLE for each question.
- Each question is worth 1 mark.
- Calculators are allowed.

Time allowed: 15 minutes — **Total marks: 10**

Marks

1 If $\frac{m}{12} = 2$, m equals — 1

(A) 2 (B) 10 (C) 14 (D) 24

2 If $2y + 7 = 15$, then y equals — 1

(A) 8 (B) 11 (C) 4 (D) 16

3 Solve for x: $\frac{x - 10}{3} = 2$ — 1

(A) –4 (B) 16 (C) 15 (D) –5

4 Solve for x: $2(3x - 4) = 34$ — 1

(A) $x = 7$ (B) $x = 21\frac{1}{3}$ (C) $x = 4\frac{1}{3}$ (D) $x = 63$

5 Solve for t: $4(t - 6) = 3(t + 2)$ — 1

(A) $t = 8$ (B) $t = -\frac{4}{7}$ (C) $t = -2\frac{4}{7}$ (D) $t = 30$

6 Solve the equation $\frac{2c + 8}{7} = 4$ — 1

(A) $c = 9\frac{1}{2}$ (B) $c = 1\frac{1}{2}$ (C) $c = 10$ (D) $c = 18$

7 If $y = mx + b$, find y when $m = -4$, $x = \frac{1}{2}$, $b = 6$ — 1

(A) $y = 1\frac{1}{2}$ (B) $y = 4$ (C) $y = -8$ (D) $y = -10\frac{1}{2}$

8 I think of a number, divide it by 5, subtract 7 and the result is 18. The number is — 1

(A) 5 (B) 18 (C) 25 (D) 125

9 The width of a rectangle is 6.5 cm and its perimeter is 27 cm. Its length is — 1

(A) 21.5 cm (B) 14 cm (C) 7 cm (D) 7.5 cm

10 If $\frac{2x - 1}{3} = \frac{x + 1}{2}$, x equals — 1

(A) 5 (B) 7 (C) –5 (D) 8

Total marks achieved for PART A

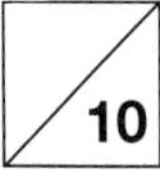

Equations

TOPIC TEST — PART B

Instructions
- This part consists of 3 questions.
- Each question part is worth 1 mark.
- Show all working.

Time allowed: 20 minutes **Total marks: 15**

Marks

1 Solve the following equations.

a $x + 4 = 16$ b $y - 8 = 12$ c $\frac{w}{15} = 3$ 3

d $2.5m = 125$ e $4n + 3.5 = 61.5$ f $\frac{a}{7} - 19 = -8$ 3

g $\frac{10 - b}{2} = -1$ h $\frac{d}{5} = \frac{5}{6}$ i $9(e + 7) = 18$ 3

j $6(x - 4) = 2(x + 1)$ k $\frac{f - 2}{3} + 5 = 9$ l $\frac{h + 6}{3} = \frac{2h + 4}{4}$ 3

2 If $S = ut + \frac{1}{2}at^2$, find S if $a = -2$, $u = 4$ and $t = 5.5$ 1

3 a Find k if the average of 5, k, 3 and 7.5 is 5.5. 1

b The width of a rectangle is 4 cm shorter than its length. Find the width of the rectangle if its perimeter is 28 cm. 1

Total marks achieved for PART B /15

CHAPTER 11
Reasoning in geometry

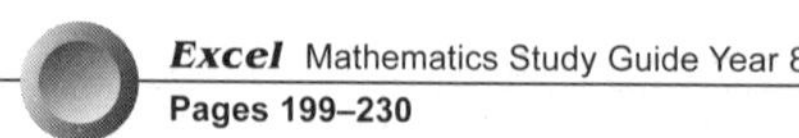

UNIT 1: Review of some basic ideas in geometry

QUESTION 1 Complete the following sentences.

a An equilateral triangle is a triangle in which ______________ sides are equal.

b An isosceles triangle is a triangle in which ______________ sides are equal.

c A scalene triangle is a triangle in which ______________ sides are equal.

d An acute-angled triangle has ______________ acute angle(s).

e An obtuse-angled triangle has ______________ obtuse angle(s).

f A right-angled triangle has ______________ right angle(s).

QUESTION 2 Complete the following statements.

a The interior angles of an equilateral triangle are ______________ .

b The base angles of an isosceles triangle are ______________ .

c The angle sum of a triangle is ______________ .

d If a triangle has all angles of equal size, the size of each angle is ______________ .

e An exterior angle of a triangle is always equal to the sum of ____________________________ angles.

f The angle sum of a quadrilateral is ______________ .

QUESTION 3 Complete the following.

a A quadrilateral is a ____________________________ plane figure.

b A polygon is a closed figure with ____________________________ sides.

c A trapezium is a quadrilateral with one pair of opposite sides ______________ .

d A parallelogram is a quadrilateral with both pairs of opposite sides ______________ .

e A rhombus is a parallelogram in which all sides are ______________ .

f A rectangle is a parallelogram in which each angle is a ____________________________ .

QUESTION 4 Complete the following sentences.

a A square is a quadrilateral with all sides ______________ and each angle is ______________ .

b A kite is a quadrilateral with two pairs of adjacent sides ______________ and no sides ______________ .

c The angle sum of a kite is ______________ .

d Opposite angles of a parallelogram are ______________ .

e Each angle of a rectangle is equal to ______________ .

QUESTION 5 Complete the following statements.

a Vertically opposite angles are ______________ .

b When a pair of parallel lines are cut by a transversal, alternate angles are ______________ .

c When a pair of parallel lines are cut by a transversal, corresponding angles are ______________ .

d When a pair of parallel lines are cut by a transversal, co-interior angles add up to ______________ .

Reasoning in geometry

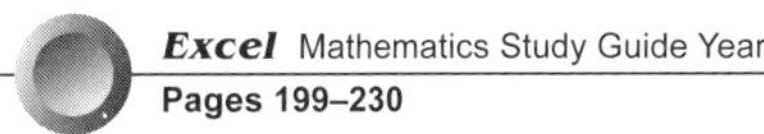

UNIT 2: Review of angles and the value of pronumerals

QUESTION **1** Find the value of the pronumeral in each of the following.

a

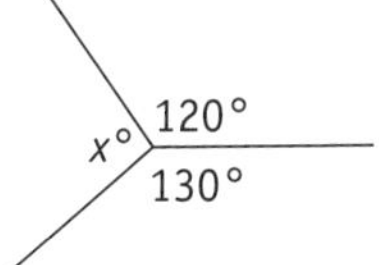

b

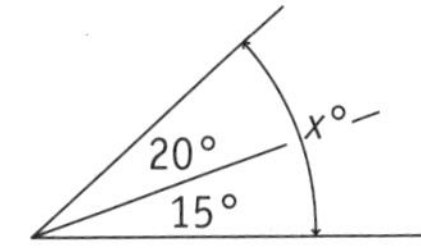

c

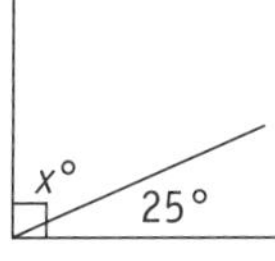

d

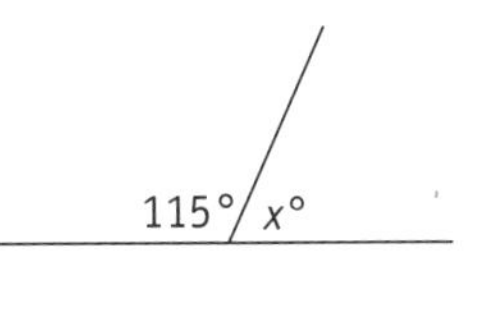

e

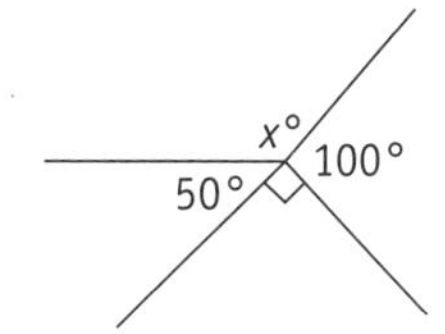

f

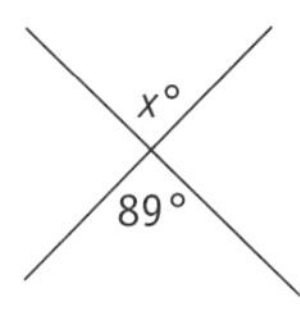

QUESTION **2** Find the value of the pronumeral and give reasons for your answer.

a

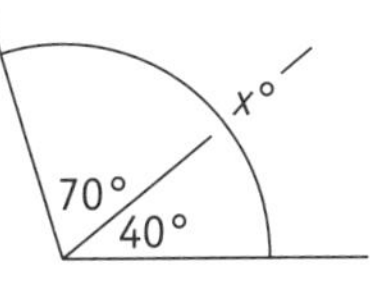

b

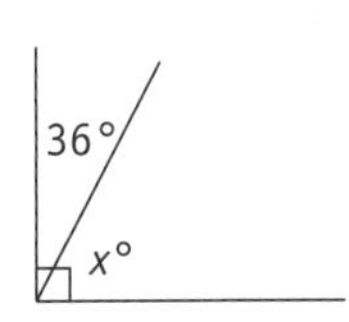

c

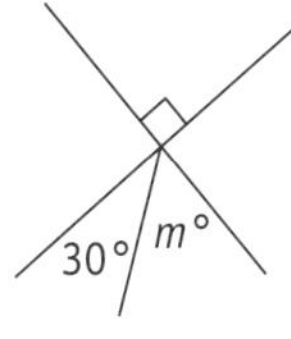

QUESTION **3** Find the value of the pronumeral, giving reasons.

a

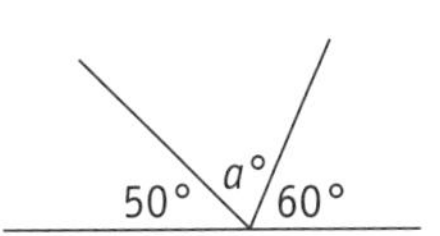

b

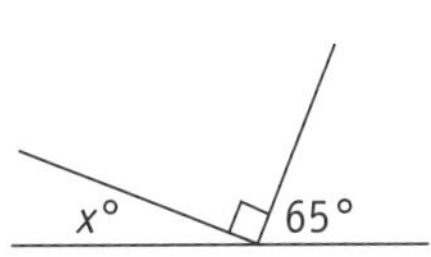

c

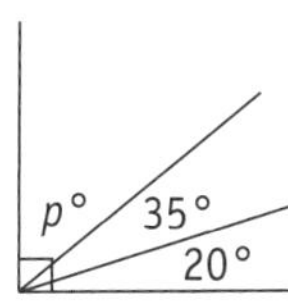

d

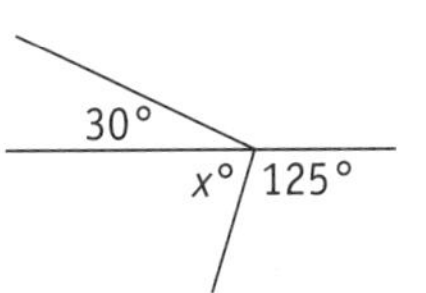

e

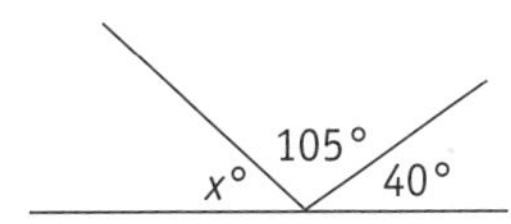

f

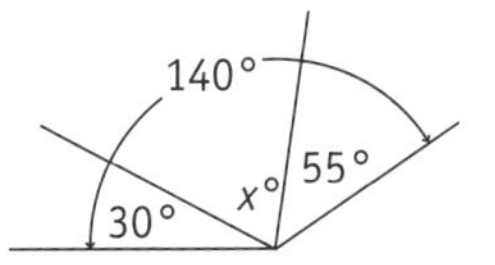

Reasoning in geometry

UNIT 3: Review of parallel lines and angles

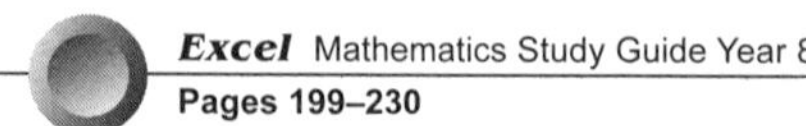

QUESTION 1 Find the value of the pronumerals in each of the following.

a

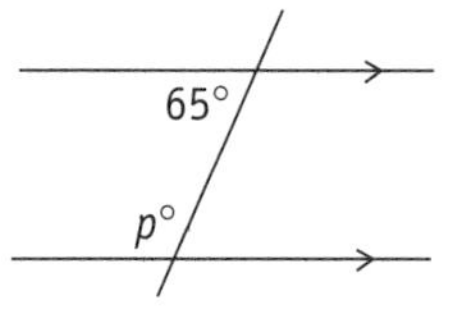

b

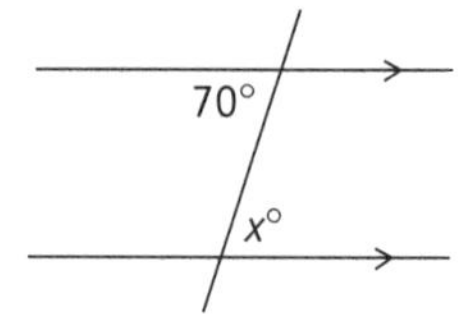

c

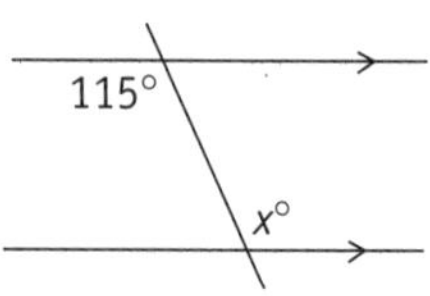

d

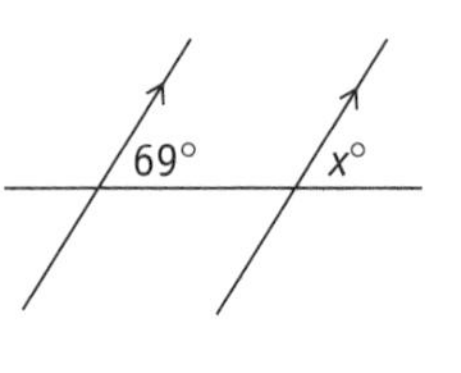

e

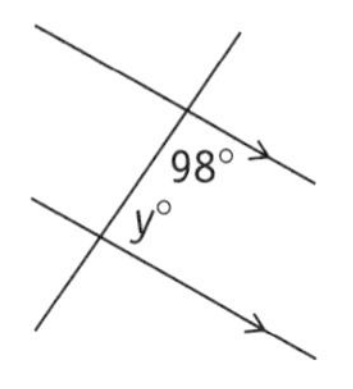

f

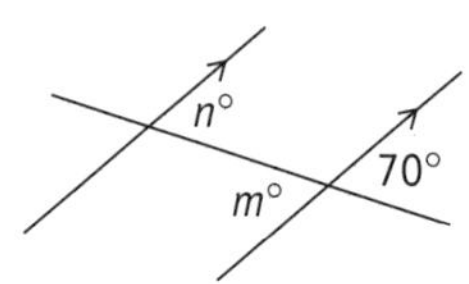

QUESTION 2 Find the value of the pronumeral, giving reasons.

a

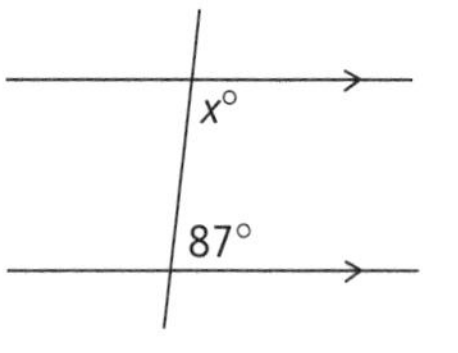

b

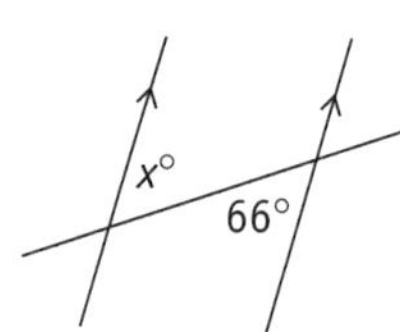

c

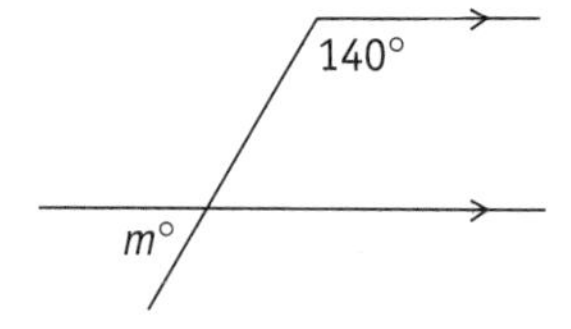

QUESTION 3 Find the value of the pronumeral, giving reasons.

a

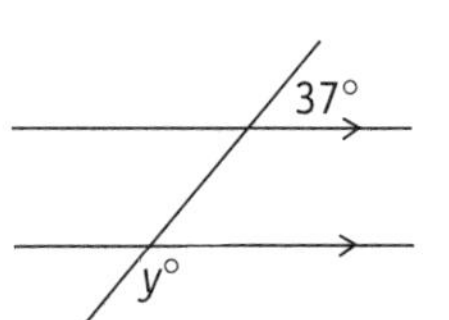

b

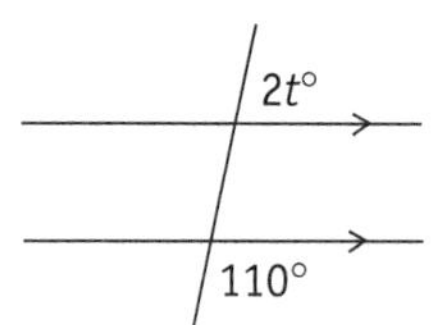

c

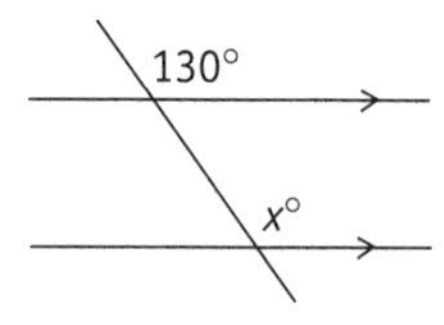

d

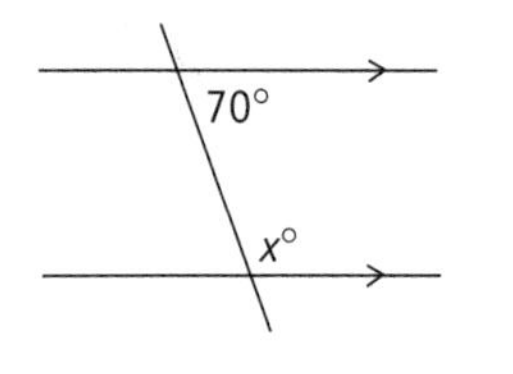

e

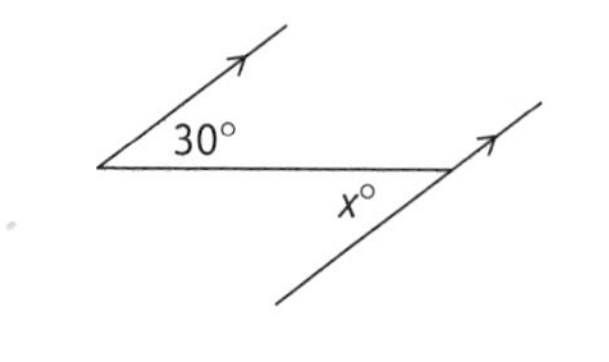

f

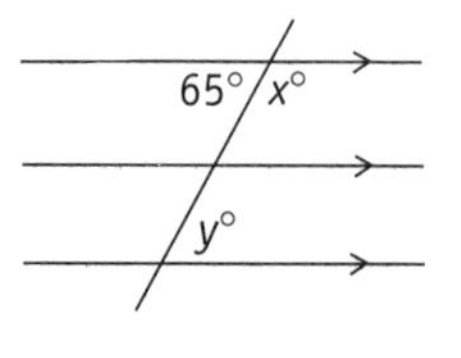

Reasoning in geometry

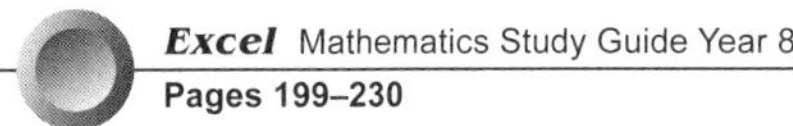

UNIT 4: Review of the angle sum of a triangle

QUESTION **1** Find the value of each pronumeral.

a

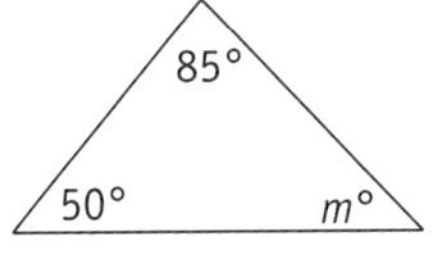

b

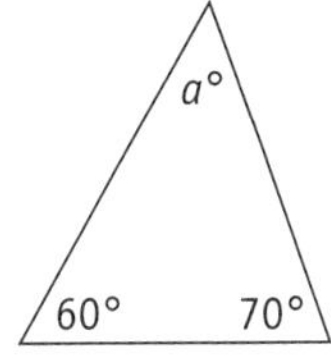

c

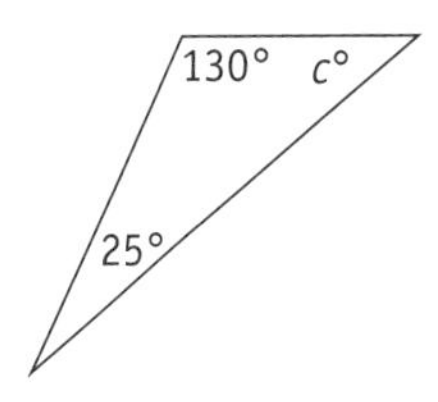

d

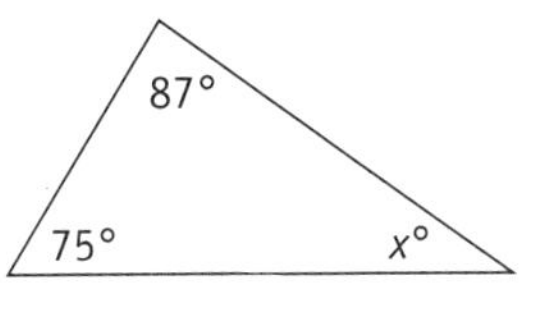

e

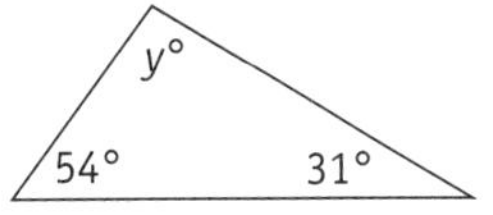

f

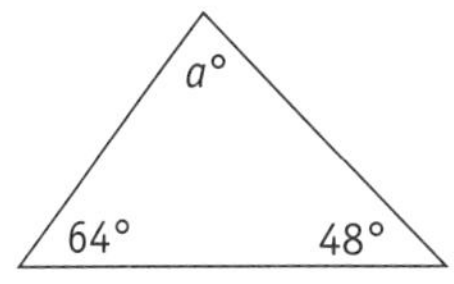

QUESTION **2** Find the value of each pronumeral.

a

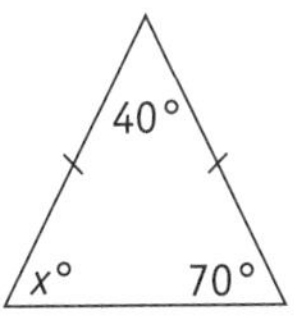

b

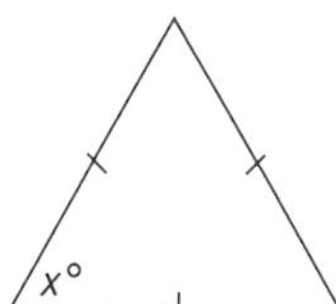

c

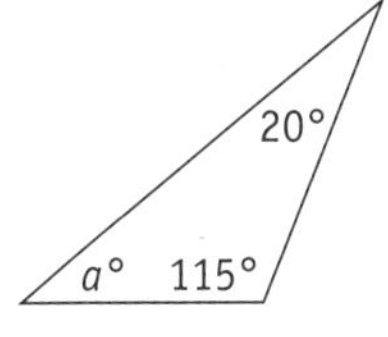

QUESTION **3** Find the value of each pronumeral.

a

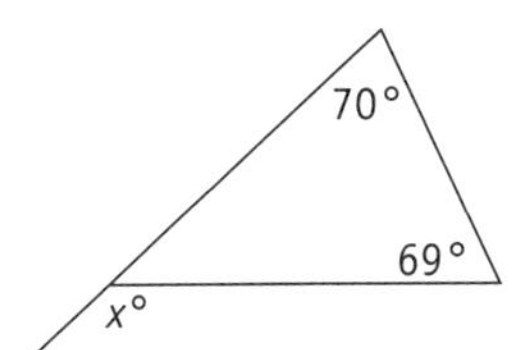

b

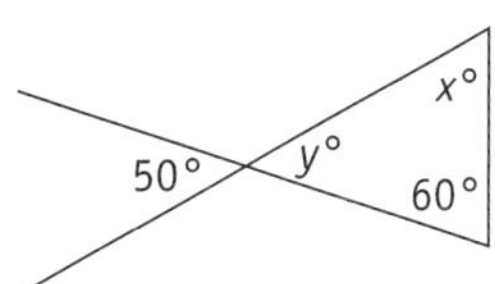

c

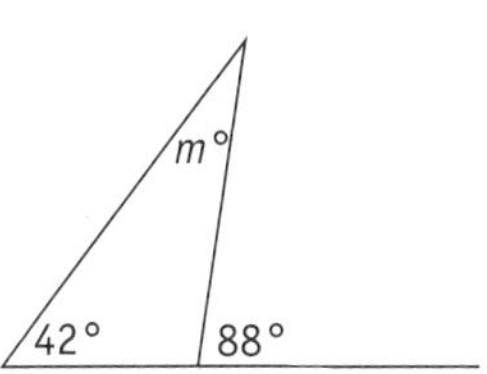

d

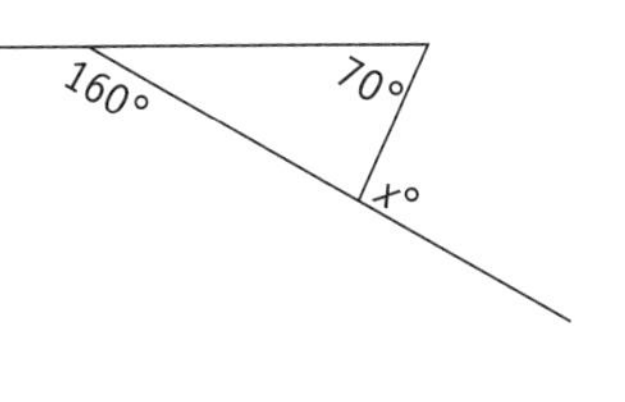

e

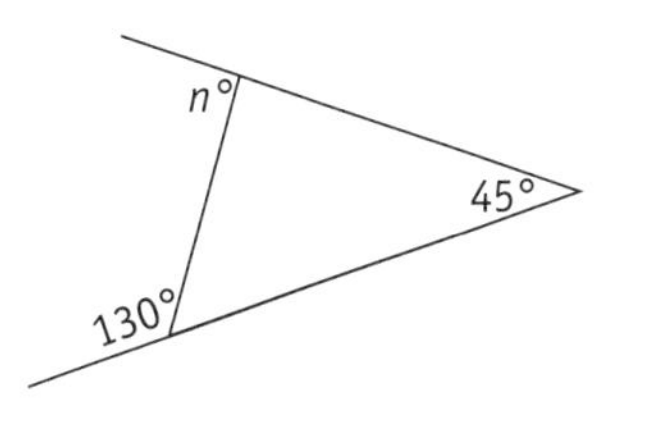

f

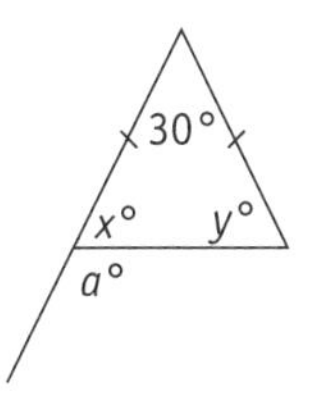

Reasoning in geometry

UNIT 5: Review of the angle sum of a quadrilateral

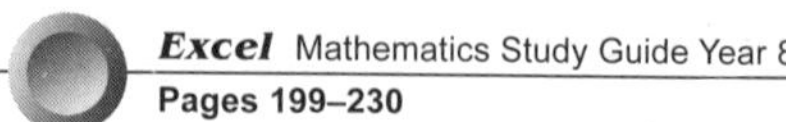

QUESTION **1** Find the value of each pronumeral.

a 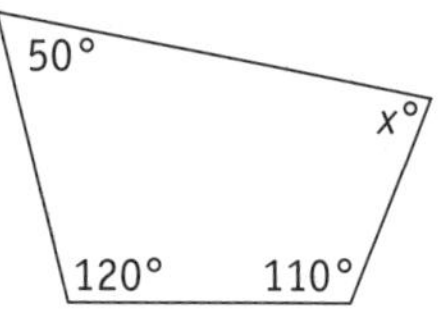

b 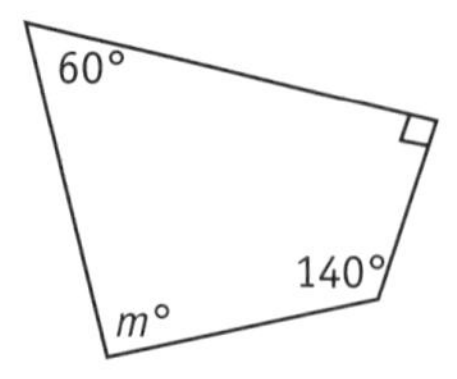

c 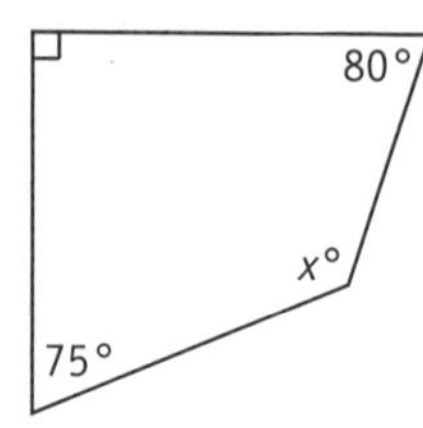

d 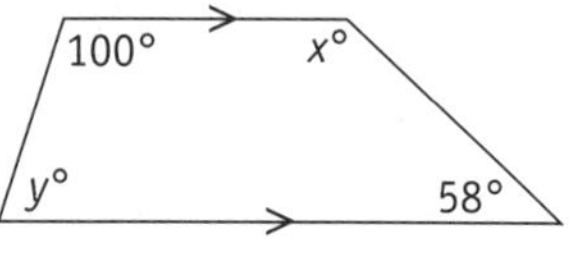

e 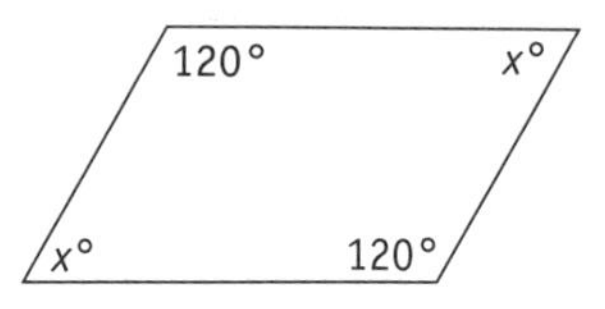

f 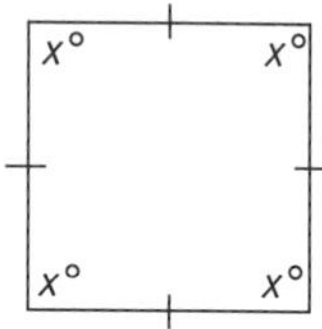

QUESTION **2** Find the unknown angles.

a 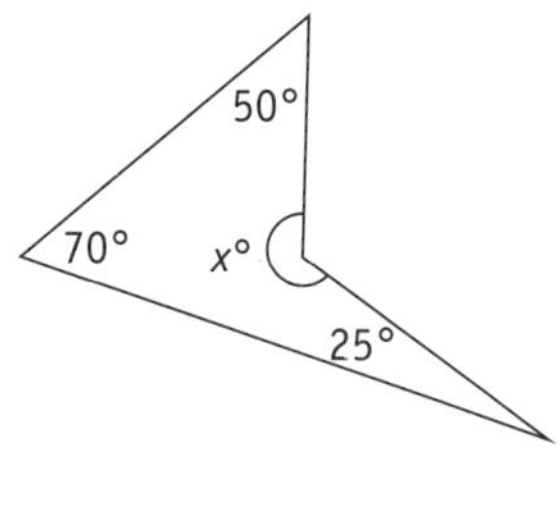

b 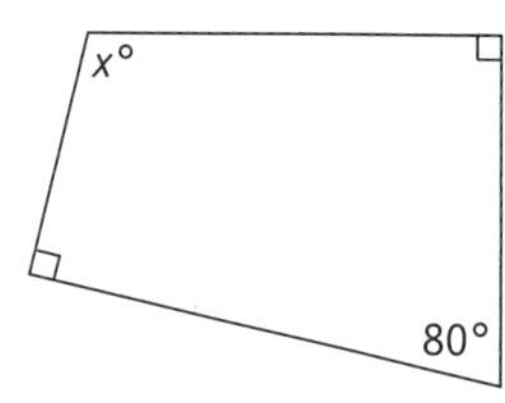

c 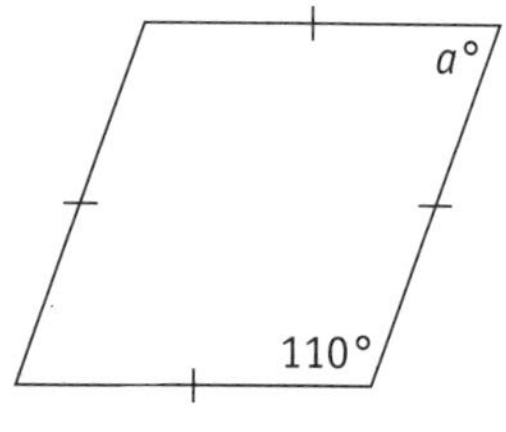

QUESTION **3** Find the value of the following pronumerals.

a 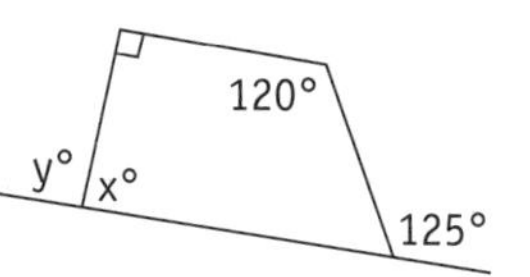

b 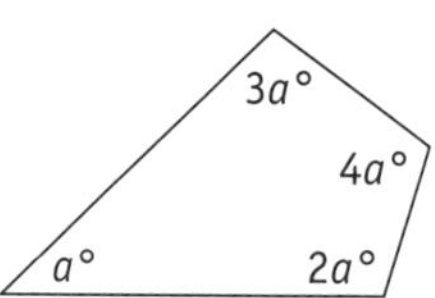

c 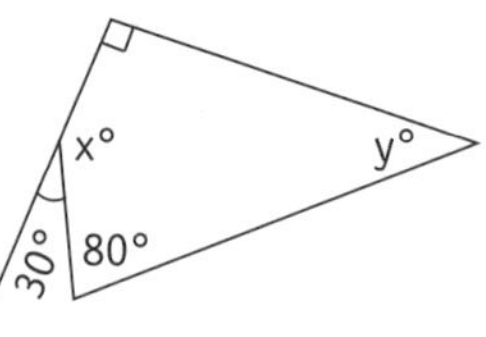

d 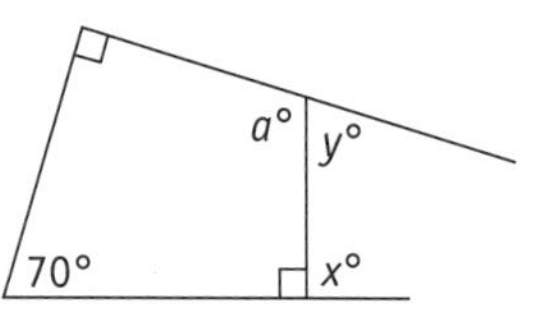

e 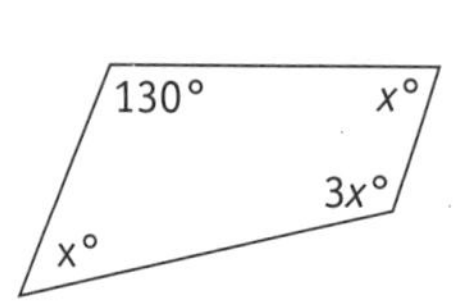

f 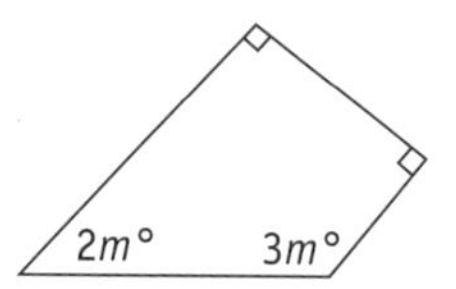

Reasoning in geometry

UNIT 6: Transformations

QUESTION 1 Complete the following sentences.

a When a shape is moved from one position to another, a ______________________ takes place.

b In translation, the shape is moved by ____ .

c In rotation, the shape is moved by ____ .

d In reflection the shape is moved by ____ .

QUESTION 2 Name the transformation used when each of the following shapes is moved to its new position.

a

____

b

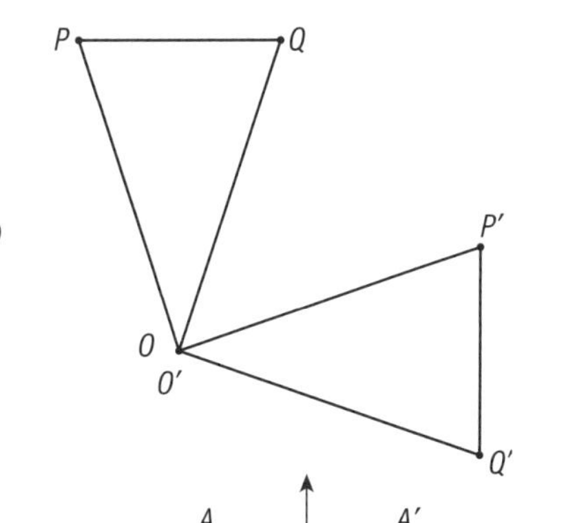

____

c

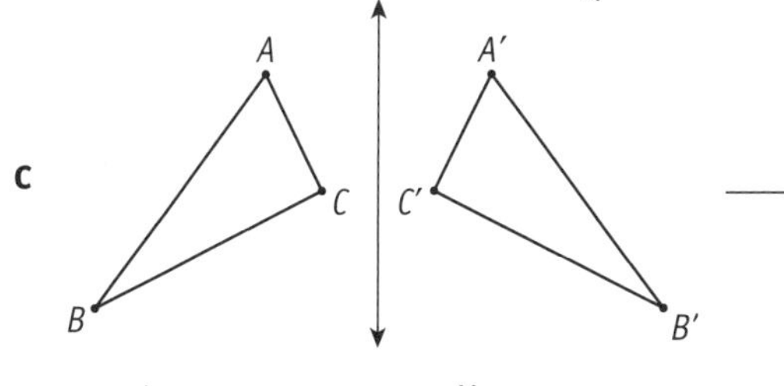

____

d

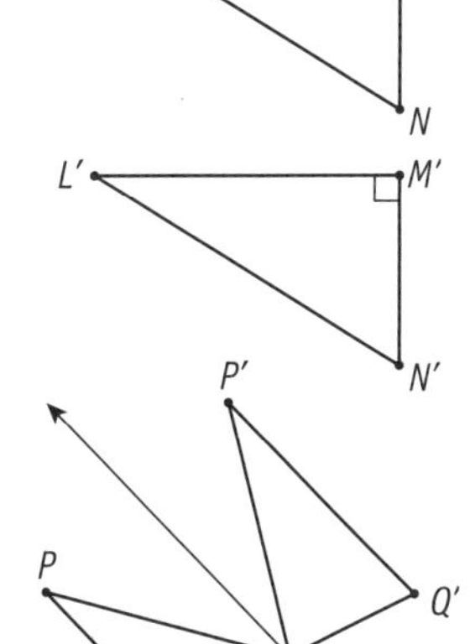

____

e

P, Q, R, P′, Q′, R′

____

Reasoning in geometry

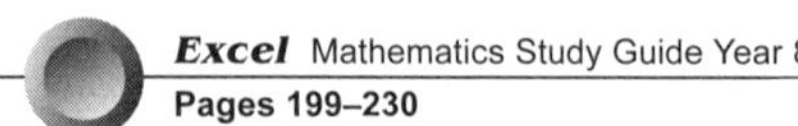

UNIT 7: Congruence of plane shapes using transformations

Question 1 Complete the following.

a The symbol for congruence is ________________.

b Congruent shapes have the ________________ shape and the ________________ size.

c Two plane shapes are called congruent if one shape fits exactly on top of the other by a ________________.

d When writing a congruence statement for two shapes the vertices must be given in ________________ order.

Question 2 Name the transformation that has been used in each pair of congruent triangles below.

a

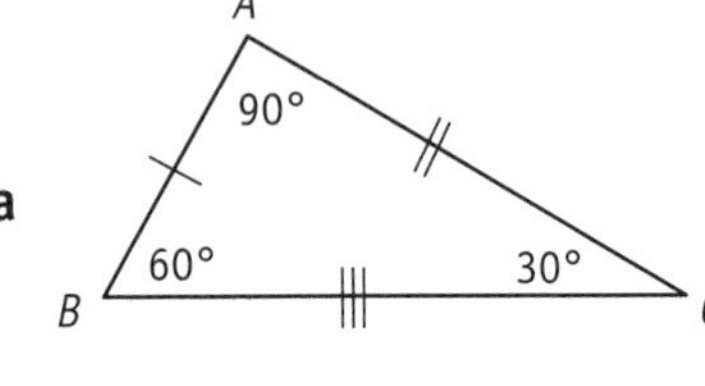

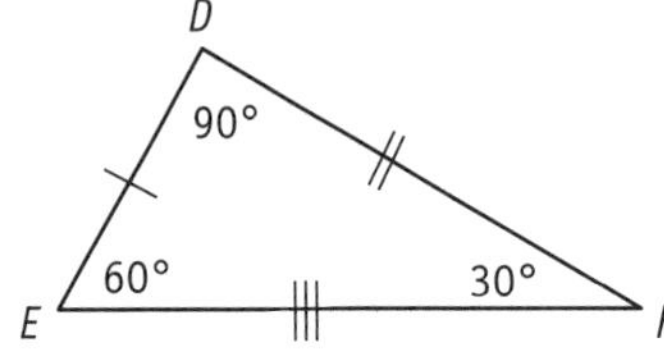

____

b

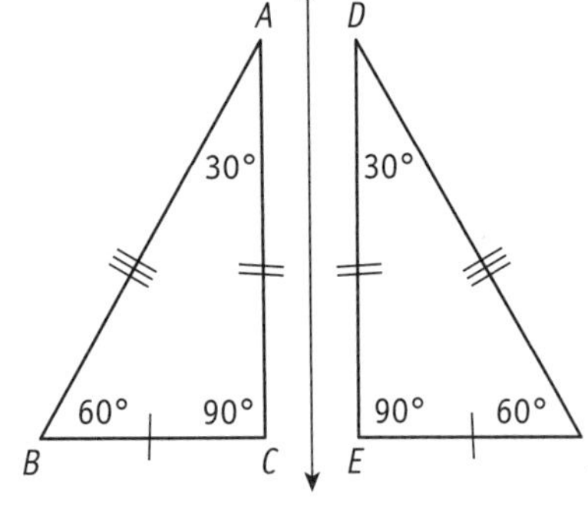

____

c

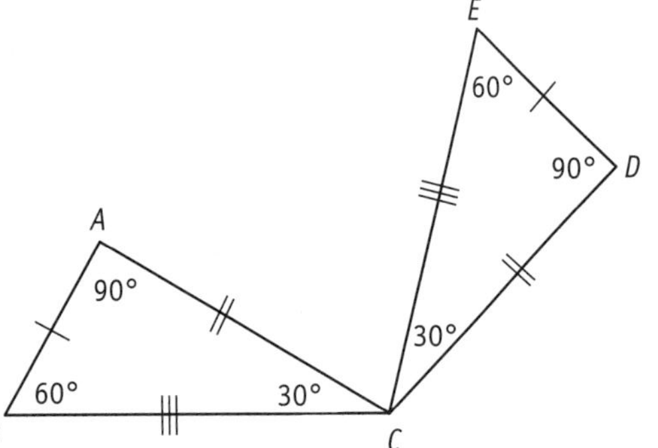

____

Question 3 For each pair of congruent triangles below, write the congruence statement.

a

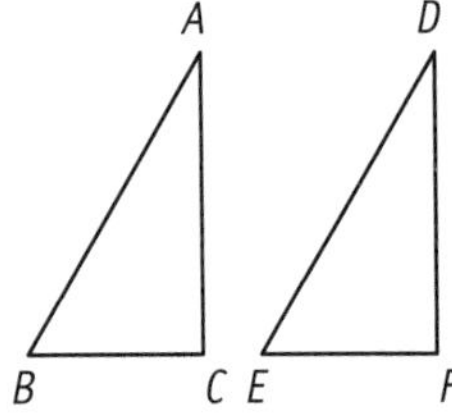

b

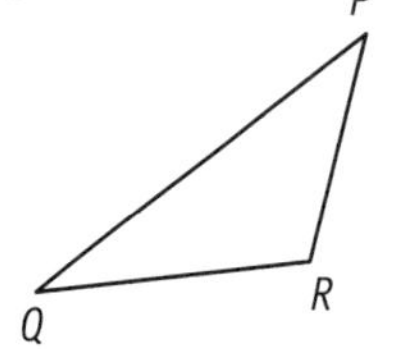

c

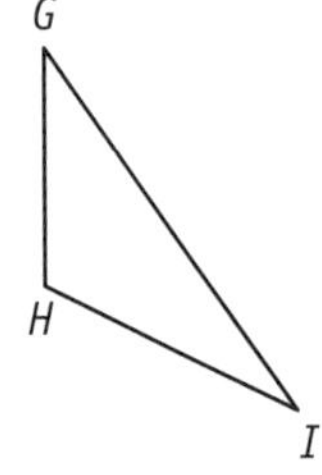

________________________ ________________________ ________________________

Reasoning in geometry

UNIT 8: Recognising congruent figures

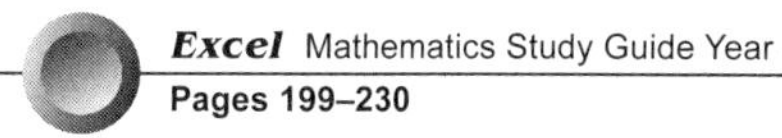
Excel Mathematics Study Guide Year 8
Pages 199–230

QUESTION 1 Complete the following sentences.

a Congruent figures have exactly the ________________ size and the ________________ shape.

b In congruent figures:

i matching sides are ____________________.

ii matching angles are ____________________.

iii matching areas are ____________________.

iv the symbol for congruence is ________________.

QUESTION 2 Name the congruent faces in each prism.

a

A B C D E F G H

____

____

b

P Q R S T U

____

____

QUESTION 3 Name the pairs of congruent figures.

A B C D E F G H I J K L M N

____

QUESTION 4 The following pair of figures are congruent.

A B C D E F G H

a Name all pairs of matching angles. ______________________________________

b Name all pairs of matching sides. ______________________________________

Reasoning in geometry

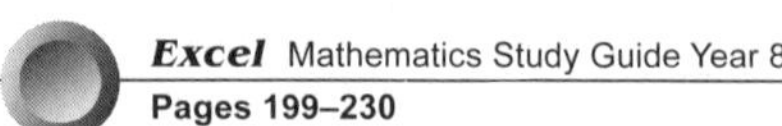

UNIT 9: Naming congruent triangles

QUESTION 1 Two pairs of congruent triangles are drawn below.

a

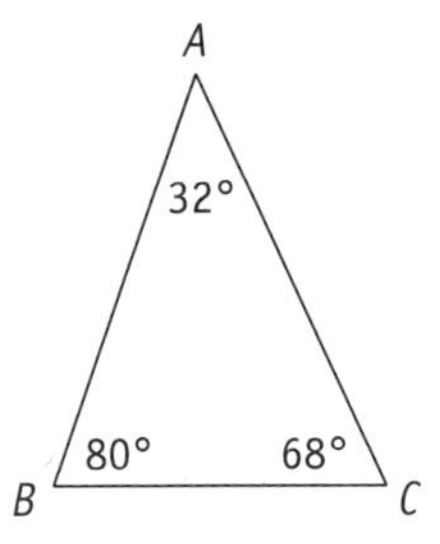

i Write all pairs of matching sides.

__________, __________, __________

ii Name these congruent triangles, giving the vertices in the matching order.

____

iii Write the congruence statement.

____

b

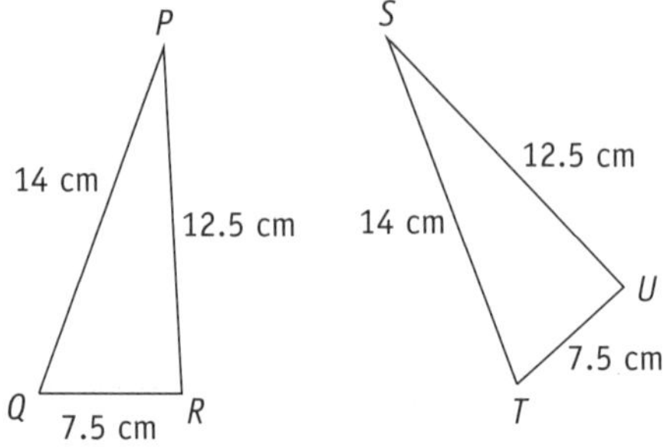

i Write all pairs of matching angles.

__________, __________, __________

ii Name the congruent triangles giving the vertices in the matching order.

____

iii Write the congruence statement.

____

QUESTION 2 In each of the following quadrilaterals, a diagonal is drawn to divide it into two congruent triangles. Write: **a** all pairs of equal sides **b** all pairs of equal angles. **c** Identify the pair of congruent triangles. **d** Complete the congruence statement.

i *ABCD* is a rectangle.

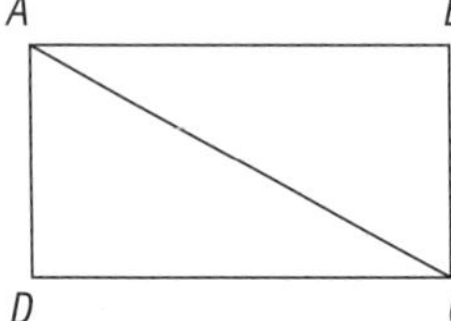

a ______________________________

b ______________________________

c ______________________________

d ______________________________

ii *EFGH* is a parallelogram.

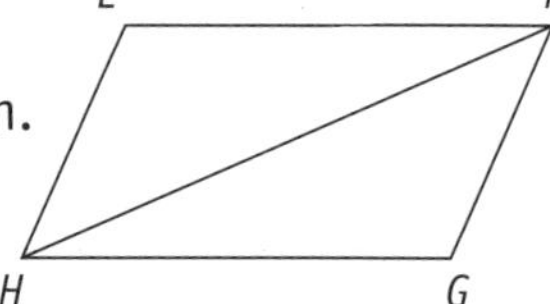

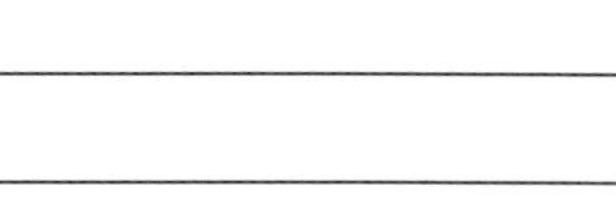

a ______________________________

b ______________________________

c ______________________________

d ______________________________

iii *PQRS* is a kite.

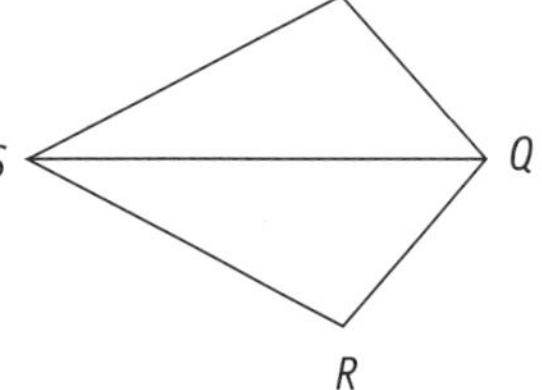

a ______________________________

b ______________________________

c ______________________________

d ______________________________

iv *TUVW* is a square.

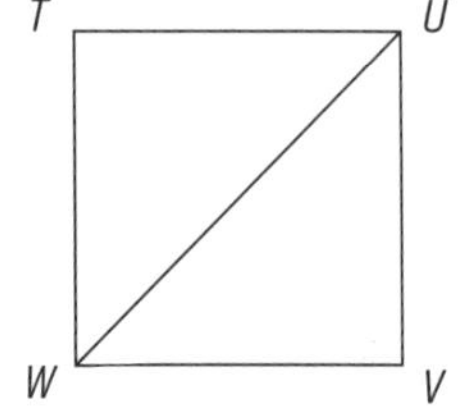

a ______________________________

b ______________________________

c ______________________________

d ______________________________

Reasoning in geometry

UNIT 10: Tests for congruent triangles

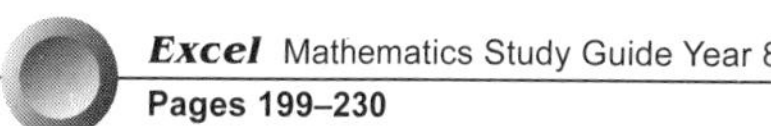

QUESTION **1** Answer the following questions.

a The symbol for congruent triangles is ______________________________.

b Two triangles are congruent if three sides of one triangle are equal to __________ of the other triangle.

c Two triangles are congruent if two angles and a side of one triangle are equal to ____________________ ______________________________ of the other triangle.

d Two triangles are congruent if two sides and the included angle of one triangle are equal to __________ ______________________________ of the other triangle.

e Two right-angled triangles are congruent if the hypotenuse and one side of one triangle are equal to ______________________________ of the other triangle.

QUESTION **2** In each pair of triangles write the congruence test that would be used to prove that the triangles are congruent.

a

____

b

____

c

____

d

____

QUESTION **3** Triangle *XYZ* is an equilateral triangle.

a Name the common side in $\triangle XYW$ and $\triangle XZW$.

____

b Are the triangles congruent? ______________________________

c If they are, which congruence test would be used to prove it? ______________________________

Reasoning in geometry

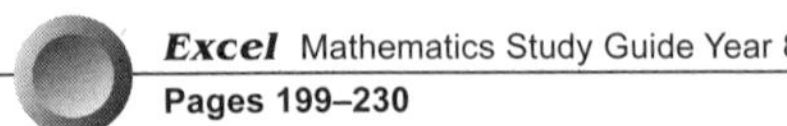
Excel Mathematics Study Guide Year 8
Pages 199–230

UNIT 11: Conditions for congruent triangles

QUESTION 1 For the following diagrams: **i** name the two congruent triangles **ii** state the congruence test **iii** write the congruence statement.

a

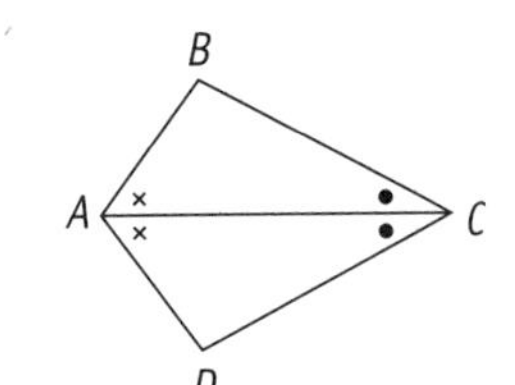

i ______

ii ______

iii ______

b

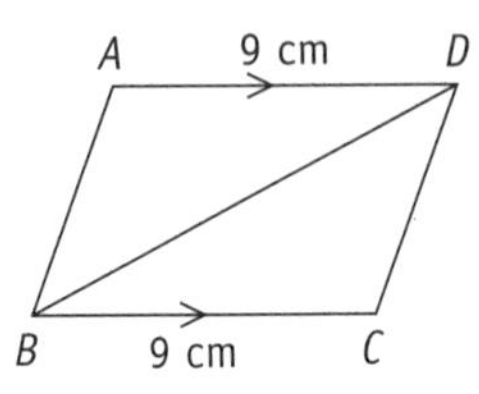

i ______

ii ______

iii ______

c

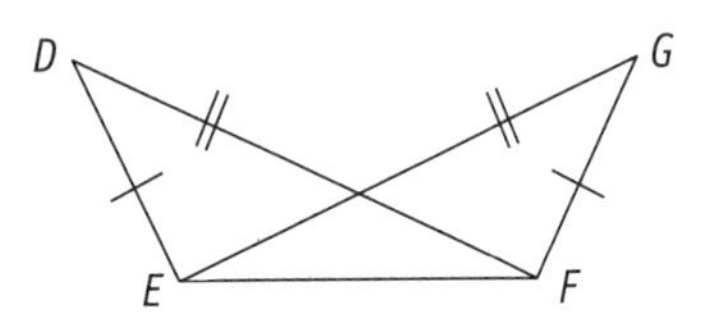

i ______

ii ______

iii ______

d

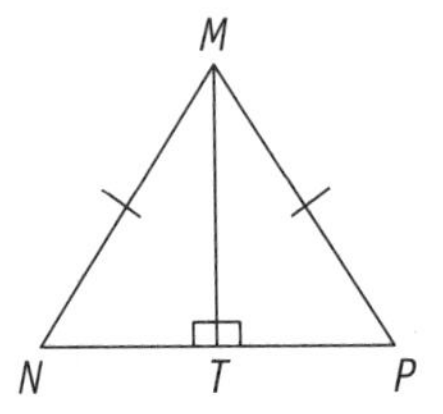

i ______

ii ______

iii ______

QUESTION 2 For the following pairs of triangles: **i** Name the two congruent triangles **ii** state the congruence test **iii** write the congruence statement **iv** find the value of the pronumerals.

a

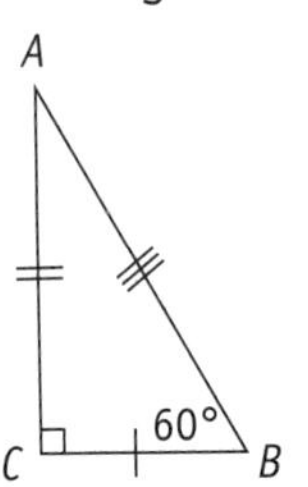

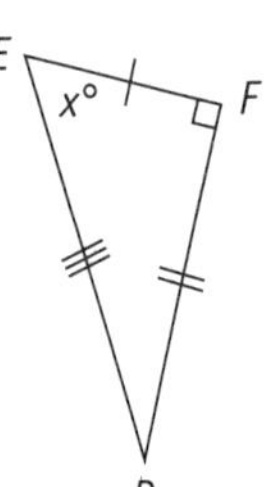

i ______

ii ______

iii ______

iv ______

b

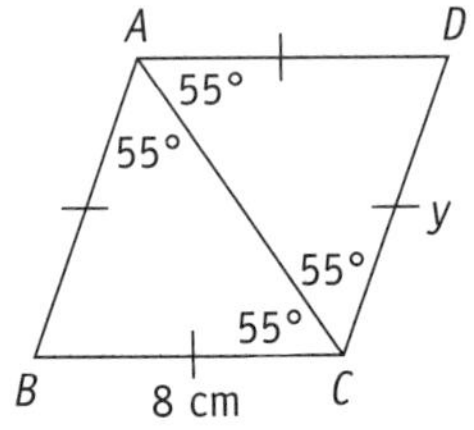

i ______

ii ______

iii ______

iv ______

c

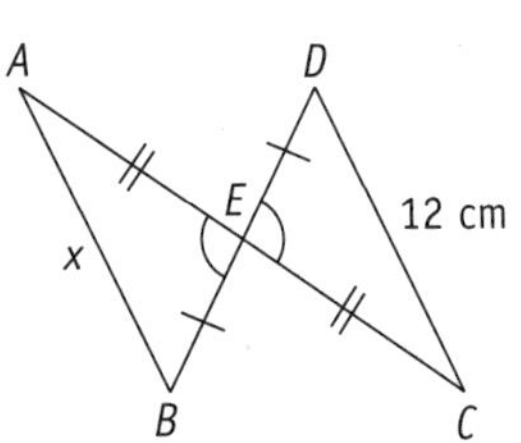

i ______

ii ______

iii ______

iv ______

d

i ______

ii ______

iii ______

iv ______

Reasoning in geometry

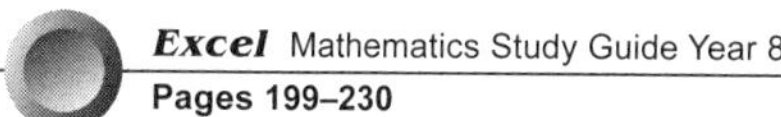

UNIT 12: Triangle congruence tests and numerical problems

QUESTION **1** Find the value of each pronumeral and justify your answer.

a

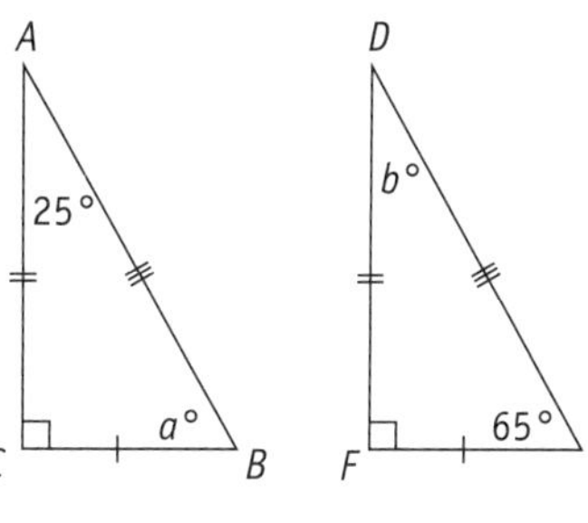

b

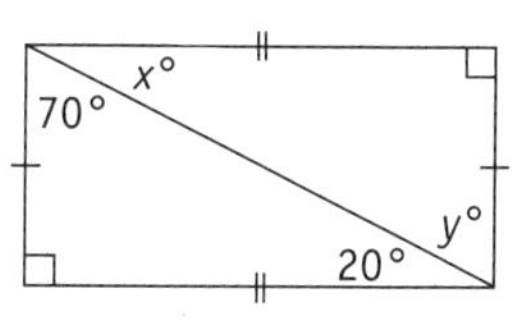

c

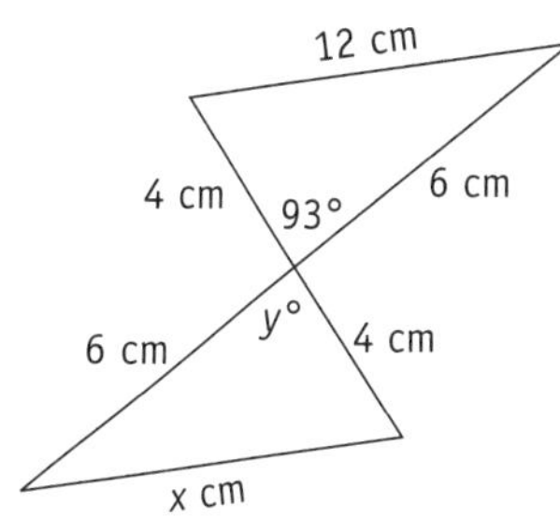

QUESTION **2** Find the value of each pronumeral.

a

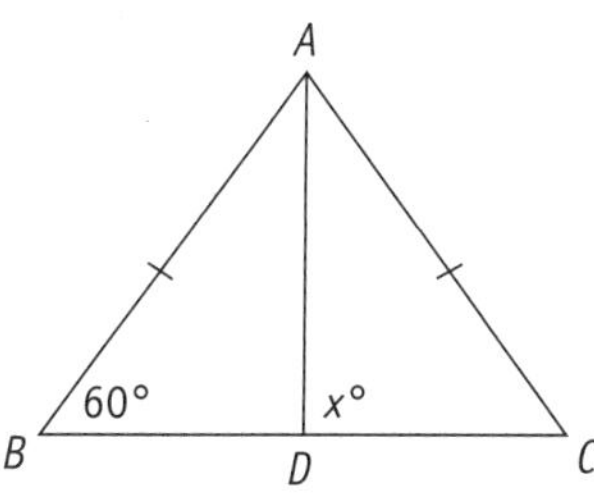

b

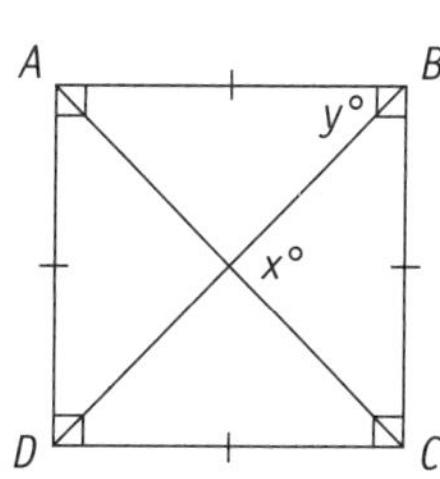

c

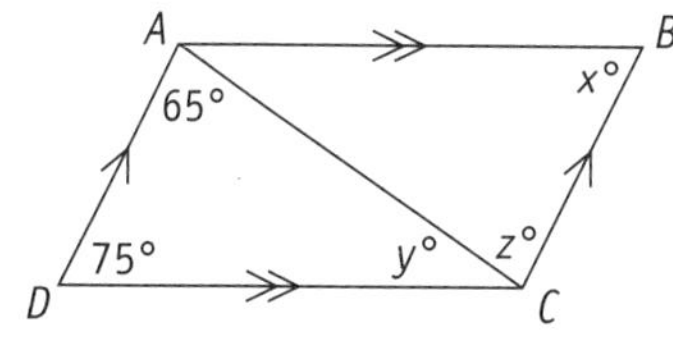

QUESTION **3** Find the value of each pronumeral and justify your answer.

a

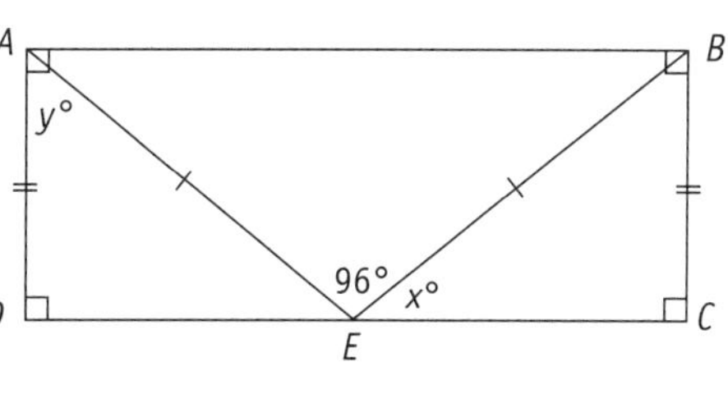

b

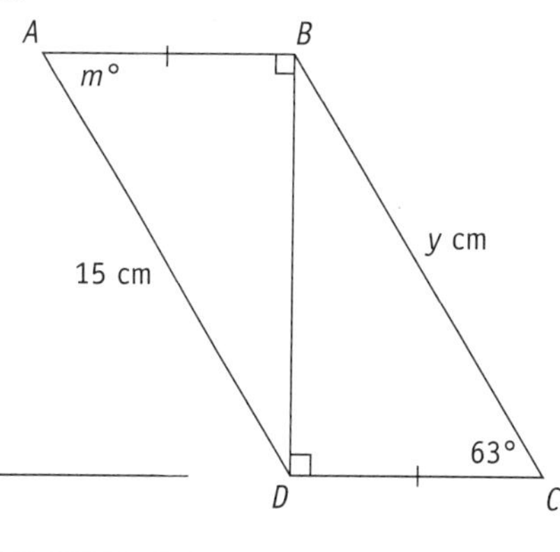

Reasoning in geometry

UNIT 13: Applying properties to solve numerical problems

QUESTION 1 Find the value of the pronumerals in these quadrilaterals.

a

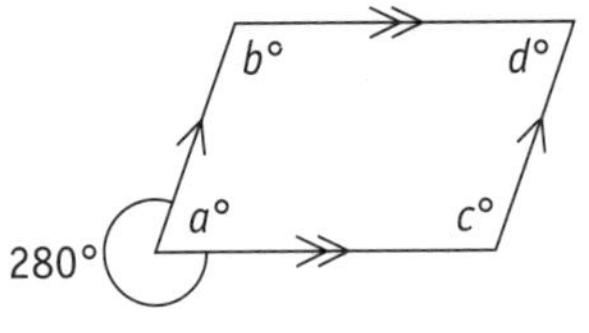

b

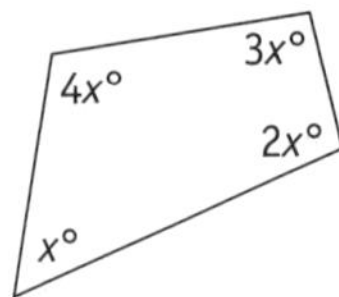

c

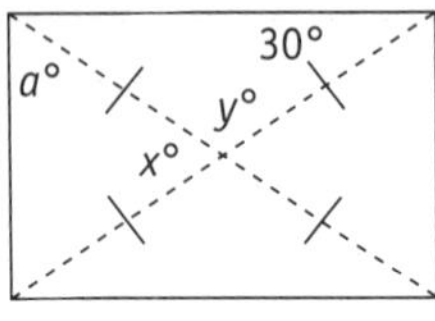

QUESTION 2 Find the value of each pronumeral in the following quadrilaterals, giving reasons.

a

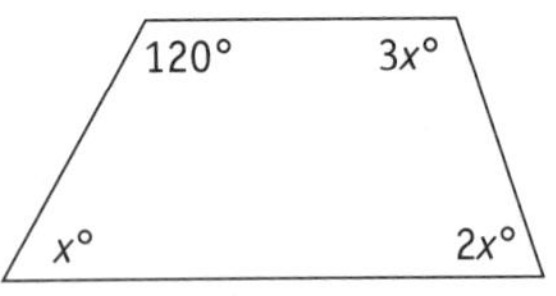

b

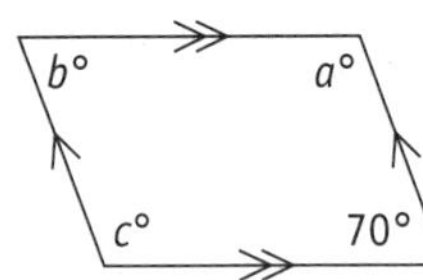

c

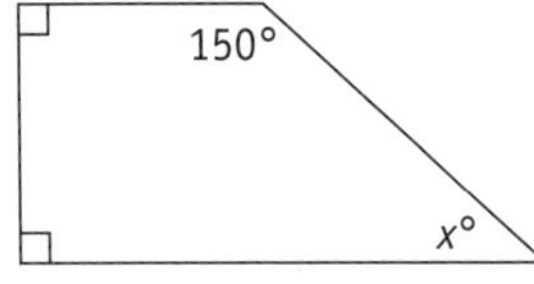

QUESTION 3 Find the value of each pronumeral in the following quadrilaterals.

a

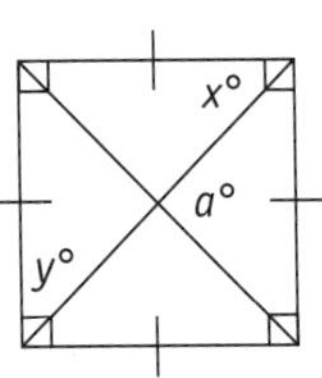

b

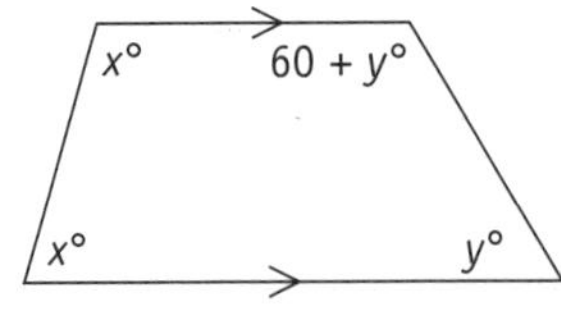

c

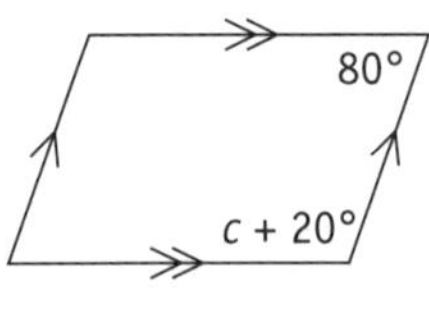

Reasoning in geometry

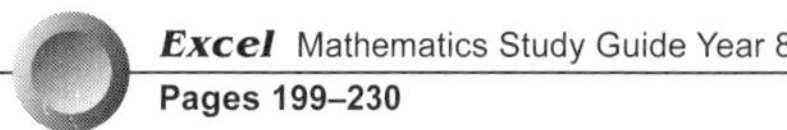

UNIT 14: More numerical problems

QUESTION 1 For each of the following quadrilaterals, find the value of the pronumeral and then find the size of each angle.

a

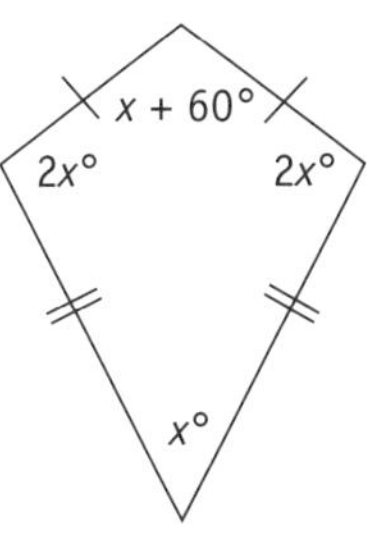

b

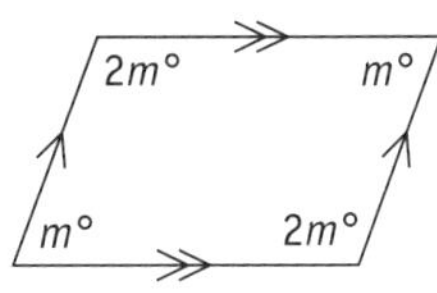

c

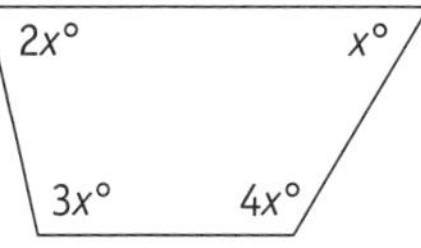

QUESTION 2 Find the value of each pronumeral in the following quadrilaterals.

a

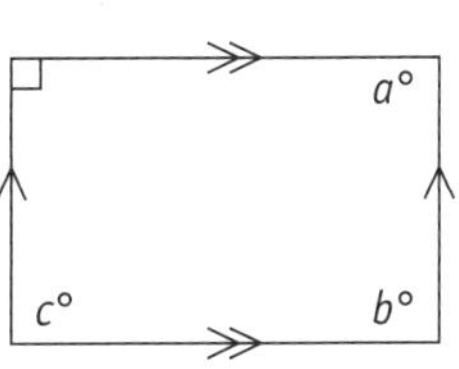

b

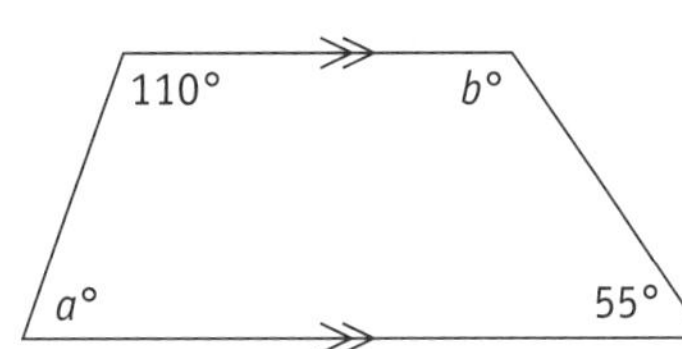

c

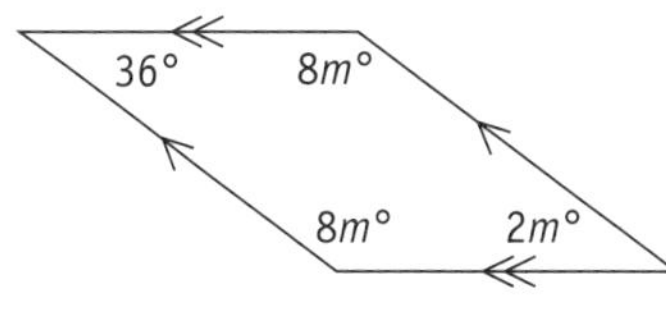

d

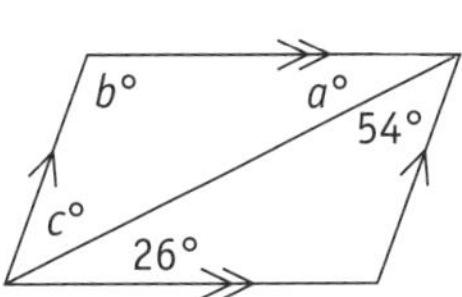

e

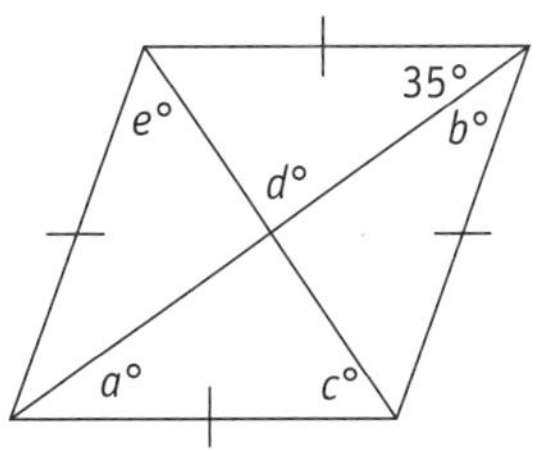

f

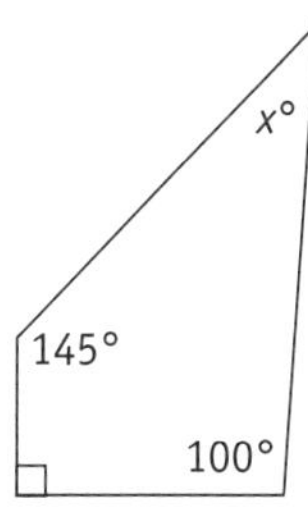

Reasoning in geometry

TOPIC TEST — PART A

Instructions
- This part consists of 10 multiple-choice questions.
- Fill in only ONE CIRCLE for each question.
- Each question is worth 1 mark.

Time allowed: 15 minutes **Total marks: 10**

Marks

1 The angles marked on the pair of parallel lines shown are

Ⓐ alternate. Ⓑ corresponding.
Ⓒ co-interior. Ⓓ complementary. 1

2 The four triangles made by the diagonals of a square are

Ⓐ equilateral only. Ⓑ isosceles only.
Ⓒ equilateral and congruent. Ⓓ isosceles and congruent. 1

3 Figure ABC has been moved to a new position by a

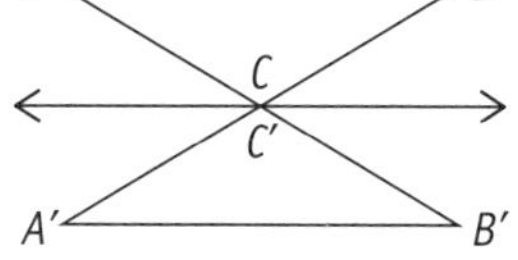

Ⓐ translation. Ⓑ rotation.
Ⓒ reflection. Ⓓ none of these. 1

4 In congruent triangles the lengths of the matching sides are

Ⓐ equal. Ⓑ different. Ⓒ in ratio 1:2. Ⓓ none of these. 1

5 When two triangles are congruent, their matching angles are

Ⓐ different. Ⓑ the same size. Ⓒ in ratio 1:2. Ⓓ none of these. 1

6 Two triangles are congruent if they are of the

Ⓐ same shape. Ⓑ same area. Ⓒ same shape and size. Ⓓ none of these. 1

7 Which congruence test proves $\triangle IJK \equiv \triangle ILK$?

Ⓐ SSS Ⓑ SAS Ⓒ AAS Ⓓ RHS 1

8 A diagonal of a rhombus divides the rhombus into two triangles that are

Ⓐ equilateral. Ⓑ isosceles only.
Ⓒ congruent only. Ⓓ isosceles and congruent.

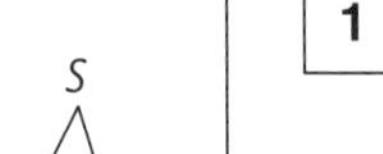

9 Which congruence test proves $\triangle PTQ \equiv \triangle STR$?

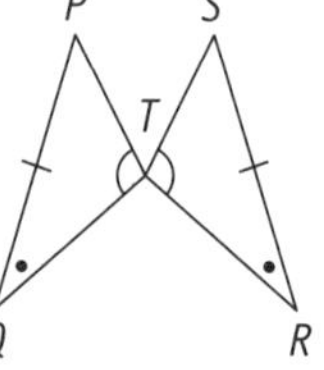

Ⓐ SSS Ⓑ SAS Ⓒ AAS Ⓓ RHS

10 The shorter diagonal of a kite divides it into two triangles that are both

Ⓐ isosceles. Ⓑ equilateral. Ⓒ congruent. Ⓓ none of these. 1

Total marks achieved for PART A

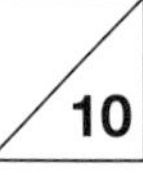

Reasoning in geometry

TOPIC TEST — PART B

Instructions
- This part consists of 4 questions.
- Each question part is worth 1 mark.
- Write only the answer in the answer column.

Time allowed: 15 minutes — **Total marks: 15**

Questions	Answers	Marks
1 For the diagram to the right:		
a Name the pairs of equal sides.	__________	1
b Name the pair of equal angles.	__________	1
c State the congruence test.	__________	1
d Write the congruence statement.	__________	1
2 Find the value of the pronumeral for each of the following:		
a	__________	1
b	__________	1
c	__________	1
d	__________	1
3 For $\triangle DEF$ and $\triangle GFE$ shown:		
a State the congruence test.	__________	1
b Write the congruence statement.	__________	1
c Find the value of the pronumeral.	__________	1
d Name another pair of congruent triangles in this figure.	__________	1
4 For the diagram to the right:		
a State the congruence test.	__________	1
b Write the congruence statement.	__________	1
c Find the value of x.	__________	1

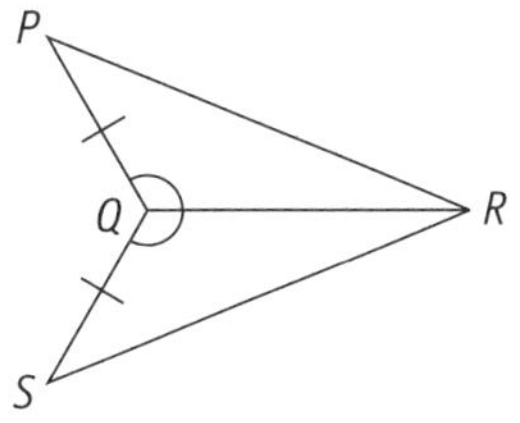

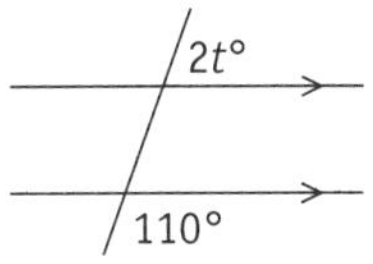

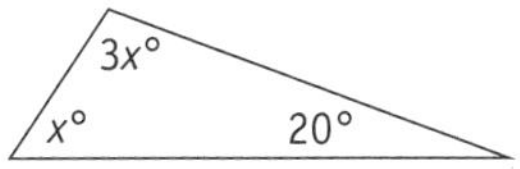

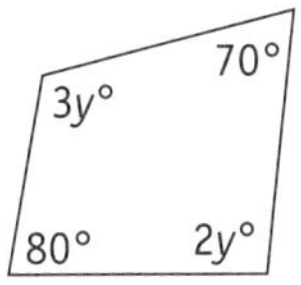

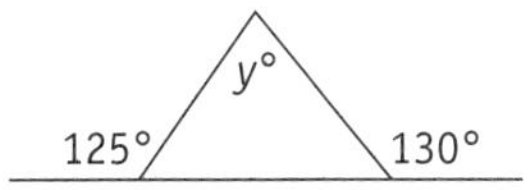

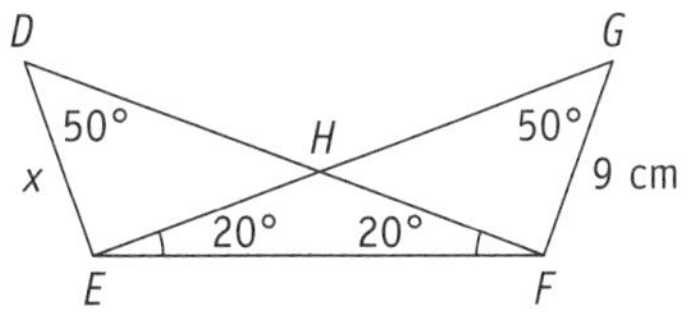

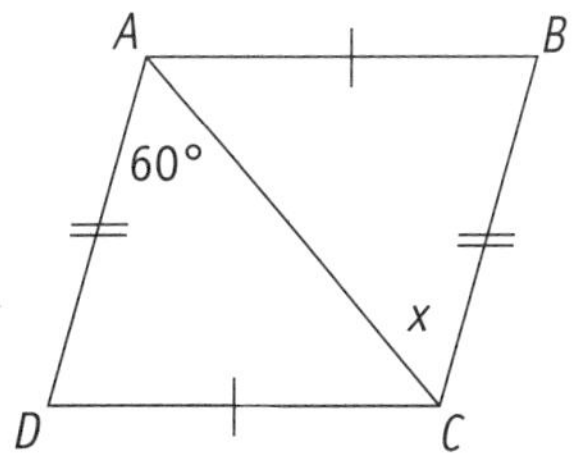

Total marks achieved for PART B ___ / 15

CHAPTER 12

Probability

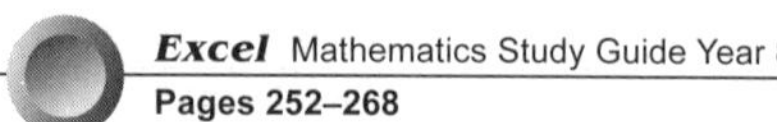

Excel Mathematics Study Guide Year 8
Pages 252–268

UNIT 1: Basic probability

QUESTION 1 A card is drawn at random from a normal pack of 52 cards. Find the probability that the card is:

a a diamond ______ **b** a red card ______ **c** a king ______

d not a club ______ **e** a red ace ______ **f** a black card ______

QUESTION 2 From the letters of the word FREQUENCY, one letter is selected at random. What is the probability that the letter is:

a a vowel? ______ **b** a consonant? ______

c the letter R? ______

QUESTION 3 A die is thrown once. Find the probability that the number is:

a a six ______ **b** an even number ______

c a number less than 5 ______ **d** seven ______

e a prime number ______ **f** a square number ______

QUESTION 4 A bag contains 8 white, 5 yellow and 7 blue balls. If a ball is drawn at random, find the probability that it is:

a white ______ **b** yellow ______ **c** blue ______

d not white ______ **e** red ______ **f** either white or blue ______

QUESTION 5 A three-digit number is to be formed from the digits 4, 5 and 6 written on separate cards. What is the probability that the number formed is:

a odd? ______ **b** even? ______

c less than 600? ______ **d** divisible by 3? ______

e divisible by 5? ______ **f** greater than 600? ______

QUESTION 6 The numbers 1 to 9 are written on separate cards. One card is chosen at random. What is the probability that:

a the number is even? ______ **b** the number is odd? ______

c it is 5? ______ **d** it is 10? ______

e it is a prime number? ______ **f** it is divisible by 3? ______

QUESTION 7 A letter is chosen from the word EXCELLENT. What is the probability that the letter is:

a a vowel? ______ **b** a consonant? ______

c the letter L? ______ **d** the letter N? ______

e the letters E or L? ______ **f** the letters T or X? ______

Probability

UNIT 2: Experimental probability

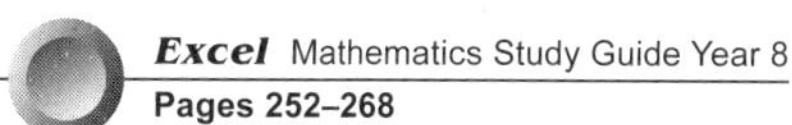

QUESTION **1** Jeanne tossed a coin many times and the results were tabulated.

	Heads	Tails
Frequency	43	57

a How many times did Jeanne toss the coin? ______________________

b What is the relative frequency of tossing heads? ______________________

c What is the probability of tossing heads? ______________________

d What is the relative frequency of tossing tails? ______________________

e What is the probability of tossing tails? ______________________

f What is the sum of the relative frequencies? ______________________

g How many tails do you expect to get in 200 tosses of a coin? ______________________

QUESTION **2** Michelle rolled a die many times and recorded the results.

a Complete the table showing the relative frequencies as fractions in simplest form.

Number	**Frequency**	**Relative frequency**
1	10	
2	16	
3	20	
4	12	
5	9	
6	13	

b From Michelle's experiment, find the probability of rolling the following.

i 4 ______________

ii an even number ______________

iii 2 or 3 ______________

iv 4, 5 or 6 ______________

v 5 (as a percentage) ______________

vi an odd number (as a percentage) ______________

vii 6 (as a decimal) ______________

viii 3 (as a decimal) ______________

ix 2 (as a fraction) ______________

x 1 (as a fraction) ______________

UNIT 3: Theoretical probability

Excel Mathematics Study Guide Year 8
Pages 252–268

QUESTION 1 Complete the following sentence.

The probability of any event is always in the range from ____________ to ____________.

QUESTION 2 A bag contains 6 red marbles and 2 green marbles. If one marble is drawn at random, what is the probability, as a decimal, that it is:

a red? ______________ **b** green? ______________ **c** white? ______________

QUESTION 3 A raffle ticket is drawn from a box containing 100 tickets numbered from 1 to 100. Find the percentage chance that the number of the ticket is:

a divisible by 10 ______________________

b less than 10 ______________________

c greater than 10 ______________________

d a multiple of 5 ______________________

e greater than 90 ______________________

f a number containing the digit 9 ____________

QUESTION 4 A spinner used in a game is in the shape of a pentagon and has an equal chance of landing on any of its sides. The sides are numbered 1, 2, 3, 4 and 5. What is the probability, as a percentage, that the spinner lands on:

a 5? ______________________

b an even number? ______________________

c an odd number? ______________________

d a number greater than 3? ________________

QUESTION 5 A bag holds 9 brown, 7 yellow and 4 white golf balls. If a ball is selected at random from the bag, what is the probability (as a fraction in its simplest form) that the ball is:

a brown? ______________________

b yellow? ______________________

c white? ______________________

d brown or yellow? ______________________

e pink? ______________________

f brown, yellow or white? ________________

g yellow or white? ______________________

h white or brown? ______________________

i not white? ______________________

j neither yellow nor brown nor white? __________

Probability

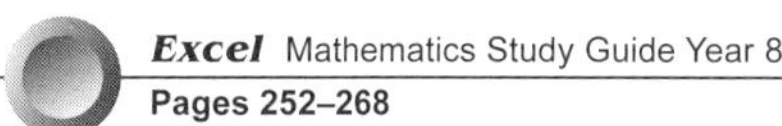

UNIT 4: Complementary events

QUESTION 1 Write the complement of each event.

a tossing a head with a coin ______

b rolling an odd number with a die ______

c selecting a diamond from a pack of cards ______

d winning a game of tennis ______

e spinning a number less than 10 on a wheel ______

f If $P(A) + P(B) = 1$, what can we say about the events A and B? ______

QUESTION 2 If the probability of having a sunny day, today is $\frac{2}{3}$, what is the probability of the day not being sunny?

______

QUESTION 3 A card is drawn at random from a normal pack of 52 cards. Find the probability that the card is:

a a spade ______

b not a spade ______

c a black card ______

d not a black card ______

e a queen ______

f not a queen ______

QUESTION 4 A letter is chosen from the word HAPPINESS. What is the probability that the letter is:

a a vowel? ______

b not a vowel? ______

c a letter N? ______

d not N? ______

e the letter P or S? ______

f not the letter P or S? ______

QUESTION 5 The numbers 1 to 7 are written on separate cards. One card is chosen at random. What is the probability that:

a the number is odd? ______

b the number is not odd? ______

c a prime number? ______

d not a prime number? ______

QUESTION 6 A bag contains 6 red, 5 blue and 4 white balls. If a ball is drawn at random, what is the probability that it is:

a red? ______

b not red? ______

c white? ______

d not white? ______

e blue? ______

f not blue? ______

g either red or white? ______

h either red or blue? ______

Probability

UNIT 5: Two dice rolled simultaneously

QUESTION **1** Two dice are rolled simultaneously. Complete the table showing all the possible numbers on each die. The first column has been done for you.

1st die

36 sample points	1	2	3	4	5	6
1	1, 1					
2	1, 2					
3	1, 3					
4	1, 4					
5	1, 5					
6	1, 6					

2nd die

QUESTION **2** Use the above table to find the probability of the event in each case listed below.

a a double six ______

b any double ______

c the sum of the two numbers rolled equaling 6 ______

d the two numbers being even ______

e the sum of the two numbers equaling 9 ______

f a total of 10 ______

g a sum greater than 8 ______

h a sum of either 5 or 7 ______

i a sum of less than 5 ______

j at least one five on the uppermost face of a die ______

k the sum of the numbers being greater than twelve ______

QUESTION **3** Suppose we wish to throw a total of five. Which is a better chance — rolling one die or rolling two dice?

______

______

______

QUESTION **4** What is the probability of throwing odd numbers on the uppermost faces of both dice in one roll of a pair of dice?

______

______

______

UNIT 6: Venn diagrams

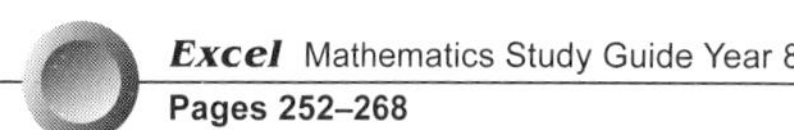

QUESTION **1** In a class of 28 students, 16 play soccer, 10 play tennis and 7 play neither.

a Draw a Venn diagram to show this information.

b How many students play both sports? ______

If one student is selected at random,

c What is the probability that he or she plays only soccer? ______

d What is the probability that this student plays either tennis or soccer (but not both)? ______

QUESTION **2** There are 115 year 8 students in a school. 80 of them study French, 50 of them study German and 5 study neither.

a Draw a Venn diagram to show this information.

b If one student is selected at random, what is the probability that the student studies French? ______

c What is the probability that the student studies German? ______

d What is the probability that the student studies neither? ______

QUESTION **3** In a survey, people were asked whether they owned a car or a motor bike. The results are shown on the Venn diagram given below.

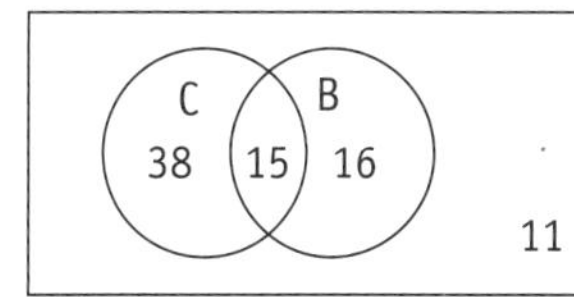

a How many people owned only a car? ______

b How many people owned both a car and a motor bike? ______

c How many people owned neither a car nor a motor bike? ______

d How many people were surveyed? ______

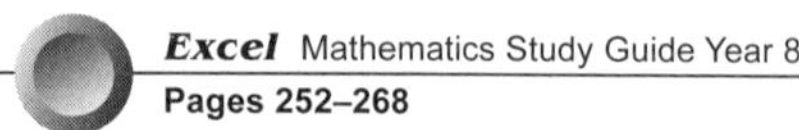
Excel Mathematics Study Guide Year 8
Pages 252–268

UNIT 7: Using language of *at least*, exclusive *or*, inclusive *or*, *and*

QUESTION 1 In a family of three children, find the probability that there is at least one girl.

____

____

QUESTION 2 When a coin is tossed four times, find the probability of getting at least one head.

____

____

QUESTION 3 From a pack of cards numbered 1 to 10, one card is selected at random. Find the probability that the number drawn is:

a odd ______________________________

b greater than 5 ______________________________

c less than 4 ______________________________

d greater than 5 or less than 4 ______________________________

e odd or greater than 5 ______________________________

f odd and greater than 5 ______________________________

g odd and less than 5 ______________________________

h greater than 5 and less than 4 ______________________________

i less than 5 or divisible by 6 ______________________________

j less than 5 or divisible by 3 ______________________________

QUESTION 4 From a pack of 52 playing cards, one card is drawn at random. Find the probability that it is:

a a black card or ten ______________________________

b a black ten or a king ______________________________

c a spade or a ten ______________________________

QUESTION 5 Two dice are thrown. Find the probability of:

a a double or a total of 10 ______________________________

b a double or a total of 15 ______________________________

Probability

UNIT 8: Two-way tables

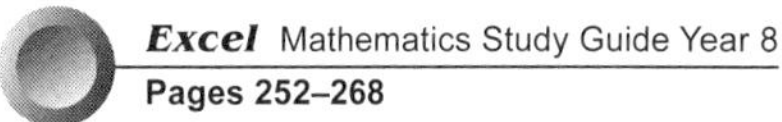
Excel Mathematics Study Guide Year 8
Pages 252–268

QUESTION **1** A school has a population of 900 students. The two-way table given below shows whether or not each student is in the school choir.

	Boys	Girls	**Total**
In the school choir	350	225	
Not in the school choir		175	**325**
Total	**500**		

a Fill in the missing boxes in the table.

b If one student is selected at random, what is the probability that:

i the student is in the school choir? ____________

ii the student is a girl and not in the school choir? ____________

iii the student is a boy and in the school choir? ____________

QUESTION **2** The two-way table below shows the people in a movie session. A person is randomly selected from the audience and is given a prize. What is the probability that the person receiving the prize is:

a a male? ____________

b a female? ____________

c a child? ____________

d an adult female? ____________

e not a female child? ____________

f not an adult? ____________

	Male	Female	**Total**
Adult	120	80	**200**
Child	60	40	**100**
Total	**180**	**120**	**300**

QUESTION **3** A two-way table is given below. What is the probability that a person selected at random from this group:

a likes milk? ____________

b likes toast but not milk? ____________

c likes milk but not toast? ____________

d doesn't like milk? ____________

e does not like milk and does not like toast? ____________

f likes milk and likes toast? ____________

	Like toast	Don't like toast	**Total**
Like milk	120	30	**150**
Don't like milk	80	20	**100**
Total	**200**	**50**	**250**

Probability

TOPIC TEST PART A

Instructions
- This part consists of 10 multiple-choice questions.
- Fill in only ONE CIRCLE for each question.
- Each question is worth 1 mark.
- Calculators are NOT allowed.

Time allowed: 15 minutes **Total marks: 10**

Marks

1 From a pack of 52 cards, one card is drawn at random. Find the probability of not drawing a king. 1

(A) $\frac{1}{13}$ (B) $\frac{12}{13}$ (C) $\frac{1}{4}$ (D) $\frac{3}{4}$

2 A bag contains 4 white, 3 red and 2 black balls. If a ball is drawn at random, find the probability that it is not white. 1

(A) $\frac{3}{9}$ (B) $\frac{4}{9}$ (C) $\frac{2}{9}$ (D) $\frac{5}{9}$

3 Two dice are thrown. Find the probability that the sum is less than 5. 1

(A) $\frac{1}{4}$ (B) $\frac{1}{12}$ (C) $\frac{1}{6}$ (D) $\frac{1}{3}$

4 Consider the word MISCELLANEOUS. What is the relative frequency of the letter 'S'? 1

(A) 1 (B) 2 (C) $\frac{1}{13}$ (D) $\frac{2}{13}$

5 In a single throw of two dice, find the probability of totalling 8 or 11. 1

(A) $\frac{5}{36}$ (B) $\frac{1}{18}$ (C) $\frac{7}{36}$ (D) $\frac{1}{12}$

6 In a family of three children, the probability that there are three boys is 1

(A) $\frac{1}{3}$ (B) $\frac{3}{8}$ (C) $\frac{1}{8}$ (D) $\frac{1}{4}$

7 From a pack of 52 playing cards, a card is chosen at random. What is the probability that it is a red Queen or a 4? 1

(A) $\frac{3}{26}$ (B) $\frac{5}{52}$ (C) $\frac{2}{13}$ (D) $\frac{3}{52}$

8 A drawer contains 13 black socks, 14 white socks and 8 grey socks. If one sock is selected at random, what is the probability that it is not black? 1

(A) $\frac{13}{35}$ (B) $\frac{14}{35}$ (C) $\frac{8}{35}$ (D) $\frac{22}{35}$

9 In a single throw of two dice, find the probability of throwing a double. 1

(A) $\frac{1}{6}$ (B) $\frac{2}{3}$ (C) $\frac{1}{2}$ (D) $\frac{3}{4}$

10 The probability of it raining today is $\frac{1}{5}$. What is the probability of it not raining today? 1

(A) $\frac{1}{5}$ (B) $\frac{2}{5}$ (C) $\frac{3}{5}$ (D) $\frac{4}{5}$

Total marks achieved for PART A 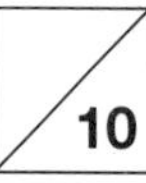

Probability

TOPIC TEST — PART B

Instructions
- This part consists of 3 questions.
- Each question part is worth 1 mark.
- Write only the answer in the answer column.
- Calculators are allowed.

Time allowed: 15 minutes — **Total marks: 15**

Questions	Answers	Marks

1 This Venn diagram displays Year 8 students' mode of transport to school one morning. Use it to answer the following questions.

Bus 42 | 25 | Train 13 | 15

- **a** How many students travelled on a bus? ______ 1
- **b** How many Year 8 students attended school that morning? ______ 1
- **c** If one student was selected at random, what is the probability he or she caught both a bus and train to school that day? ______ 1
- **d** What is the probability that a student did not travel by bus to school that day? ______ 1

2 A three-digit number is to be formed randomly from the digits 4, 5 and 6 that are written on separate cards. What is the probability that the number will:

- **a** be even? ______ 1
- **b** be odd? ______ 1
- **c** be greater than 600? ______ 1
- **d** be less than 600? ______ 1
- **e** be divisible by 3? ______ 1
- **f** be less than 400? ______ 1
- **g** be divisible by 5? ______ 1

3 This two-way table shows a group of Year 8 students at a particular school.

	Male	Female	**Total**
Play an instrument	32		**54**
Don't play an instrument	15	21	
Total	**47**	**43**	**90**

- **a** Calculate how many female students play an instrument. ______ 1
- **b** Find out how many students do not play an instrument. ______ 1
- **c** If one student is selected at random, what is the probability that it is a female who doesn't play an instrument? ______ 1
- **d** If a male student is selected at random, what is the probability that he plays an instrument? ______ 1

Total marks achieved for PART B

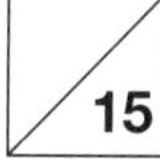

Chapter 13

Statistics

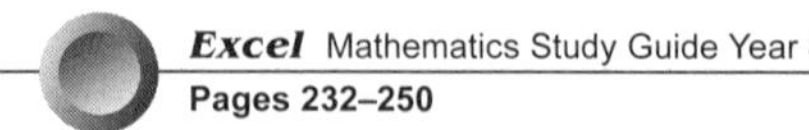
Excel Mathematics Study Guide Year 8
Pages 232–250

UNIT 1: Collecting data

Question 1 Complete the following.

a When a surveyor collects information about a whole class of objects or the total number of people the process is called a ____________________.

b When each and every member of the population is surveyed it is called ____________________ .

c When only a part of the population is surveyed it is called a ____________________ .

Question 2 State whether a census or a sample should be used in each of the following.

a The average number of lollies in a box of lollies. ____________________

b The total number of books available for sale in a bookshop. ____________________

c The number of students who travel to school by car. ____________________

d The number of people who bought 2 tickets for a movie. ____________________

e All the students and teachers in a high school who voted for the selection of a new school captain.

____

f The yearly exam results of all the students in a high school. ____________________

Question 3 State whether a sample or a census should be used in the following.

a The total number of people working in a hotel. ____________________

b The number of adults who went to a hall to attend a meeting. ____________________

c The total number of students in a high school. ____________________

____

d The number of students who were selected to participate in a swimming carnival. ____________________

e The total number of teachers who work in the state of NSW. ____________________

f The number of people who travel to work by bus. ____________________

Statistics

UNIT 2: Types of sampling

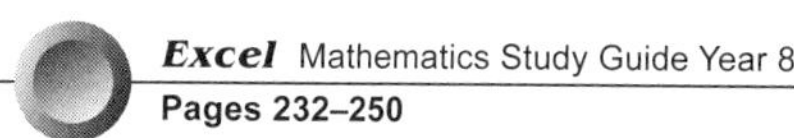

Question 1 Complete the following.

a A random sample may be defined as a sample in which each member of the population has the same ____________________ of selection.

b A systematic random sample is a sample where those chosen for the sample are chosen in a ____________________ way.

c In a stratified sample, the population is divided into different ____________________ and a sample is taken from each group.

Question 2 State whether the samples below are chosen by random sampling, systematic sampling or stratified sampling.

a Asking 85 men and 70 women in a company of 850 men and 700 women about work practices.

____

b Checking every 15th computer on a production line for quality and functioning.

____

c Choosing two teams of students for a soccer match by drawing names from a hat.

____

Question 3 Which sampling method would be most suitable for each of the following surveys?

a Checking the quality and weight of biscuits produced in a factory.

____

b Surveying students in each grade of a high school to determine whether a new school uniform is needed.

____

c Determining the preferred television show of households in a suburb.

____

d Selecting students from a class to check if homework has been completed.

____

e Measuring the pollution level of beaches along a State coastline.

____

Statistics

UNIT 3: Classification of data

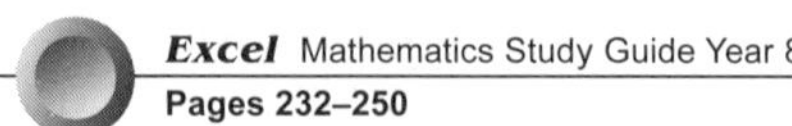
Excel Mathematics Study Guide Year 8
Pages 232–250

QUESTION **1** Answer the following questions.

a The information that can be collected in a survey is called ______________________.

b When the whole population is surveyed, it is called a ______________________.

c When a part of the whole population is surveyed, it is called a ______________________.

d The data that cannot be measured is ______________________ data and it is also called ______________________ data.

e The data that can be measured is ______________________ data and it is also called ______________________ data.

f The data that has only exact values and is obtained by counting is called ______________________ data.

g The data that has any values within a certain range and is generally obtained by measurement is called ______________________ data.

QUESTION **2** State whether the following data is numerical or categorical.

a The height of a student ______________________

b The number of rooms in a motel ______________________

c The number of days in a year ______________________

d The colour of your hair ______________________

e The length of a car ______________________

f The breed of a cat that you have ______________________

g Room temperature ______________________

h The number of people sitting in a park ______________________

QUESTION **3** State whether the data is discrete or continuous.

a The number of cars parked in a street ______________________

b The height of the Centre Point Tower ______________________

c The number of people in a restaurant ______________________

d The number of books in a bookshop ______________________

e The width of a wall ______________________

f The number of patients in a hospital ______________________

g The weight of a baby ______________________

h The weight of a car ______________________

i The number of lollies in a packet ______________________

j The body temperature of a person ______________________

k The length of a boat ______________________

UNIT 4: Frequency histograms and polygons

Question 1 For the set of scores given:

a complete the frequency distribution table.
b draw a frequency histogram.
c draw a frequency polygon.

5	10	8	5	4	7	7	10	7	5
9	6	7	9	10	4	7	10	8	6
8	11	9	9	8	9	10	11	9	7
7	9	8	10	8	9	8	9	9	8

Score (x)	Tally	Frequency (f)

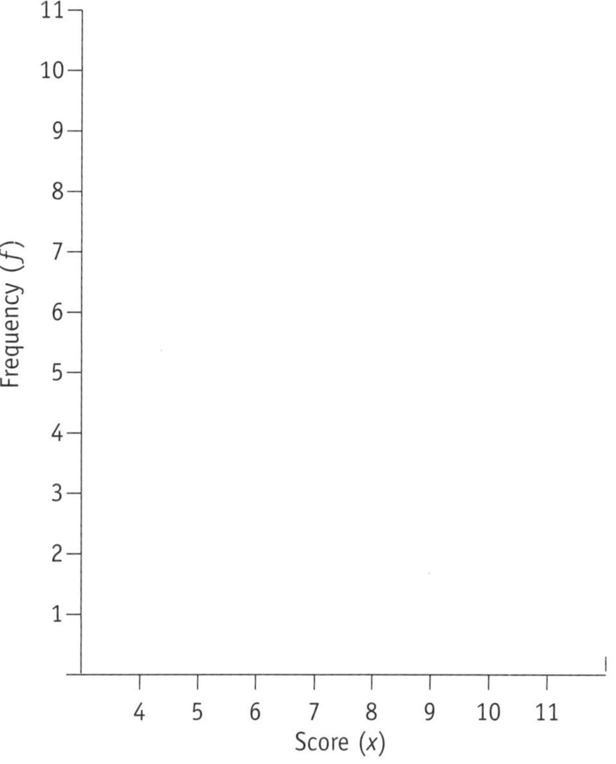

Question 2 From the following distribution table, draw a frequency histogram and frequency polygon.

Score (x)	8	9	10	11	12	13	14	15
Frequency (f)	2	4	3	7	11	8	5	2

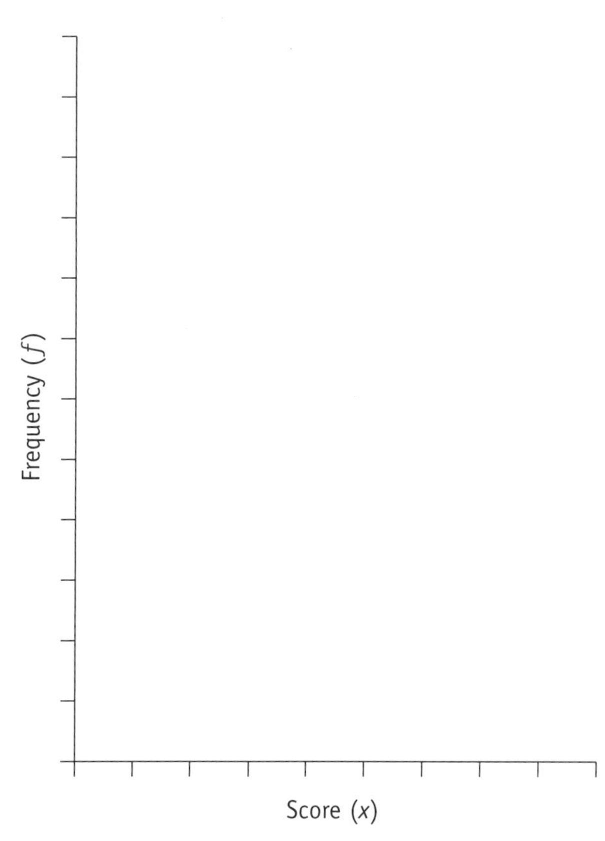

UNIT 5: Cumulative frequency histograms and polygons

QUESTION 1 A class of 25 students sat for a test. The results are shown in the frequency table.

a Complete the frequency distribution table.

b Draw a cumulative frequency histogram and polygon.

Score	Frequency	Cumulative frequency
5	1	
6	2	
7	3	
8	6	
9	5	
10	4	
11	3	
12	1	

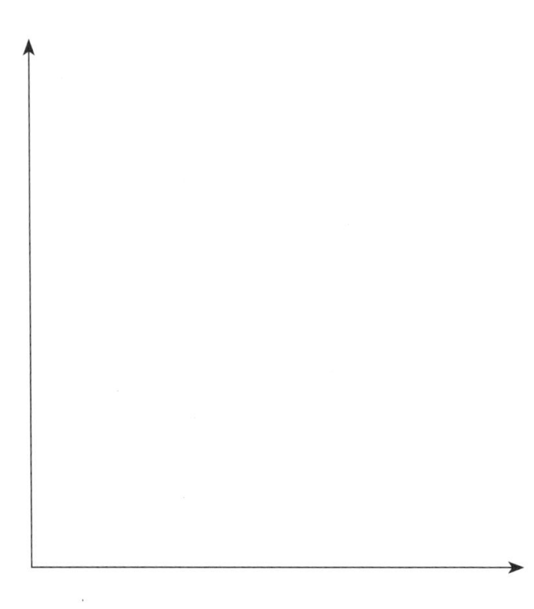

QUESTION 2 A class of 20 students obtained the following results in a class test.

8 4 10 9 10 5 6 8 6 8
12 11 10 9 6 7 10 12 10 5

a Complete the frequency distribution table.

b Draw a cumulative frequency histogram and polygon.

Score (x)	Tally	Frequency (f)	Cumulative frequency
4			
5			
6			
7			
8			
9			
10			
11			
12			

Statistics

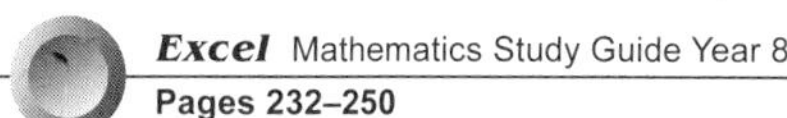

UNIT 6: Frequency histograms and polygons with grouped data

QUESTION 1 The percentage results of 50 students in a Mathematics exam are given below.

87	88	74	67	80	70	76	77	82	82
77	73	68	88	58	83	66	87	79	63
73	85	86	79	74	76	84	86	67	78
62	89	90	67	57	84	66	80	85	81
85	59	60	84	85	68	79	57	78	75

a Complete the grouped frequency distribution table.

Class	Class centre (c.c.)	Tally	Frequency (f)	Cumulative frequency (cf)
57–61	59			

b Label the axes and draw a grouped frequency histogram and polygon.

c Label the axes and draw a grouped cumulative frequency histogram and polygon.

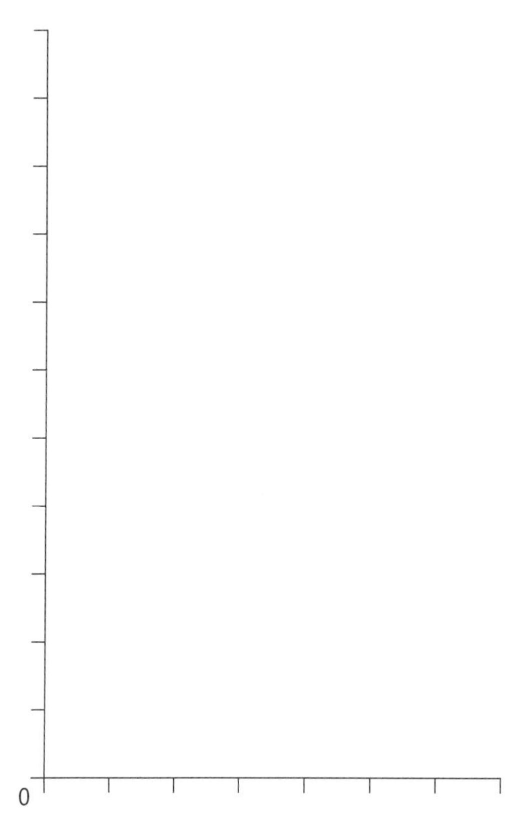

UNIT 7: Mean

Excel Mathematics Study Guide Year 8
Pages 232–250

Question 1 Find the mean of each of the following sets of scores.

a 10, 11, 12 ______

b 6, 8, 12, 16 ______

c 12, 13, 14, 15 ______

d 11, 15, 19, 23 ______

e 9, 11, 14, 16 ______

f 17, 18, 21, 22, 24, 30 ______

Question 2 Find the mean of each of the following sets of scores.

a 5, 5, 7, 7, 7, 7, 9, 9, 9, 9

b 11, 11, 12, 12, 12, 11, 11, 11, 11

c 12, 12, 15, 15, 15, 16, 16, 16, 16

d 6, 6, 6, 6, 6, 7, 7, 8, 8, 9

e 8, 9, 10, 11, 10, 9, 8, 7

f 16, 15, 17, 14, 13, 14, 16, 15

Question 3 Find the mean of each of the following sets of scores.

a 15, 15, 15, 18, 18, 20, 20, 20, 20 ______

b 10, 10, 12, 12, 12, 13, 14, 16 ______

c 7, 7, 7, 9, 9, 9, 9, 9, 10, 10, 10, 10 ______

d 8, 8, 8, 9, 9, 9, 10, 10, 10, 10, 11, 11, 11 ______

Question 4 Use a calculator to find the mean for each set of scores.

a

x	f	fx
1	2	
2	1	
3	3	
4	4	
5	5	
6	5	

b

x	f	fx
3	3	
4	2	
5	4	
6	2	
7	1	
8	3	

c

x	f	fx
7	4	
8	1	
9	3	
10	2	
11	6	
12	5	

Statistics

UNIT 8: Median and Mode

Excel Mathematics Study Guide Year 8
Pages 232–250

QUESTION **1** Find the median of the following sets of scores.

a 4, 5, 6, 8, 6 ____________________

b 9, 12, 16, 11, 9 ____________________

c 15, 18, 22, 15, 17, 16, 15 ____________________

d 47, 57, 37, 27, 57, 77, 67 ____________________

e 8, 11, 16, 13, 12, 13, 16, 11, 8, 7 ____________________

f 6, 12, 5, 9, 12, 9, 7, 10, 11, 12, ____________________

QUESTION **2** What is the mode of the following sets of scores?

a 3, 3, 4, 9, 10, 4, 3, 6, 3, 4 ____________________

b 8, 9, 7, 8, 10, 8, 5, 6, 8, 7, 8 ____________________

c 11, 10, 11, 14, 11, 12, 11, 11 ____________________

d 37, 32, 33, 37, 38, 37, 37, 39, 37 ____________________

e 3, 4, 5, 5, 6, 5, 6, 5, 7, 6, 5, 5, 5, 4, 5 ____________________

f 7, 8, 9, 7, 10, 7, 8, 7, 8, 7, 9, 7, 5, 7 ____________________

QUESTION **3** Complete each table, then find the median and mode.

a

Score	Frequency	Cumulative frequency
1	3	
2	6	
3	7	
4	5	
5	8	

Median = __________ Mode = __________

b

Score	Frequency	Cumulative frequency
18	6	
19	1	
20	5	
21	3	
22	3	

Median = __________ Mode = __________

c

Score	Frequency	Cumulative frequency
0	5	
1	2	
2	6	
3	9	
4	2	
5	1	

Median = __________ Mode = __________

d

Score	Frequency	Cumulative frequency
10	12	
11	10	
12	16	
13	14	
14	8	
15	6	

Median = __________ Mode = __________

UNIT 9: Using the mean, mode and median

Excel Mathematics Study Guide Year 8
Pages 232–250

QUESTION **1** An extension class has eight students. The class sat for a test and the following marks resulted. 7, 93, 95, 96, 96, 99, 90, 94

a Find:

i the median **ii** the mean **iii** the mode

b Edwin scored 93. "I did well in the test," Edwin told his mother. "I was way above average." Do you agree with his statement? Briefly comment.

QUESTION **2** When talking about real-estate, people in the industry and the media refer to the median house price. Why is the median a better means of describing the data than the mean or the mode?

QUESTION **3** A shop sells sportswear. The table shows the number of each size of shorts sold over the previous month.

Size	8	10	12	14	16	18	20	22	24
Number sold	2	13	28	42	35	26	23	19	21

a Find the mean size of shorts.

b What is the modal size?

c What is the median?

d The shop owner is most interested in the modal size of shorts. Why do you think she would find that important?

UNIT 10: Range

QUESTION 1 Find the range of the following sets of scores.

a 2, 4, 8, 9, 10 ________

b 13, 16, 18, 22 ________

c 68, 70, 70, 83, 85, 89, 90, 90 ________

d 13, 3, 21, 80, 22, 9, 37, 42, 50 ________

e 116, 163, 134, 127, 129, 130, 117 ________

f 63, 69, 72, 78, 76, 65, 38, 73, 69 ________

QUESTION 2 Find the range for the scores in each dot plot.

a

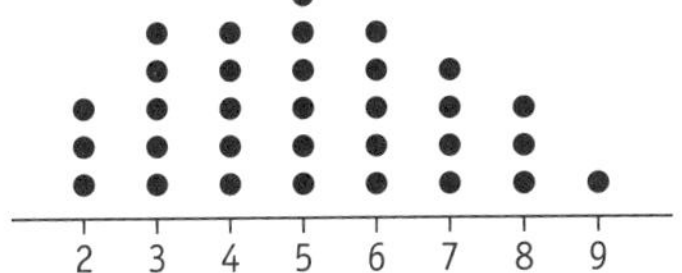

____

b

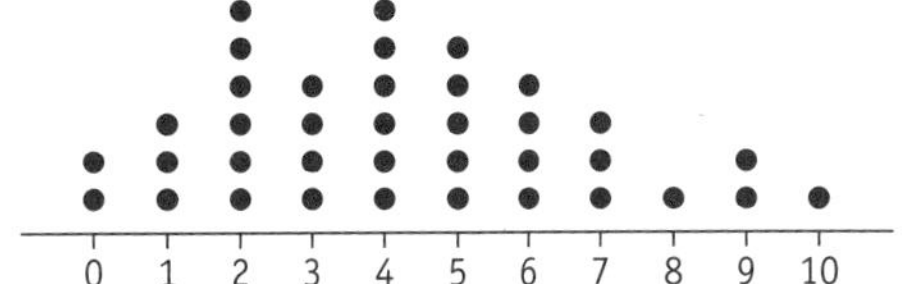

____

QUESTION 3 Find the range for the scores in each stem-and-leaf plot.

a

1	4 5 8
2	0 1 5 7 9
3	1 1 5 6
4	3 5 7 9
5	0 7
6	1

____

b

10	0 2 4 6
11	1 3 5 5 6
12	3 4 6 8 9 9
13	0 1 2 5
14	2 4 8
15	1 3

____

QUESTION 4 Find the range for the following sets of scores.

a

Score	4	5	6	7	8	9	10	11	12
Frequency	0	2	5	8	6	4	14	9	7

____

b

Score	0	1	2	3	4	5	6	7	8
Cumulative Frequency	1	2	5	8	12	12	14	17	20

____

QUESTION 5 Answer the following questions.

a If the range of a set of scores is 14 and the highest score is 29, what is the lowest score?

____

b The range of marks in a Maths test is 43. If the lowest mark was 32, what was the highest mark?

____

Statistics

TOPIC TEST — PART A

Instructions
- This part consists of 10 multiple-choice questions.
- Fill in only ONE CIRCLE for each question.
- Each question is worth 1 mark.

Time allowed: 15 minutes — **Total marks: 10**

Marks

1 Checking every 30th box of chocolates on a production line to ensure quality and correct measurements is an example of

Ⓐ stratified sampling. Ⓑ random sampling.
Ⓒ systematic sampling. Ⓓ a survey. 1

2 The number of chairs in a classroom is an example of which kind of data?

Ⓐ numerical discrete Ⓑ numerical continuous
Ⓒ categorical Ⓓ census 1

3 If a score of 3 is added to the data in the dot plot, the median

Ⓐ increases by 1. Ⓑ remains the same.
Ⓒ decreases by 1. Ⓓ decreases by 2. 1

1 2 3 4 5

4 The average height of five students is 162 cm. If another student who is 168 cm joins the group, the new average height is

Ⓐ 162.5 cm Ⓑ 163 cm Ⓒ 164 cm Ⓓ 165 cm 1

5 The median of the numbers 8, 3, 5, 7, 10, 8, 3 is

Ⓐ 3 Ⓑ 5 Ⓒ 7 Ⓓ 8 1

6 For the set of scores 1, 2, 3, 4, 4, 4, 5, 5, which measure is **not** 4?

Ⓐ Mean Ⓑ Mode Ⓒ Median Ⓓ Range 1

7 The mean of the numbers 4, 8 and x is the same as the mean of the numbers 4, 6, 8 and 10. Find the value of x.

Ⓐ 6 Ⓑ 7 Ⓒ 8 Ⓓ 9 1

QUESTIONS **8–9** refer to the set of scores

Score	1	2	3	4	5
Frequency	4	5	7	6	3

8 The mode is

Ⓐ 2 Ⓑ 3 Ⓒ 4 Ⓓ 5 1

9 The mean to the nearest whole number is

Ⓐ 2 Ⓑ 3 Ⓒ 4 Ⓓ 5 1

10 For the set of scores 5, 7, 3, 7, 6, 7, 9, 7, 8, 7, 7, 5, 3, find the difference between the median and the range.

Ⓐ 0 Ⓑ 1 Ⓒ 2 Ⓓ 3 1

Total marks achieved for PART A /10

Statistics

TOPIC TEST PART B

Instructions
- This part consists of 3 questions.
- Write only the answer in the answer column.
- Calculators are allowed.

Time allowed: 15 minutes **Total marks: 15**

Questions	Answers	Marks
1 a If a survey collects data from every person or thing in a group, the survey is called a	________	1
b Choosing a number of people proportional to the total they represent in a subset of a population is which type of sampling? ...	________	1
c 'Which day of the week a restaurant has the highest number of patrons' is an example of which type of data?	________	1
d The mean of a set of scores is 86. If another score of 86 is added to the set, which measure(s) (mean, mode, median or range) will definitely not change?	________	1
2 For the given set of scores 3, 7, 3, 8, 3, 5, 3, 9, 2, 1, 8, 3, 3, 5, 3, find:		
a the mean	________	1
b the median	________	1
c the difference between the range and the mode	________	1
3 For the given set of scores, find:		
a the mean	________	1
b the range	________	1
c the mode	________	1
d Draw a frequency histogram.		2
e Draw a frequency polygon.		2
f Find the median.	________	1

Score (x)	Frequency (f)
2	3
3	5
4	2
5	9
6	2
7	3

Frequency (f)

Score (x)

Total marks achieved for PART B

Exam Paper 1

Instructions for all parts
- Attempt all questions.
- Calculators are NOT allowed.

Time allowed: 45 min **Total marks: 50**

EXAM PAPER 1 PART A

Fill in only one circle for each question.

Marks

1 If $x - 10 = -2$ then x equals 1

Ⓐ −12 Ⓑ −8 Ⓒ 8 Ⓓ 12

2 In a group of 35 children the ratio of boys to girls is 5:2. How many children are girls? 1

Ⓐ 14 Ⓑ 10 Ⓒ 21 Ⓓ 25

3 If $a = 5 - 3b^2$ and $b = 2$ then a is equal to 1

Ⓐ −31 Ⓑ −7 Ⓒ 17 Ⓓ 41

4 Convert 1.4 square metres to square centimetres. 1

Ⓐ 14 Ⓑ 140 Ⓒ 1400 Ⓓ 14 000

5 $1 + \frac{7}{100} + \frac{3}{1000} =$ 1

Ⓐ 1.73 Ⓑ 0.173 Ⓒ 1.073 Ⓓ 1.0703

6 A rectangle has perimeter 20 cm and area 21 cm^2. What are its dimensions in cm? 1

Ⓐ 1 and 20 Ⓑ 4 and 6 Ⓒ 9 and 2 Ⓓ 3 and 7

7 3.25 km = 1

Ⓐ 325 m Ⓑ 32 500 mm Ⓒ 3250 m Ⓓ 32 500 m

8 $3(2x - 5) - 2(4 - x)$ simplifies to 1

Ⓐ $8x - 23$ Ⓑ $7x - 13$ Ⓒ $5x - 13$ Ⓓ $4x - 2$

9 $(-3)^2 + (-2)^3$ equals 1

Ⓐ −17 Ⓑ −12 Ⓒ 1 Ⓓ 17

10 If 120 is 20% of a number, then the number is 1

Ⓐ 24 Ⓑ 96 Ⓒ 480 Ⓓ 600

Total marks achieved for PART A 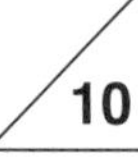

EXAM PAPER 1 PART B

Write only the answer in the answer column. For any working use the question column.

Questions	Answers	Marks
Evaluate:		
1 $8 \times 12 + 13 \times 8 =$		1
2 $-7 - (-5) =$		1
3 $\frac{1}{9}$ is an example of what kind of decimal?		1
4 Find 25% of \$800		1
5 Answer in index form: $2^3 \times 2^2 \times 2^4 =$		1
6 Simplify the following. 30 : 48		1
7 How many centimetres are in 1 kilometre?		1
8 Solve: $m - \frac{m}{5} = 12$		1
9 $-6a - 2b + 5a =$		1
10 Find the range for the set of scores 3, 8, 12, 15, 32, 41, 85		1
11 Solve $\frac{8y + 24}{8} = 8$		1
12 Find the value of x in the diagram given opposite ($x°$, $70°$, $25°$)		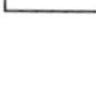 1
13 If a movie started at 4.20 pm and finished at 7.05 pm, how long was the movie?		 1
14 Simplify 60 cents : \$15		 1
15 Find the value of $a^2\sqrt{b}$ when $a = 8$ and $b = 25$		1

Total marks achieved for PART B /15

EXAM PAPER 1 PART C

Show all working for each question.

Marks

1 Simplify the following.

a $2a + 7a \times 3 - 4a$ __________ **b** $7a \times 3xy + 5 \times 6yx$ __________ 4

c $-7x \times -x$ __________ **d** $9mn \times m \times m$ __________ 2

2 a Find $2\frac{2}{7} \times 7$ __________ **b** Increase \$900 by 20% __________ 4

3 A greengrocer bought 200 kg of oranges for \$150 and sold them for \$1.35 per kg. Find:

a his profit __________

b his percentage profit on the cost price __________ 4

4 Write an expression to describe the perimeter and area of each shape.

a (rectangle with sides m and n) Perimeter = __________ Area = __________ 2

b (right-angled triangle with sides $3z$, $2y$ and x) Perimeter = __________ Area = __________ 2

5 Goods costing \$$C$ have a selling price of \$$S$. Write a formula for the profit \$$P$.

__________ 1

6 Does the point (0,1) lie on the line $y = 2x - 1$? __________ 1

7 Name the parts of the circle indicated: 2

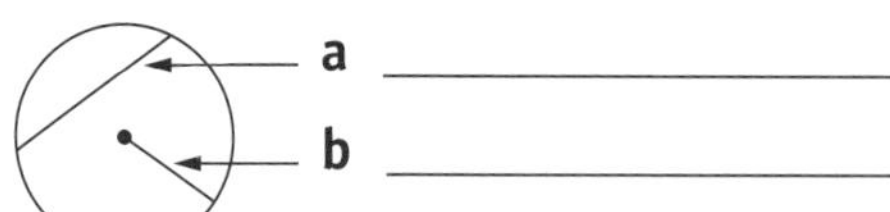

a __________

b __________

8 What transformation is shown below?

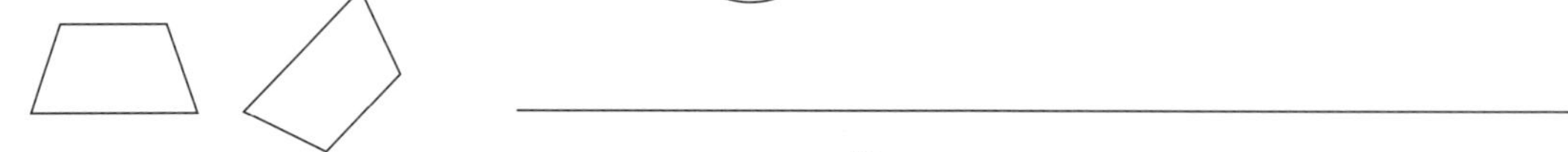

__________ 1

9 Find the value of a and b, giving reasons.

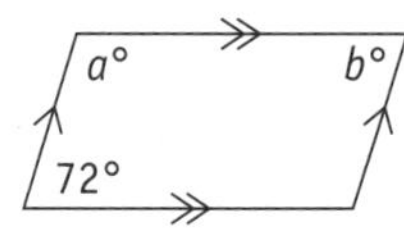

__________ 2

Total marks achieved for PART C /25

Exam Paper 2

Instructions for all parts
- Attempt all questions.
- Calculators are allowed.

Time allowed: 45 min **Total marks: 50**

EXAM PAPER 2 PART A

Fill in only one circle for each question.

		Marks
1	$a(a - 5) =$ Ⓐ $a^2 - 5$ Ⓑ $a^2 - 5a$ Ⓒ $-4a$ Ⓓ $-5a^2$	1
2	The sum of the smallest and the largest of the numbers 0.5129, 0.9, 0.89 and 0.289 is Ⓐ 1.189 Ⓑ 0.8019 Ⓒ 1.428 Ⓓ 1.179	1
3	The angles of a triangle are in the ratio 1 : 2 : 3. Find the smallest angle. Ⓐ 30° Ⓑ 45° Ⓒ 60° Ⓓ 72°	1
4	$(0.2)^2 =$ Ⓐ 0.04 Ⓑ 0.4 Ⓒ 2.0 Ⓓ 0.02	1
5	The congruency test for $\triangle ABC$ and $\triangle ADC$ is Ⓐ RHS Ⓑ AAS Ⓒ SAS Ⓓ SSS 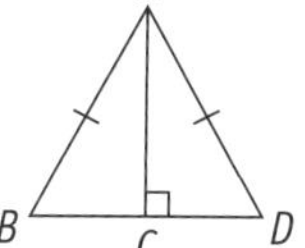 	1
6	0.3% equals Ⓐ $\frac{1}{3}$ Ⓑ $\frac{3}{10}$ Ⓒ $\frac{3}{100}$ Ⓓ $\frac{3}{1000}$	1
7	Increase 250 by 6% Ⓐ 265 Ⓑ 256 Ⓒ 400 Ⓓ none of these	1
8	The perimeter of this shape is given by Ⓐ $\pi + 8$ Ⓑ $4\pi + 4$ Ⓒ $2\pi + 8$ Ⓓ $2\pi + 4$ 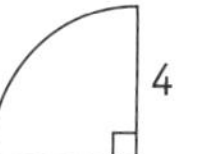 	1
9	As a decimal, $8\frac{1}{2}\%$ is Ⓐ 0.085 Ⓑ 0.85 Ⓒ 8.005 Ⓓ 8.5	1
10	Find the area, in square metres, of a rectangle 2 m long and 90 cm wide. Ⓐ 0.18 m^2 Ⓑ 1.8 m^2 Ⓒ 18 m^2 Ⓓ 180 m^2	1
11	'Favourite day of the week' classifies as which type of data? Ⓐ Numerical Ⓑ Categorical Ⓒ Discrete Ⓓ Continuous	1

Total marks achieved for PART A

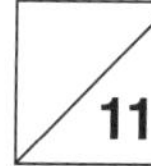

EXAM PAPER 2 PART B

Write only the answer in the answer column. For any working use the question column.

Questions	Answers	Marks
1 Find the average of $6a$ and $2a$		1
2 What is the y-intercept for the line $y = 3x + 4$?		1
3 Find the percentage rate of commission when a car sales assistant earns \$1560 on the sale of a car worth \$60 800		1
4 How many minutes are there in 270 seconds?		1
5 Factorise $9a - 9$		1
6 Simplify $\frac{8x + 16}{8}$		1
7 If 7 tickets cost \$12.60, how much would 2 tickets cost?		1
8 Solve $\frac{2x}{3} = \frac{8}{9}$		1
9 Simplify $(-4) \times (-6) \times (-1)$		1
10 Write the following as a rate in simplest form: 255 km in 3 hours.		1
11 Express $\frac{3}{8}$ as a decimal.		1
12 From a pack of 52 cards, a card is drawn at random. What is the probability that it is not a spade?		1
13 How many 375 mL containers can be filled with 9000 cm^3 of water?		1
14 Simplify $4 - 2(3a - 1)$		1
15 Find $\frac{\sqrt{4036}}{185 + 3.548}$ correct to two decimal places.		1

Total marks achieved for PART B $\frac{\quad}{15}$

EXAM PAPER 2 PART C

Show all working for each question.

Marks

1 Find the area of these shapes to 1 decimal place where necessary.

a 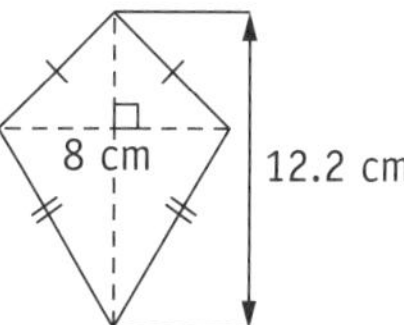

b 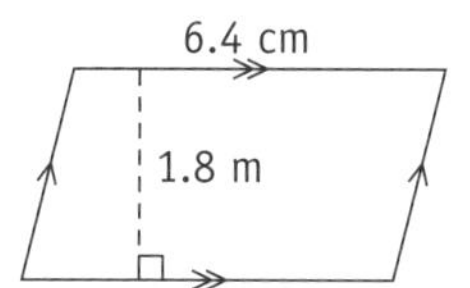

c d 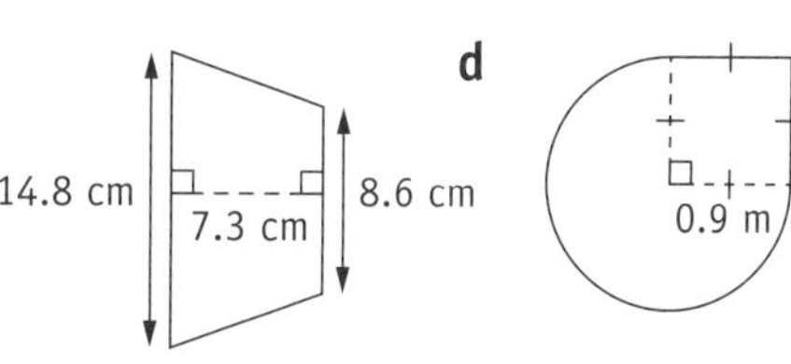

_______ _______ _______ _______ 4

2 Expand and simplify.

a $5a - 2(a - 6) =$ _______ b $4x(x - 2) - 3(5x - 1) =$ _______ 4

3 Find the value of the pronumeral, giving reasons. All measurements are in cm.

a 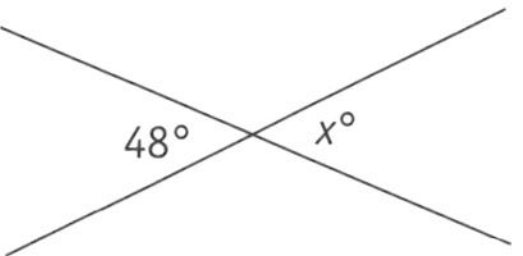

b 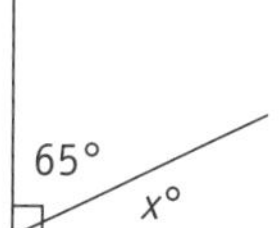

c 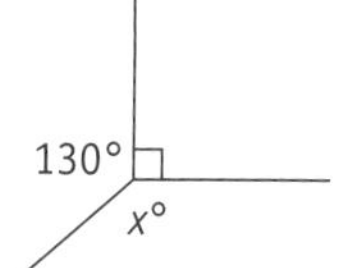

_______ _______ _______ 6

4 Nicole receives a gross pay of $680 for a 40 hour week. Calculate Nicole's hourly rate of pay.

_______ 1

5 Computer disks are $3 each. A 10% discount is given for buying a box of 20 disks. Find the cost of a box of disks. _______ 2

6 Simplify the following.

a  $8 : 1\frac{1}{2}$ _______ b $8x^0 + (8x)^0$ _______

c $\sqrt{64} + \sqrt[3]{125}$ _______ d $2.5 : 1.5$ _______ 4

7 At a school, the ratio of boys to girls is $5 : 4$. What fraction of students are girls? _______ 1

8 A bag contains red, black and green marbles in the ratio $3 : 5 : 6$. If there are 168 marbles in total, find the number of green marbles. _______ 2

Total marks achieved for PART C /24

Exam Paper 3

Instructions for all parts
- Attempt all questions.
- Calculators are allowed.

Time allowed: 45 min **Total marks: 50**

EXAM PAPER 3 PART A

Fill in only one circle for each question.

Marks

1 Which number is the largest? 1

(A) 0.907 (B) 0.0907 (C) 0.970 (D) 0.9007

2 If $2x - 5 = 21$, what is the value of x? 1

(A) 6 (B) 8 (C) 13 (D) 26

3 Each student in year 8 was asked how long they spent doing homework the previous week. This survey was 1

(A) a sample (B) a census (C) systematic (D) stratified

4 A car was bought for $6000 and sold at a profit of 30%. What was the selling price? 1

(A) $1800 (B) $4200 (C) $6200 (D) $7800

5 How much larger is the area of a circle with radius 8 cm than a circle with a radius of 6 cm? 1

(A) 2π (B) 4π (C) 14π (D) 28π

6 $a \times a \times a \times a \times a =$ 1

(A) a^5 (B) $5a$ (C) 5^a (D) $a + 5$

7 $-6a + 3a =$ 1

(A) $3a$ (B) $9a$ (C) $-3a$ (D) -3

8 $a \times a + a \times a =$ 1

(A) $4a$ (B) a^4 (C) $2a^4$ (D) $2a^2$

9 Five more than p, can be written as 1

(A) $5 > p$ (B) $p < 5$ (C) $5 - p$ (D) $p + 5$

10 The volume of this solid is 1

(A) 42 cm³ (B) 70 cm³

(C) 84 cm³ (D) 140 cm³

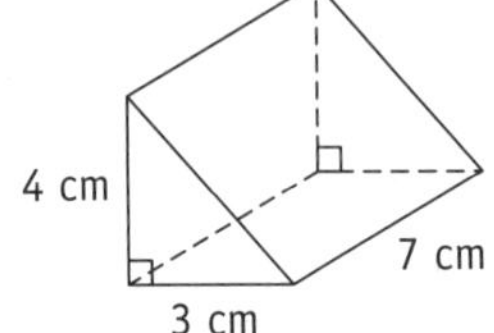

Total marks achieved for PART A /10

EXAM PAPER 3 PART B

Write only the answer in the answer column. For any working use the question column.

Questions	Answers	Marks
1 Simplify the following.		
a $(-9) \times (-5)$		1
b $\frac{1}{7} - \frac{1}{3}$		1
c $8.7 - 0.3$		1
d $\frac{1}{2} - 0.35$		1
2 Find the value of $x^2 + 1$ when $x = \frac{1}{3}$		1
3 $\frac{2x + 3}{4.5} = 10$		1
4 How many litres in 3527 mL? Leave answer in decimal form.		1
5 A die is rolled. What is the probability that the number rolled is not a six?		1
6 Convert:		
a 11.3 km to metres		1
b 0.2 kg to milligrams		1
c 345 000 000 mL to kilolitres		1
d 62 080 mcg to milligrams		1
7 For the scores in the table shown, find the:		

Score (x)	0	1	2	3	4
Cumulative Frequency (cf)	4	8	13	17	20

Questions	Answers	Marks
a mode		1
b median		1
c mean		1

Total marks achieved for PART B  15

EXAM PAPER 3 PART C

Show all working for each question.

Marks

1 Express 600 mL as a percentage of 2 L. ______________________ 1

2 A retailer buys an armchair for \$350 and sells it at a profit of 20% of the cost price.
Find the selling price. ______________________ 

3 At a farm the ratio of pigs to goats is 5:2. If there are 35 pigs, find the number of goats.

______________________ 

4 A man runs at a speed of 18 km/h for 30 minutes and then walks at a speed of 6 km/h for 1 hour. Find:

a the total distance covered ______________________ 2

b the average speed for the whole journey ______________________ 1

5 If $x = 2$, $y = -2$ and $z = \frac{1}{6}$, evaluate.

a $x - y =$ __________ **b** $\frac{1}{2}x^2 - y^2 =$ __________ **c** $\frac{xyz}{y - z} =$ __________ 3

6 Simplify.

a $a \times a \times a + a \times a \times a =$ __________ **b** $-5(2a + 3b) + a =$ __________ 2

7 For the diagram given, find.

a the value of c __________

b the value of b __________

c the value of a __________

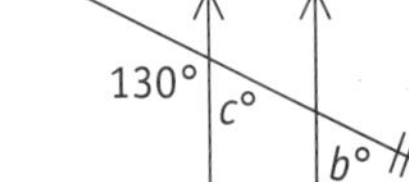

8 In the diagram, AOB is a sector of the circle with centre O. Using $\pi \approx \frac{22}{7}$,

a find the area of the sector AOB __________ **b** find the perimeter of the sector __________ 4

9 Simplify the following.

a $\frac{0.2 \times 6 \times 4}{0.32 \times 6 \times 2} =$ __________ **b** $10 - 2[3 + 6 \times (-2)] \div 3$ __________

c $72 \div 3 \times 2 \times 4 =$ __________ 

10 Solve.

a $3(x + 1) = 15$ __________ **b** $3(x + 2) = x + 10$ __________ 4

Total marks achieved for PART C /25

Exam Paper 4

Instructions for all parts
- Attempt all questions.
- Calculators are allowed.

Time allowed: 45 min **Total marks: 50**

EXAM PAPER 4 PART A

Fill in only one circle for each question.

		Marks
1	A square has a perimeter of 32 cm. Its area is Ⓐ 32 cm^2 Ⓑ 48 cm^2 Ⓒ 64 cm^2 Ⓓ 72 cm^2	1
2	The mode of the set of scores 2, 3, 3, 4, 5, 3, 8, 3, 6, 4 is Ⓐ 2 Ⓑ 3 Ⓒ 4 Ⓓ 5	1
3	The factors of $6y - 3$ are Ⓐ $6(y - 1)$ Ⓑ $y(6 - y)$ Ⓒ $3(2y - 1)$ Ⓓ $3y$	1
4	By what number must (–3 + 7 – 6) be multiplied by to give an answer of 6? Ⓐ –3 Ⓑ 3 Ⓒ 4 Ⓓ –4	1
5	$\frac{1}{2}$ of $\frac{1}{3}$ equals Ⓐ $\frac{1}{2}+\frac{1}{3}$ Ⓑ $\frac{1}{2}\times\frac{1}{3}$ Ⓒ $\frac{1}{2}-\frac{1}{3}$ Ⓓ $\frac{1}{2}\div\frac{1}{3}$	1
6	If $\frac{y}{3} = -6$ then y equals Ⓐ –2 Ⓑ 2 Ⓒ –18 Ⓓ 18	1
7	If $2x + 7 = 1$ then x equals Ⓐ –3 Ⓑ –4 Ⓒ 3 Ⓓ 4	1
8	When simplified, the expression $4a + 5b - 3a - 7b$ equals Ⓐ $7a + 12b$ Ⓑ $7a - 2b$ Ⓒ $a + 12b$ Ⓓ $a - 2b$	1
9	$\frac{2}{5}$ as a percentage is Ⓐ 25% Ⓑ 250% Ⓒ 40% Ⓓ 4%	1
10	Jasmine draws a sketch of a house using a scale of 1 cm to 50 cm. How long is the line representing a wall of length 8 m? Ⓐ 4 cm Ⓑ 8 cm Ⓒ 12 cm Ⓓ 16 cm	1

Total marks achieved for PART A 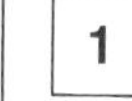 /10

EXAM PAPER 4 PART B

Write only the answer in the answer column. For any working use the question column.

Questions	Answers	Marks
1 Write 12.05 am in 24-hour time		1
2 Write 19:52 in 12-hour time		1
3 $(-6) \times (-2) \times (-1) =$		1
4 Find the value of x 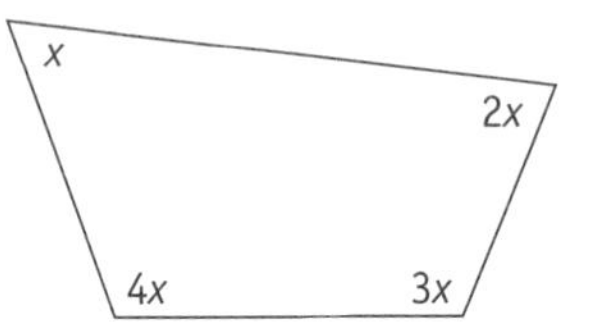		1
5 If $d = 7$ find $3d^2$		1
6 $\frac{5}{6} \times 2\frac{1}{2} =$		1
7 $7.8 \times 10^4 - 3.9 \times 10^2 =$		1
8 Find $\left(5\frac{5}{6}\right)^3$ to 3 significant figures		1
9 Write $(8^4)^2$ in simplest index form.		1
10 True or false: 'Favourite fruit' is an example of categorical data.		1
11 Expand $5a(4b - a)$		1
12 Factorise $9a + 24$		1
13 Divide \$200 in the ratio 2 : 3		1
14 Decrease \$42 by 30%		1
15 For the formula $v = u + at$, find the value of v when $u = 3$, $a = 2$ and $t = 4$		1

Total marks achieved for PART B ___ / 15

EXAM PAPER 4 PART C

Show all working for each question.

Marks

1 For $\triangle ABC$ and $\triangle DCB$

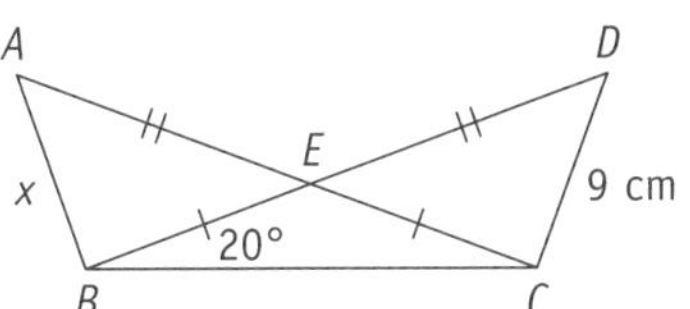

a Find $\angle BCA$, giving reasons __________ 1

b What test proves $\triangle ABC$ and $\triangle DCB$ are congruent? __________ 1

2 Find the area of each shape correct to 1 decimal place.

a

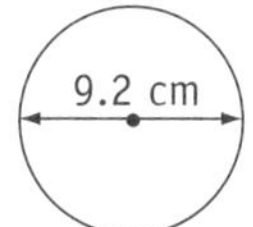

__________ 2

b

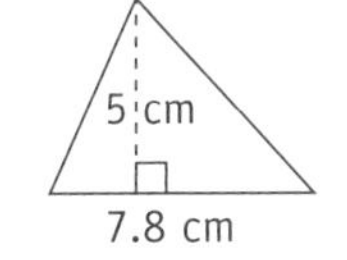

__________ 2

c

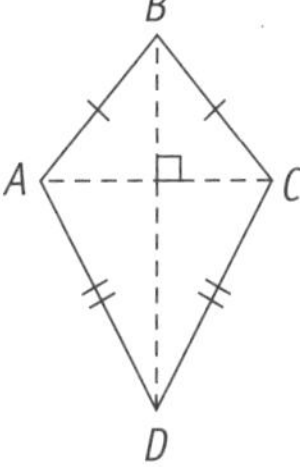

$AC = 10.4$ cm $BD = 15.8$ cm

__________ 2

d

8.2 cm
6 cm
11 cm
17 cm

__________ 2

3 A rectangular swimming pool is 4 m wide and 10 m long. At one end it is 1.3 m deep and at the other end it is 2.5 m deep.

a What is the area of the trapezoidal cross-section? __________ 2

b What is the pool's volume when filled to the top? __________ 1

c How many litres of water can it hold? __________ 1

d Evaporation over summer causes the water level to drop 9 cm from the top. How many kilolitres of water has it lost? __________ 2

Continued on the next page

EXAM PAPER 4 PART C

Show all working for each question.

Marks

4 The histogram shows the lateness of a class of students for a year.

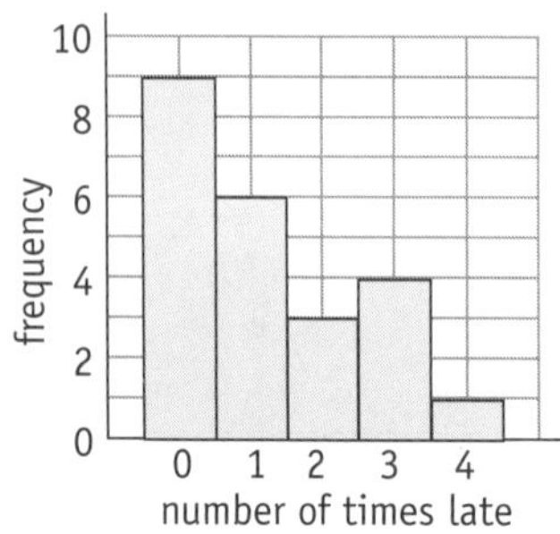

a How many were never late? ______

b How many were late more than twice? ______

c Find the range. ______

d Find the mode. ______

e Find the median. ______

f Find the mean. ______ 6

5 There are 24 people in a Food Technology class. During a lesson, 12 students make nachos, 14 make enchiladas and 3 make neither.

a Complete the Venn Diagram to show this information.

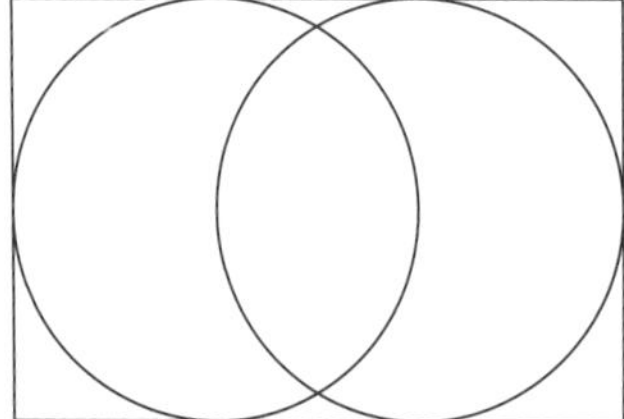

2

b What is the probability that a student selected at random would have made enchiladas only?

______ 1

Total marks achieved for PART C /25

Answers

Chapter 1 – Rational numbers

Page 1 **1 a** 100 **b** 8300 **c** 1000 **d** 600 **e** 100 **f** 100 **g** 900 **h** 300 **i** 800 **j** 6400 **k** 6100 **l** 8500 **2 a** 9000 **b** 1000 **c** 2000 **d** 5000 **e** 8000 **f** 10 000 **g** 10 000 **h** 1000 **i** 6000 **j** 6000 **k** 1000 **l** 8000 **3 a** 8.32 **b** 0.93 **c** 72.36 **d** 92.61 **e** 0.22 **f** 65.12 **g** 53.81 **h** 36.21 **i** 92.99 **j** 98.31 **k** 48.12 **l** 63.13 **4 a** 1.8 **b** 9.4 **c** 15.1 **d** 39.2 **e** 63.1 **f** 83.3 **g** 93.9 **h** 58.2 **i** 96.2 **j** 51.4 **k** 68.9 **l** 53.0 **5 a** 92 **b** 370 **c** 5800 **d** 6600 **e** 9400 **f** 57 **g** 910 **h** 510 **i** 920 **j** 1900 **k** 830 **l** 910 **6 a** 9360 **b** 5690 **c** 81 300 **d** 56.1 **e** 83.8 **f** 98.7 **g** 67.8 **h** 916 **i** 86.3 **j** 110 **k** 72.6 **l** 982

Page 2 **1 a** 19 **b** 32 **c** 58 **d** 60 **e** 82 **f** 180 **g** 60 **h** 496 **i** 60 **j** 15 **k** 478 **l** 60 **2 a** 453.1 **b** 49.2 **c** 64 **d** 35 **e** 64.512 **f** 118.4 **g** 20.32 **h** 27.65 **3 a** 19.522 **b** 16 **c** 5.2 **d** 14.06 **e** 9.402 **f** 16.9 **g** 1612.85 **h** 9.546 **i** -5.2 **j** 268.3 **k** 261.744 **l** 230 **4 a** $4\frac{2}{3}$ **b** 10 **c** $8\frac{2}{3}$ **d** $\frac{1}{34}$ **e** $48\frac{3}{8}$ **f** $6\frac{1}{3}$ **g** $26\frac{9}{16}$ **h** $2\frac{2}{5}$ **i** 13 **j** $3\frac{1}{8}$ **k** $8\frac{8}{9}$ **l** $4\frac{4}{45}$ **5 a** 34.69 **b** 14.58 **c** 16 **d** 114.6 **e** 120 **f** 33 **g** 7.5 **h** 50 **i** 47.66 **6 a** 23.205 **b** 277.98 **c** 9.3 **d** 37.6 **e** 19.008 **f** 6.24 **g** 0.17578125 **h** 27.62 **i** 885.5

Page 3 **1 a** 9 **b** 49 **c** 81 **d** 25 **e** 121 **f** 169 **g** 1681 **h** 2304 **i** 3969 **j** 3364 **k** 2.89 **l** 8335.69 **2 a** 8 **b** 125 **c** 512 **d** 9261 **e** 314 432 **f** 373 248 **g** 1024 **h** 1296 **i** 324 **j** 256 **k** 729 **l** 39.0625 **3 a** 9.261 **b** 75.69 **c** 857.375 **d** 11 698.5856 **e** 328.509 **f** 112 151.3121 **g** 50 625 **h** 4032.25 **i** 85.766 121 **j** 5112.25 **k** 6331.625 **l** 3769.96 **4 a** 6.3 **b** 2.1 **c** 4521.2 **d** 74.6 **e** 65.6 **f** 4173.3 **g** 753.6 **h** 380.3 **i** 731.2 **j** 46.7 **k** 3336.2 **l** 64.4 **5 a** $6\frac{1}{4}$ **b** $\frac{1}{27}$ **c** $3\frac{1}{16}$ **d** $18\frac{7}{9}$ **e** $190\frac{7}{64}$ **f** $3164\frac{1}{16}$ **g** $12\frac{19}{27}$ **h** $12\frac{24}{25}$ **i** $34\frac{1}{36}$ **j** $11\frac{25}{64}$ **k** $52\frac{47}{64}$ **l** $18\frac{1}{16}$

Page 4 **1 a** 2 **b** 7 **c** 9 **d** 4 **e** 10 **f** 15 **g** 11 **h** 30 **i** 12 **j** 13 **k** 5 **l** 1 **2 a** 3 **b** 14 **c** 5 **d** 2 **e** 27 **f** 44 **g** 4 **h** 18 **i** 1.81 (2 d.p.) **j** 5 **k** 20 **l** 3.86 (2 d.p.) **3 a** 71.31 **b** 19.48 **c** 2.08 **d** 10.82 **e** 4.52 **f** 4.53 **g** 9.31 **h** 4.92 **i** 4.94 **j** 13.47 **k** 2.28 **l** 13.12 **4 a** 4.2 **b** 5.0 **c** 4.6 **d** −0.4 **e** 5.2 **f** 31.3 **g** 14.4 **h** 27.6 **i** 13.0 **j** 1.9 **k** 38.1 **l** 19.3 **5 a** 3.623 **b** 2.050 **c** 3.640 **d** 3.252 **e** 2.566 **f** 5.439 **g** 3.081 **h** 2.776 **i** −0.013 **j** 0.237 **k** 2.136 **l** 1.504 **m** 1.950 **n** 2.743 **o** 2.561

Page 5 **1** \$47.25 **2** 3.51 m **3** 68.0625 m^2 **4** 42.875 cm^3 **5** \$405.60 **6** \$1223.75 **7** \$36.40 **8** 9.21 **9** 2653.56 cm^3 **10** 29 cm **11** 70.98 **12** $3\frac{1}{2}$ **13** 3.03 **14** \$61.80 **15** 21

Page 6 **1 a** 0.1 **b** 0.5 **c** 0.9 **d** 0.03 **e** 0.07 **f** 0.11 **g** 0.002 **h** 0.035 **i** 0.163 **j** 0.0834 **k** 0.00564 **l** 0.000021 **2 a** 6.5 **b** 9.7 **c** 23.18 **d** 39.07 **e** 27.03 **f** 91.081 **g** 29.003 **h** 52.17 **i** 63.013 **j** 83.02 **k** 94.012 **l** 69.0035 **3 a** 0.3 **b** 0.12 **c** 0.015 **d** 0.64 **e** 0.127 **f** 0.0635 **g** 0.625 **h** 0.6 **i** 0.36 **j** 0.125 **k** 0.128 **l** 1.125 **4 a** 0.1 **b** 0.27 **c** 0.2 **d** 3.64 **e** 0.072 **f** 1.45 **g** 0.17 **h** 0.608 **i** 0.48 **j** $0.022\dot{6}$ **k** 0.144 **l** $0.4\dot{6}$ **5 a** 2.4 **b** 6.38 **c** 0.539 **d** 0.0249 **e** 0.00865 **f** 0.86 **g** 0.55 **h** 0.096 **i** 0.074 **j** 0.475 **k** 0.288 **l** 0.70375

Page 7 **1 a** $\frac{3}{10}$ **b** $\frac{7}{10}$ **c** $\frac{1}{2}$ **d** $\frac{81}{100}$ **e** $\frac{6}{25}$ **f** $\frac{19}{50}$ **g** $\frac{1}{8}$ **h** $\frac{23}{40}$ **i** $\frac{17}{20}$ **j** $\frac{109}{200}$ **k** $\frac{43}{50}$ **l** $\frac{193}{1000}$ **2 a** $\frac{1}{50}$ **b** $\frac{1}{25}$ **c** $\frac{1}{20}$ **d** $\frac{3}{1000}$ **e** $\frac{1}{125}$ **f** $\frac{1}{40}$ **g** $\frac{7}{1000}$ **h** $\frac{1}{200}$ **i** $\frac{3}{200}$ **j** $\frac{161}{200}$ **k** $\frac{38}{125}$ **l** $\frac{177}{250}$ **3 a** $1\frac{1}{5}$ **b** $3\frac{2}{5}$ **c** $8\frac{9}{10}$ **d** $2\frac{1}{4}$ **e** $3\frac{16}{25}$ **f** $5\frac{3}{4}$ **g** $8\frac{3}{10}$ **h** $9\frac{2}{5}$ **i** $7\frac{1}{5}$ **j** $6\frac{1}{10}$ **k** $5\frac{21}{50}$ **l** $8\frac{93}{100}$ **4 a** $6\frac{1}{8}$ **b** $8\frac{171}{500}$ **c** $9\frac{9}{20}$ **d** $10\frac{13}{40}$ **e** $29\frac{9}{25}$ **f** $41\frac{7}{40}$ **g** $18\frac{63}{100}$ **h** $128\frac{12}{25}$ **i** $526\frac{1}{2}$ **j** $16\frac{801}{1000}$ **k** $19\frac{1}{2}$ **l** $428\frac{11}{20}$ **5 a** $5\frac{16}{25}$ **b** $9\frac{47}{200}$ **c** $60\frac{181}{250}$ **d** $83\frac{3}{4}$ **e** $83\frac{21}{50}$ **f** $81\frac{12}{25}$ **g** $94\frac{1}{4}$ **h** $49\frac{123}{1000}$ **i** $53\frac{231}{1000}$ **j** $51\frac{1}{8}$ **k** $86\frac{257}{500}$ **l** $32\frac{519}{1000}$

Page 8 **1 a** T **b** T **c** T **d** R **e** T **f** T **g** T **h** T **i** R **j** R **k** T **l** R **2 a** $0.\dot{3}$ **b** $0.4\dot{3}$ **c** $0.\dot{1}3\dot{5}$ **d** $0.\dot{7}$ **e** $0.80\dot{7}\dot{8}$ **f** $0.2\dot{3}$ **g** $0.\dot{2}\dot{5}$ **h** $0.\dot{3}\dot{6}$ **i** $0.\dot{1}$ **j** $0.\dot{5}0\dot{5}$ **k** $0.2\dot{7}$ **l** $0.\dot{7}\dot{2}$ **3 a** 0.222 … **b** 5.333 … **c** 9.555 … **d** 0.2323 … **e** 0.861111 … **f** 0.49595 … **g** 0.5454 … **h** 0.4545 … **i** 0.2777 … **j** 0.7333 … **k** 0.375375 … **l** 0.259393 … **4 a** $0.\dot{6}$ **b** $0.\dot{2}\dot{6}$ **c** $1.\dot{5}$ **d** $0.\dot{2}$ **e** $0.0\dot{1}\dot{5}$ **f** $9.\dot{8}$ **g** $5.\dot{3}\dot{6}$ **h** $0.\dot{2}\dot{8}$ **i** $7.\dot{9}\dot{3}$ **j** $1.5\dot{3}$ **k** $0.\dot{1}$ **l** $0.04\dot{8}$ **5 a** $0.0\dot{5}$ **b** $0.0\dot{6}$ **c** $0.\dot{2}\dot{7}$ **d** $0.0\dot{3}\dot{6}$ **e** $0.8\dot{3}$ **f** $0.0\dot{1}$ **g** $1.\dot{2}$ **h** $0.0\dot{5}\dot{4}$ **i** $2.\dot{7}$ **j** $0.\dot{7}\dot{2}$ **k** $0.0\dot{4}\dot{8}$ **l** $0.00\dot{6}\dot{2}$ **6 a** $0.\dot{1}\dot{8}$ **b** $0.\dot{3}$ **c** $0.8\dot{3}$ **d** $0.\dot{7}$ **e** $0.5\dot{3}$ **f** $0.\dot{1}8\dot{5}$ **g** $0.1\dot{4}\dot{5}$ **h** $0.1\dot{0}\dot{6}$ **i** $1.\dot{2}$ **j** $3.\dot{5}7142\dot{8}$ **k** $4.\dot{8}4615\dot{3}$ **l** $0.\dot{8}5714\dot{2}$

Page 9 **1 a** rational **b** rational **c** irrational **d** irrational **e** irrational **f** rational **g** rational **h** rational **i** irrational **j** rational **k** rational **l** irrational **2 a** rational **b** rational **c** rational **d** neither **e** rational **f** irrational **g** rational **h** rational **i** irrational **j** rational **k** rational **l** irrational

Page 10 **1 a** 1:2 **b** 1:3 **c** 1:3 **d** 1:6 **e** 1:7 **f** 1:12 **g** 1:9 **h** 1:8 **i** 1:9 **j** 1:4 **k** 1:5 **l** 1:9 **2 a** 5:13 **b** 1:3 **c** 1:3 **d** 2:19 **e** 1:13 **f** 2:21 **g** 2:9 **h** 1:13 **i** 3:14 **j** 2:11 **k** 2:15 **l** 4:9 **3 a** 1:3:5 **b** 3:1 **c** 9:16 **d** 3:4 **e** 1:5 **f** 5:1 **g** 1:3 **h** 5:7 **i** 6:13 **j** 5:6 **k** 12:1 **l** 1:5 **4 a** 4:21 **b** 4:15 **c** 1:6 **d** 2:5 **e** 3:4 **f** 1:7 **g** 8:25 **h** 2:3 **i** 5:2 **j** 4:5 **k** 10:1 **l** 5:4 **5 a** 6:25 **b** 1:15 **c** 5:18 **d** 3:1 **e** 7:18 **f** 8:1 **g** 5:7 **h** 7:8

Answers

PAGE 11 1 a 20 kg, 160 kg b $10, $35 c $168, $672 d $300 e $800 2 a 25 cm b 15 c 40°, 60°, 80° d 16:25 e 9 3 a 28 m b $36, $60 c 3:10 d $1200, $1600, $2000

PAGE 12 1 a 80 km b $20 c $5\frac{1}{3}$ L d $7.95 2 a 1 b 960 c 3600 d $300 e 1800 f 0.6° 3 a 80 km/h b 108 km c 48.6 L 4 a 2.5 km/min b $0.02\dot{7}$ km/s c 136 5 a 168.75 km/h b 2.58 m/y c 37

PAGE 13 1 a 4800 b 100 c $1020.8\dot{3}$ d 3600 e 0.275 f 2400 g 2 h 300 i 0.45 j 2160 k 540 000 l 14 100 2 a 3480 b 208 800 c 208.8 d 5011.2 3 $1833.\dot{3}$ 4 a 2700 b 162 000 c 162 d 0.162 5 a $26.\dot{6}$ b 108 c $33.\dot{3}$ d 1200

PAGE 14 1 a 1:1500 b 1:200 c 1:30 000 d 1:5000 e 3:500 f 1:40 g 1:30 000 h 3:10 i 1:8000 2 a 1 m b 4 m c 6 m d 50 cm e 70 cm f 2 km g 60 cm h 80 cm i 1.5 km 3 a 4 m b 100 m c 8 km d 75 m e 9.3 km f 23.45 km g 9 m h 15.2 km i 48.25 km 4 a 6 cm b 7 cm c 13.8 cm d 0.5 cm e 50 cm f 39.65 cm 5 a 4.8 m × 5.6 m b 0.58 m

PAGE 15 1 a 3: 4 b 4:7 2 a 4:5 b 5:4 3 80 km/h 4 $240, $320 5 a 4:9 b 3:8 6 120 km/h 7 30 cents 8 3:4 9 $4000 10 30°, 60° 11 25 kg/s

PAGE 16 1 C 2 A 3 C 4 D 5 C 6 B 7 C 8 C 9 B 10 B

PAGE 17 1 $34.75 2 51 3 2004 4 0.06 5 1:6 6 65 7 50 8 12:7 9 $33\frac{1}{3}$ m/s 10 13 h 20 min 11 256 m, 384 m 12 22.8 L 13 40° 14 98 15 3 and 4

CHAPTER 2 – Integers

PAGE 18 1 a 3 km east b 5 km east c 6 km east d 1 km west e 7 km east f 9 km west

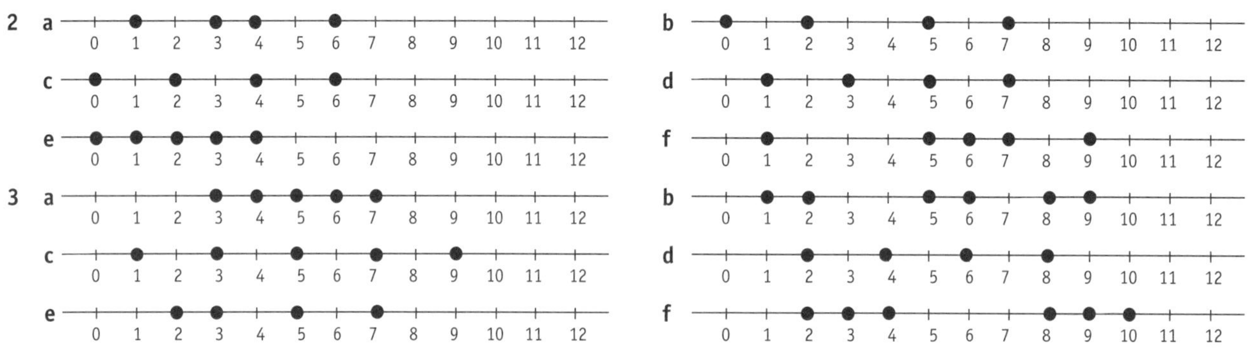

PAGE 19 1 a 7 b 7 c 11 d 6 e 7 f 11 g 4 h 7 i 4 j 14 2 a 6 b −1 c −8 d −2 e 1 f −7 g −3 h 8 i −6 j −4 k −6 l 8 m −8 n −7 o 2 3 a −3 b 11 c 9 d 7 e 7 f 6 g 5 h 4 i −4 j −4 k −2 l −7 4 a 4, 2, −1, −2, −7, −9 b 5, 3, −2, −6, −8, −9 c 9, 8, 3, 2, 1, −4 d 6, 3, 2, −1, −3, −5 e 8, 5, 2, −2, −4, −6 5 a > b > c > d < e < f > g < h < i < j > k < l > m < n < o <

PAGE 20 1 a 13 b 3 c 8 d 8 e 9 f 11 g 11 h 6 i 11 2 a 0 b −5 c −4 d −1 e 4 f 2 g 2 h 6 i 0 3 a −8 b −12 c −4 d −7 e −17 f −13 g −7 h −18 i 8 4 a −5 b −8 c −8 d −14 e −5 f −15 g −13 h −18 i −12 j −29 k −20 l −17 5 a 13 b 12 c 9 d −10 e −14 f −25 g 2 h −3 i −15 j 2 k 4 l −4 6 a −2 b 5 c 7 d −2 e −6 f −15 g 3 h 12 i −9 j −11 k −9 l 21 m 8 n −9 o −14

PAGE 21 1 a 3 b 3 c 7 d 2 e 12 f 27 g 3 h 15 i 11 j 6 k 25 l 10 2 a 8 b 24 c 37 d 13 e 37 f 21 g 15 h 20 i 28 j 20 k 16 l 45 3 a −22 b −14 c −19 d −18 e −25 f −21 g −25 h −15 i −34 j −10 k −37 l −49 4 a −2 b 3 c −5 d 1 e −6 f −3 g 3 h −5 i −6 j 3 k −7 l 7 5 a 2 b −57 c −123 d −5 e 64 f −60 g −34 h −29 i 53 j −21 k 24 l 28 6 a 5 b −33 c 8 d −1 e 25 f 17 g −19 h 42 i 17 j −6 k −5 l −15

PAGE 22 1 a 5 b 120 c 126 d 72 e 14 f 54 g 121 h 65 i 90 j 336 k 119 l 84 2 a −8 b −3 c −45 d −18 e −60 f −90 g −42 h −32 i −48 j −105 k −110 l −105 3 a −30 b −48 c −63 d −150 e −22 f −12 g −65 h −154 i −120 j −54 k −20 l −84 4 a 70 b 4 c 112 d 22 e 108 f 39 g 20 h 12 i 55 j 30 k 12 l 42 5 a −12 b 100 c −54 d −15 e 144 f −75 g −48 h 154 i −72 j −48 k −72 l 175 6 a 6: 48, −36, 60; −8: −64, 48, −80; 3: 24, −18, 30 b 4: −32, 36, −20; −2: 16, −18, 10; 5: −40, 45, −25 c −7: −28, 42, −56; 12: 48, −72, 96; −15: −60, 90, −120

PAGE 23 1 a −5 b −4 c −7 d −3 e −6 f −7 g −3 h −16 i −13 j −3 k −6 l −2 2 a −3 b −2 c −3 d −3 e −13 f −3 g −3 h −3 i −3 j −3 k −2 l −9 3 a 5 b 5 c 2 d 3 e 9 f 2 g 6 h 19 i 3 j 2 k 8 l 13 4 a $-1\frac{2}{3}$ b $3\frac{3}{5}$ c −3 d $-\frac{3}{5}$ e $2\frac{7}{9}$ f 4 g $-8\frac{1}{3}$ h $-\frac{2}{3}$ i $-1\frac{2}{3}$ j 20 k 1 l 4 5 a 10 b −12 c −3 d −36 e −8 f −49 g 4 h 16 i 11 j 42 k −9 l −4 6 a 19 b −6 c −9 d −23 e 5 f 4 g 17 h 7 i −17 j 8 k −2 l −8

Answers

PAGE 24 **1 a** −1 **b** 39 **c** 22 **d** −14 **e** 44 **f** −18 **g** 3 **h** −56 **i** 43 **j** −27 **k** −6 **l** −40 **2 a** 7 **b** −9 **c** 18 **d** −18 **e** 19 **f** −30 **g** 27 **h** 18 **i** −31 **j** −22 **k** −41 **l** −49 **3 a** −7 **b** −64 **c** 6 **d** 8 **e** 6 **f** 100 **g** 7 **h** −9 **i** 60 **j** 72 **k** −16 **l** −72 **4 a** −40 **b** 30 **c** −19 **d** −30 **e** 24 **f** −4 **g** 4 **h** 60 **i** 90 **j** 6 **k** 12 **l** −2
5 a −8 **b** −24 **c** 6 **d** −54 **e** −54 **f** −3 **g** 1 **h** 9 **i** −10 **j** −3 **k** −13 **l** −6
6 a −3: −18, 24, −30, 36, 9, −15; 5: 30, −40, 50, −60, −15, 25; −9: −54, 72, −90, 108, 27, −45
b 4: −20, −12, 8, 24, −32, 36; −8: 40, 24, −16, −48, 64, −72; 7: −35, −21, 14, 42, −56, 63

PAGE 25 **1 a** 8 **b** 25 **c** −27 **d** 9 **e** 16 **f** 1 **g** 125 **h** 64 **i** −1 **j** 64 **k** −8 **l** 1 **m** 32 **n** 81 **o** 16 **p** 81 **2 a** 1 **b** 625 **c** −243 **d** −1 **e** −32 **f** 32 **g** −1 **h** −1000 **i** −64 **j** −216 **k** 64 **l** 108 **m** 81 **n** 343 **o** 100 **p** −125 **3 a** −72 **b** 85 **c** −63 **d** 41 **e** 125 **f** 2 **g** −28 **h** −49 **i** −150 **j** 4 **k** 82 **l** −288 **m** 441 **n** 208 **o** 100 **p** −4 **q** −11 **r** −16 **4 a** −5 **b** −128 **c** −81 **d** 21 **e** 2500 **f** 8 **g** 2100 **h** −15 **i** −100 000 **j** −6000 **k** 24 **l** −116 **m** 63 **n** 176 **o** −241 **p** 1 **q** −964 **r** 5000

PAGE 26 **1 a** −21 **b** −1 **c** −24 **d** −13 **e** 14 **f** −48 **g** 2 **h** −10 **i** 4 **j** −15 **k** −48 **l** 4 **2 a** −4 **b** 4 **c** −1 **d** −5 **e** 5 **f** 14 **g** −113 **h** −4 **i** 0 **j** −34 **k** −18 **l** −3 **3 a** 3 **b** 0 **c** 14 **d** −2 **e** −28 **f** −26 **g** −96 **h** 66 **i** 1 **j** −9 **k** 72 **l** −12 **4 a** 20 **b** 20 **c** 34 **d** −48 **e** 64 **f** −54 **g** 54 **h** 110 **i** 240 **j** 21 **k** 12 **l** −72 **5 a** −40 **b** −22 **c** −33 **d** −42 **e** 14 **f** −4 **g** 32 **h** 0 **i** −110 **j** −1 **k** 60 **l** 97 **6 a** 1200 **b** 280 **c** 61 **d** 375 **e** −9 **f** 4 **g** 4 **h** 10 **i** 20 **j** 4 **k** −7 **l** 1

PAGE 27 **1** 4 km north **2** −$200 **3** −7 and −8 **4** −6 and 4 **5** −7 **6** loss of $12 000 **7** −10°C **8** gain of 1 kg **9** −5 and −6 **10** −36 **11** 47 **12** −40 **13** −1 **14** −7, −3, 9 **15** −90

PAGE 28 **1** A **2** A **3** B **4** A **5** C **6** B **7** B **8** A **9** C **10** B

PAGE 29 **1** −3 **2** 17 **3** 49 **4** −18 **5** −17 **6** −80 **7** 32 **8** $-3\frac{1}{2}$ **9** −36 **10** 1 **11** −63 **12** −6 **13** −11 **14** −14 **15** −7

CHAPTER 3 – Indices

PAGE 30 **1 a** $4\times4\times4$ **b** $8\times8\times8\times8\times8\times8$ **c** $5\times5\times5$ **d** $6\times6\times6\times6$ **e** $2\times2\times2\times2$ **f** 8×8 **g** $3\times3\times3\times3\times3$ **h** $3\times3\times3\times3\times3\times3$ **i** $9\times9\times9$ **2 a** $1.5\times1.5\times1.5$ **b** $3\times3\times5\times5\times5\times5\times5$ **c** $2\times2\times2\times2\times3\times3\times3\times3\times3\times3$ **d** $3\times3\times3\times3\times4\times4$ **e** $7\times7\times7\times7\times7$ **f** $\frac{3}{4}\times\frac{3}{4}\times\frac{3}{4}\times\frac{3}{4}\times\frac{3}{4}$ **g** $5\times5\times8\times8\times8\times8\times8\times8$ **h** $9\times9\times9\times9\times9$ **i** $-5\times-5\times-5$ **j** $-\frac{2}{3}\times-\frac{2}{3}\times-\frac{2}{3}\times-\frac{2}{3}\times-\frac{2}{3}$ **3 a** $6\times6\times6$ **b** $2\times2\times2\times2\times2\times2$ **c** $7\times7\times7\times7$ **d** $3\times3\times4\times4\times4$ **e** $2\times2\times2\times2\times2\times3\times3\times3\times3$ **f** $5\times5\times3\times3\times3\times4\times4\times4\times4$ **g** $2\times2\times2\times3\times3\times3\times3\times4\times4\times4\times4\times4$ **h** $5\times5\times5\times6\times6\times6\times6\times6$ **i** $-3\times2\times2\times2\times5\times5\times5\times5\times5$ **j** $17\times3\times3\times3\times5\times5\times5\times5\times5\times5\times5\times5\times5$ **4 a** 3^6 **b** 5^4 **c** 2^8 **d** 9^5 **e** 7^6 **f** 4^3 **g** $(-3)^4$ **h** $\left(\frac{1}{2}\right)^4$ **i** $(1.8)^5$ **j** $(3.5)^4$ **5 a** $2^3\times3^3$ **b** $5^3\times6^2$ **c** $4^4\times6^3$ **d** $2^4\times8^2$ **e** $4^5\times5^2$ **f** $2^5\times4^2$ **g** $7^5\times3^3$ **h** $7^3\times5^3$ **i** $5^3\times2^3$ **j** $4^3\times5^4$ **6 a** 7776 **b** 81 **c** 1.331 **d** 625 **e** 343 **f** 6561 **g** 8 **h** 512 **i** 10 000 000

PAGE 31 **1 a** 3^{12} **b** 6^{10} **c** 7^8 **d** 5^{12} **e** 2^{10} **f** 6^{10} **g** 10^{12} **h** 11^9 **i** 7^{13} **j** 6^{18} **k** 10^{21} **l** 13^{29} **2 a** 5^3 **b** 2^5 **c** 6^4 **d** 7^3 **e** 6^4 **f** 3^7 **g** 11^4 **h** 10^3 **i** 5^5 **j** 3^6 **k** 5^7 **l** 2^8 **3 a** 5^5 **b** 6^{12} **c** 6^{11} **d** 3^{14} **e** 5^9 **f** 10^{12} **g** 2^{12} **h** 11^{12} **i** 3^{14} **j** 7^{10} **k** 3^{21} **l** 2^{22} **4 a** 13^7 **b** 2^9 **c** 2^{11} **d** 6^{11} **e** 3^{11} **f** 6^{10} **g** $3^5\,5^7$ **h** 11^{10} **i** 7^5 **j** 7^5 **k** 5^8 **l** 10^{10} **5 a** 6^5 **b** 10^8 **c** 2^5 **d** 5^7 **e** 6^5 **f** 7^{12} **g** 2^{14} **h** 3^{15} **i** 6^{10} **j** 7^4 **k** 5^9 **l** 11^{15} **6 a** 7^3 **b** 6^3 **c** 5^7 **d** 9^2 **e** 3^3 **f** 4 **g** 8^7 **h** 2 **i** 3^4 **j** 4^6 **k** 9^3 **l** 2^2

PAGE 32 **1 a** 10^4 **b** 6^3 **c** 11^{13} **d** 3^3 **e** 5^{13} **f** 13^5 **g** 2^{23} **h** 3^{15} **i** 7^{11} **j** 5^7 **k** 2^6 **l** 5^9 **2 a** 3^5 **b** 11^5 **c** 7^4 **d** 6^2 **e** 6^4 **f** 3^7 **g** 5^4 **h** 10^3 **i** 5^3 **j** 13^2 **k** 5^5 **l** 2^4 **3 a** 3^7 **b** 10^6 **c** 6^4 **d** 5^4 **e** 3^{33} **f** 7^4 **g** 2^{32} **h** 2^3 **i** 5^{23} **j** 7^{34} **k** 11^{12} **l** 3^3 **4 a** 5^8 **b** 2^6 **c** 6^5 **d** 6^6 **e** 5^9 **f** 11^6 **g** 10^8 **h** 3^{15} **i** 13^{12} **j** 3^7 **k** 7^{14} **l** 2^{26} **5 a** 11^3 **b** 6^7 **c** 6 **d** 6^5 **e** 7^3 **f** 10 **g** 3^3 **h** 2^5 **i** $3^3\,5^6$ **j** $2^2\,5^3$ **k** $4^2\,5^5$ **l** $2^5\,7^1$ **6 a** 6^5 **b** 6^6 **c** 9^5 **d** 8^6 **e** 9^3 **f** 8^3 **g** 3^4 **h** 4^9 **i** 3^2 **j** $3^1\,4^2$ **k** 2^8 **l** $4^3\,5^6$

PAGE 33 **1 a** 11^{25} **b** 2^8 **c** 3^{32} **d** 7^{16} **e** 5^{24} **f** 3^{54} **g** 10^{24} **h** 6^{56} **i** 11^{40} **j** 3^{24} **k** 2^{54} **l** 7^{36} **2 a** 2^{16} **b** 10^{16} **c** 6^4 **d** 5^8 **e** 6^4 **f** 3^4 **g** 13^6 **h** 9^8 **i** 6^3 **j** 2^{12} **k** 2^{64} **l** 6^{16} **3 a** 2^{42} **b** 3^{56} **c** 6^{48} **d** 5^{54} **e** 7^{24} **f** 10^{64} **g** 6^{90} **h** 11^{40} **i** 2^{40} **j** 3^{81} **k** 6^{18} **l** 7^{16} **4 a** 7^{40} **b** 8^{45} **c** 3^{40} **d** 6^{144} **e** 2^{90} **f** 6^{108} **g** 5^{18} **h** 7^{80} **i** 2^{66} **j** 11^{128} **k** 3^{63} **l** 10^{42} **5 a** 6^8 **b** 2^{72} **c** 3^{32} **d** 5^{48} **e** 7^{36} **f** 6^{36} **g** 3^{96} **h** 6^{18} **i** 11^{16} **j** 2^{108} **k** 2^{96} **l** 13^{48} **6 a** 2^{32} **b** 2^{36} **c** 3^{48} **d** 6^{32} **e** 3^{96} **f** 6^{40} **g** 2^{16} **h** 2^{45} **i** 2^{36} **j** 3^{84} **k** 12^{64} **l** 3^{72}

PAGE 34 **1 a** 1 **b** 1 **c** 1 **d** 1 **e** 1 **f** 9 **g** 1 **h** 8^4 **i** 1 **j** 1 **k** 1 **l** 16 **2 a** true **b** true **c** true **d** true **e** true **f** true **g** true **h** true **i** true **j** true **k** true **l** true **3 a** 12 **b** 64 **c** 9 **d** 6 **e** 36 **f** 9 **g** 9 **h** 2 **i** 3 **j** 1 **k** 2 **l** 72 **4 a** 7 **b** 10 **c** $\frac{1}{6}$ **d** −2 **e** 9 **f** $12a^4$ **g** $28y^4$ **h** $2a$ **i** 1 **j** 1 **k** 1 **l** 14 **5 a** 2^8 **b** 3^6 **c** 4^8 **d** 10^8 **e** 8^8 **f** 5^6 **g** 8^9 **h** 5^9 **i** 6^9 **j** 3^{10} **k** 7^{10} **l** 1 **6 a** 6 **b** $\frac{9}{8}$ **c** $\frac{1}{26}$ **d** 1 **e** $\frac{1}{4}$ **f** 7 **g** $\frac{9}{8}$ **h** 2 **i** $\frac{1}{16}$ **j** 2 **k** $2a^9$ **l** 6

Answers

Page 35 1 a 9^{10} b 3^8 c 9^8 d 4^{12} e 8^9 f 8^6 g 8^{10} h 5^{10} i 6^6 2 a 4^3 b 2^2 c 6 d 6^3 e 4^5 f 8^2 g 3^4 h 6^2 i 4^3 3 a 3^{16} b 2^{16} c 9^{16} d 4^{32} e 6^{10} f 8^{12} g 5^{44} h 3^{36} i 5^{32} 4 a 4 b 1 c 8 d 6 e 2^6 f 1 g 1 h 1 i 9^2 5 a 8^{11} b 6^{11} c 2^{24} d 5^{12} e 5^5 f 6^4 g $2^6\,6^{16}$ h $4^6\,3^2$ i 8^3 j 1 k 7^3 l 9^4

Page 36 1 C 2 A 3 B 4 B 5 D 6 B 7 D 8 B 9 D 10 B

Page 37 1 a 4^6 b 6^2 c 3^{12} d 10 e 8^{12} 2 a 8^{14} b 3^4 c $2^2\,3^2$ d $\frac{1}{27}$ e 5 3 a 12^{18} b 7^{-2} c 1 d 11 e $25^5 \times 64^9$

Chapter 4 – Percentages

Page 38 1 a 100 b 100 c 100 d 100 e %, 100 2 a 78% b 22% c 37% d 19% e 87% 3 a 73% b 34% c 48% d 31% e 61% f 82% g 66% h 13% i 16% 4 a 18% b 61% c 46% d 81% e 33% f 53% 5 a 57% b 36% c 37% d 67% 6 a 78% b 11%

Page 39 1 a $\frac{1}{5}$ b $\frac{3}{10}$ c $\frac{2}{5}$ d $\frac{3}{20}$ e $\frac{1}{4}$ f $\frac{7}{20}$ g $\frac{37}{100}$ h $\frac{22}{25}$ i $\frac{93}{100}$ j $\frac{9}{50}$ k $\frac{23}{100}$ l $\frac{47}{100}$ 2 a $\frac{3}{250}$ b $\frac{9}{250}$ c $\frac{16}{125}$ d $\frac{41}{500}$ e $\frac{37}{1000}$ f $\frac{59}{1000}$ g $\frac{1}{800}$ h $\frac{1}{250}$ i $\frac{3}{2000}$ j $\frac{1}{500}$ k $\frac{7}{1000}$ l $\frac{203}{400}$ 3 a $\frac{39}{100}$ b $\frac{1}{8}$ c $\frac{161}{400}$ d $\frac{3}{80}$ e $\frac{19}{200}$ f $\frac{17}{20}$ g $\frac{77}{200}$ h $\frac{21}{50}$ i $\frac{1}{3}$ 4 a 7% b 9% c 3% d 21% e 33% f 47% g $21\frac{1}{3}$% h 16% i 35% j $7\frac{1}{2}$% k $18\frac{3}{4}$% l $18\frac{8}{9}$% 5 a 40% b $9\frac{3}{13}$% c $12\frac{1}{2}$% d $42\frac{6}{7}$% e $33\frac{1}{3}$% f 25%

Page 40 1 a 0.1 b 0.3 c 0.5 d 0.9 e 0.8 f 0.6 g 0.85 h 0.65 i 0.45 j 0.95 k 1.15 l 0.35 2 a 0.08 b 0.72 c 0.64 d 0.648 e 0.785 f 0.394 g 0.736 h 0.557 i 0.764 j 0.932 k 0.785 l 0.329 3 a 0.025 b 0.0325 c 0.152 d 0.1225 e 0.055 f 0.3825 4 a 80% b 70% c 90% d 19% e 38% f 65% g 32.5% h 13.8% i 25.7% j 25.6% k 40.6% l 54.8% 5 a 0.1% b 0.7% c 0.3% d 12.3% e 24.6% f 87.6% g 0.4% h 0.5% i 0.8% j 2.3% k 3.4% l 7.8% 6 a 125% b 287% c 391% d 256% e 867% f 522.7% g 798% h 467% i 942%

Page 41 1 a 800 b 560 c 160 d 680 e 180 f 1170 g 1120 h 240 i 300 j 1320 k 390 l 300 2 a 165 b 157.5 c 1008 d 1625 e 760 f 562.5 g 1275 h 101.25 i 101.25 j 15 k 798 l 536.25 3 a 10.4 b 117.6 c 803.84 d 9.8 e 471.04 f 616.4 g 312.8 h 432.96 i 813.4 j 808.08 k 769.76 l 924.8 4 a 32.5 b 892.8 c 34.32 d 274.7 e 145.6 f 127.71 g 284.2 h 254.6 i 253.5 j 13.94 k 859.88 l 133.4 5 a 150 b $33\frac{1}{3}$ c 41 175 d 6435 e 3775 f 3750 g 1010 h 105 i 72 400 6 a 2800 b 8.4 c 10 800 d 187.5 e 35 500 f 11 062.2 g 3062.5 h 670 i 12 000

Page 42 1 a $153 b 353.6 h c $540 d 3080 ha e $960 f 498.4 min g 2700 h 3515 L i 3930 kg j 920 t k $1386 l 8511.5 cm 2 a $171 b $28.80 c 151.20 m d 632 L e $744 f $3161.60 g $211.50 h $533.75 i $578.43 j $110 k $48 l $66 3 a $420 b 2550 L c $385 d 132 t e 9.8 m f 270 g $1032 h $1280 i $1638.40 j $1320 k $218.50 l $4625 4 a $45 000 b $525 c $19 200 d $11 908.80

Page 43 1 a 10% b 0.67% c 1.25% d 25% e 30% f 20% g 25% h 40% i 25% j 0.5682% k 25% l 4% 2 a 5% b 3% c $16\frac{2}{3}$% d $12\frac{1}{2}$% e $11\frac{1}{9}$% f $8\frac{1}{3}$% g $16\frac{2}{3}$% h $2\frac{1}{2}$% i $2\frac{7}{9}$% j 50% k 5% l 25% 3 a 5% b 2% c $2\frac{1}{12}$% d $66\frac{2}{3}$% e $4\frac{1}{6}$% f $1\frac{7}{25}$% g 4% h $2\frac{2}{9}$% i $1\frac{2}{3}$% j $\frac{2}{5}$% k 5% l 5% 4 a $11\frac{1}{9}$% b $32\frac{8}{31}$% c $2\frac{1}{2}$% d 5%

Page 44 1 a 680 b 1150 c $1800 d $1333\frac{1}{3}$ e 160 2 a $77\frac{1}{7}$ b 600 c 80 d $316\frac{2}{3}$ e $81\frac{3}{17}$ 3 a 1000 b $653\frac{1}{3}$ c 800 d $228\frac{4}{7}$ e $266\frac{2}{3}$ f 7680 g 280 h 5760 i 750 j $1236\frac{4}{11}$ k 1400 l 375 4 a $366.67 b $85 c $124\frac{1}{2}$ d $11 000

Page 45 1 $4550 2 a $6650 b $8270 3 a $8330 b $9180 4 $20 000 5 a $1004.85 b $22 381

Page 46 1 a $222.50, $667.50 b $5.70, $32.30 c $3.70, $70.30 d $32.50, $292.50 e $378, $1512 f $7.80, $57.20 2 a $144.75 b $820.25 3 a $980 b 11.73% 4 a $86 b $344 5 a $274.39 b $49.39 6 a $2748.89 b $1511.89

Page 47 1 a $500 b 35.71% c 55.56% 2 a $16 500 b $24 750 3 a $6.82 b $68.18 4 $1046.15

Page 48 1 28 2 160 g 3 1900 4 $84 5 $336 6 $794.88 7 $63 8 $6\frac{2}{3}$% 9 $147 10 60% 11 $43 000 12 $23 072 13 $426.67 14 300 15 $375

Page 49 1 D 2 C 3 B 4 B 5 C 6 C 7 C 8 D 9 C 10 C

Page 50 1 $9\frac{1}{10}$ 2 $16\frac{2}{3}$% 3 750 4 93 5 198 kg 6 817 7 60% 8 74% 9 375 10 99.36 11 12 12 $\frac{22}{25}$

Answers

13 90 **14** $\frac{33}{50}$ **15** 900

Chapter 5 – Basic algebra

Page 51 **1 a** $5a + 7b$ **b** $2x - 12$ **c** $10x$ **d** $x - y$ **e** $2y$ **2 a** $4y - 6$ **b** $3a + 2b$ **c** $2x + y + z$ **d** $8a \times 2b$ **e** $\frac{9a + 11a}{2}$ **3 a** $5x + 11$ **b** $\frac{x}{5} + 9$ **c** $x + 35$ **d** $x - 7$ **e** $x + 9$ **4 a** $x + 16$ **b** $2a + 8b$ **c** $5a + 6b + 9c$ **d** $8x + 12y + 5z$ **e** $8m + 12n + 11p$ **5 a** $8a + 7b + 6y$ **b** $16xy - 18$ **c** $\frac{5a}{7b} + 19$ **d** $\frac{x + y}{20}$ **e** $\frac{x + 16}{8}$

Page 52 **1 a** 6 **b** 49 **c** 116 **d** 124 **e** 114 **f** 154 **g** 96 **h** 85 **i** 82 **j** 141 **k** 45 **l** 81 **2 a** $8a$, $16a$ **b** $6x$, $9x$ **c** none **d** $6c$, $9c$ **e** $14y$, $6y$ **f** $5l$, $6l$ **g** $4c$, $6c$ **h** $5a$, $6a$ **i** $18y$, y **3 a** $22a$ **b** $5y$ **c** $5m$ **d** $5x$ **e** $5a$ **f** $4d$ **g** $26d$ **h** $16k$ **i** $29a$ **4 a** $8x$ **b** $9b$ **c** $73x$ **d** $12a$ **e** $43x$ **f** $12a$ **g** $6y$ **h** $6c$ **i** $19y$ **5 a** $19a$ **b** $6ab$ **c** xyz **d** $16xy$ **e** $42mnl$ **f** $8mn$ **g** $16ab$ **h** $24mn$ **i** $44ade$ **6 a** $28 \times a \times b$ **b** $8 \times a \times b \times c$ **c** $9 \times x \times y$ **d** $71 \times m \times n$ **e** $19 \times a \times b \times c$ **f** $48 \times m \times m$ **g** $12 \times x \times y \times z$ **h** $19 \times a \times a \times b$ **i** $44 \times m \times n \times p$ **j** $8 \times a \times a \times a$ **k** $6 \times a \times c \times e$ **l** $47 \times a \times a \times b \times b$ **m** $46 \times x$ **n** $28 \times x \times x \times y \times y \times y$

Page 53 **1 a** $5a$ **b** $3x + 6y$ **c** $4x + 3y$ **d** $5l + 3m$ **e** $3m + 4n$ **f** $3k + 4m$ **2 a** $20x + 9y$ **b** $19a + 37$ **c** $11a + 15b$ **d** $16x + 11y$ **e** $14m + 7n$ **f** $16a + 21b$ **3 a** $34x + 2y$ **b** $32x + 19y$ **c** $29p + 2l$ **d** $29m + 2n$ **e** $39x + 21y$ **f** $4xyz + 8x - 9y$ **g** $17a + 17b$ **h** $34a + b$ **4 a** $7m + 2n$ **b** $2p + 3q$ **c** $6x^2 + 8y^2$ **d** $2x + 5y$ **e** $11a - 13b$ **f** $5x + 4y$ **g** $10x + 9y$ **h** $2d - c$ **5 a** $-6x$ **b** $-2a^2 - 8a + 33$ **c** $-3x + 2y$ **d** $5p^3 + 2p^2 + 15$ **e** $-a + 8b$ **f** $14x - y$ **g** $13b + 4ab$ **h** $xy - 3x$ **6 a** $5xz$ **b** $23m^2n + mn^2$ **c** $24a^2$ **d** $23xyz + 8x - 4x^2$ **e** $27xyz + 8x$ **f** $6x^2 + 17y^2$ **g** $11x^2y^2 + 17x^2$ **h** $13a^3 - 2b^3$

Page 54 **1 a** $18y$ **b** $48a$ **c** $15\,m^3$ **d** $6 \times x \times x \times y$ **e** $24 \times y$ and $8 \times y$ **2 a** $115a$ **b** $135n$ **c** $2a + 7b$ **d** $14(x + 16)$ **e** $\frac{13x}{5}$ **f** abc **g** $162a$ **h** $18ab^2$ **i** $70mn$ **j** $12m + 5$ **k** $9 + 14x$ **l** $13a + 4b$ **m** $\frac{23}{xy}$ **n** $\frac{6a}{7b}$ **o** $\frac{3m}{10}$ **p** $\frac{m + n}{6}$ **q** $\frac{15x}{3a + 2}$ **r** $\frac{8x + 3}{2a}$ **3 a** $19 \times a$ **b** $13 \times y$ **c** $42 \times a$ **d** $14 \times x - 5$ **e** $8 \times m + 32$ **f** $60 - 8 \times x$ **g** $a \times b + 36$ **h** $12 \times x - 3 \times y$ **i** $a \times a \times a \times b \times b \times c \times c$ **j** $14 \times x \times y \times z$ **k** $17 \times m \times m + 42$ **l** $x \times x - y \times y$ **4 a** $6a^3b$ **b** $42a^2b^2$ **c** $5(x + 9)$ **d** $15(m - 5)$ **e** $7a(c + 2)$ **f** $28(x - 2)$ **g** $mn(p + 3)$ **h** $6(5a + 9)$ **i** $18 + (20 + 4m)$ **j** $8(6a - 2b)$ **k** $(3a + 2)(a + 5)$ **l** $(4a + 3)(2a - 5)$ **m** $7n^4$ **n** $18a^3$ **o** x^3y^3 **p** $(3m + 2)(5m - 3)$ **q** $(a + 2)(5a + 7)$ **r** $(a - 3)(4a - 7)$

Page 55 **1 a** $11x$ **b** $16x$ **c** $16x$ **d** $27x$ **e** $41x$ **f** $45x$ **g** $14xy$ **h** $20mn$ **i** $38a$ **j** $18x$ **k** $42a^2$ **l** 19p **2 a** $4a$ **b** $8a$ **c** $5a$ **d** $15x$ **e** $7m$ **f** $4y$ **g** $6x$ **h** $2xy$ **i** $22x^2$ **j** $8y$ **k** $5x$ **l** $11p$ **3 a** $12a$ **b** $8x$ **c** $14a$ **d** $14a$ **e** $20xy$ **f** $8mn$ **g** $9x^2$ **h** $13a^2$ **i** $5x$ **j** $16t$ **k** $10m$ **l** $12a$ **4 a** $3x + 5y$ **b** $9a + 11b$ **c** $5m + 2n$ **d** $4x^2$ **e** $2a + 5b$ **f** $14a - 6b$ **g** $15y - 7x$ **h** $6xy$ **5 a** $22 - 10x$ **b** $3m + 3n$ **c** $19x - 2y$ **d** $11x^2 - 7y^2$ **e** $8m + 2n$ **f** $6xy + 6yz$ **g** $9a + 6b$ **h** $3a + 6b$ **i** $6t + 3$ **j** $18ab$

Page 56 **1 a** $10a$ **b** $18b$ **c** $72y$ **d** $21y$ **e** $-6ab$ **f** $6a$ **g** $-3ab$ **h** $6ab$ **i** $18x$ **j** $-45a^2b$ **k** $16a^2$ **l** $12x^2$ **2 a** $18ab$ **b** $6x^2$ **c** $40xy$ **d** $24y^2$ **e** $27ab$ **f** $-30a^2$ **g** $48y^2$ **h** $6t^2$ **i** $54x^2y$ **j** $-30a^2b$ **3 a** $2a^2b$ **b** $90a$ **c** $30a^2bc$ **d** $a^2b^2c^2$ **e** $-54ab^2c$ **f** $42a^2p$ **g** $-40x$ **h** $6xy^2$ **4 a** $40ab$ **b** $14k^2$ **c** $-72x^2y^2$ **d** $8a^3x^2$ **e** $36m^2$ **f** $24m^2t$ **g** $3y^3$ **h** $-30x^2$ **5 a** $-8a^2b^2$ **b** $-84a^2$ **c** $-18x^2y$ **d** $30abc$ **e** $-4a^2b$ **f** $-x^2y$ **g** $25p^2$ **h** $-36a^2$

Page 57 **1 a** $3a$ **b** $3pq$ **c** $\frac{4}{5m}$ **d** $3x$ **e** $3y$ **f** $6a$ **g** $2a$ **h** $9x$ **i** $\frac{3}{y}$ **2 a** 3 **b** $4y$ **c** $3n$ **d** $8m$ **e** $-x$ **f** $8b$ **g** 4 **h** $8a$ **i** $4y$ **j** $-3b$ **k** $3b$ **l** $4x$ **3 a** $-3x$ **b** $-2b$ **c** $2y$ **d** $3x$ **e** $3y$ **f** 4 **g** $4x^2$ **h** $12y$ **i** $-5q$ **j** $10n$ **4 a** -4 **b** 17 **c** $-8x$ **d** -2 **e** $2x$ **f** $6m$ **g** $5y$ **h** $8b$ **5 a** $-4b$ **b** $4x$ **c** -1 **d** $9y$ **e** 1 **f** $\frac{ab}{c}$ **g** $\frac{3}{b}$ **h** $2x$

Page 58 **1 a** $16a^2$ **b** $60a^2c$ **c** $-27x^2$ **d** $-36a^3$ **e** $45x^3y$ **f** $15y^3$ **g** $96a^2b$ **h** $72a^3$ **2 a** 5 **b** 9 **c** $-2q$ **d** $-7x$ **e** $-4b$ **f** $-4a$ **g** $4p$ **h** $4yz$ **3 a** -9 **b** $\frac{-5b}{a}$ **c** -1 **d** $7abc$ **e** 3 **f** $7t$ **g** $4y$ **h** $-5b$ **i** $\frac{5}{2}$ **j** $-5a$ **4 a** $-15a^2$ **b** $40abc$ **c** $-56x^2$ **d** $48a^3$ **e** $-30a^2b$ **f** $-40x^2y$ **g** $36a^3b^2$ **h** $-a^4b^4$ **i** $8xy^3$ **j** $-15x^2y$ **5 a** 6 **b** 12 **c** $\frac{10}{3}$ **d** $4b^2$ **e** $6ab$ **f** $-x$

Page 59 **1 a** $120x^2$ **b** $72b$ **c** $30m^2$ **d** $180a^2b$ **e** $96a^2$ **f** 3 **g** 21 **h** $\frac{32m}{3}$ **i** $10m$ **2 a** $24x$ **b** $320x$ **c** $\frac{1}{4}$ **d** -1 **e** 4 **f** $90m^2$ **g** $160a^2b^2$ **h** 2 **i** $12a$ **3 a** 0 **b** $28x$ **c** $-24n$ **d** $115p$ **e** $-n$ **f** $-60x$ **g** $-3t$ **h** 0 **i** 3 **4 a** 4 **b** $\frac{3}{2}$ **c** 1 **d** $9y$ **e** 18 **f** 8 **g** $46x$ **h** $10a$ **i** 6

Page 60 **1 a** $3x + 3y$ **b** $10a + 15b$ **c** $8a + 12$ **d** $48a - 42b$ **e** $8m + 24p$ **f** $15a - 12m$ **g** $5x^2 + 40x$ **h** $7x - 35$ **i** $9a^2 - 18ab$ **2 a** $-3x - 6$ **b** $-10x^2 - 5x$ **c** $-4x + 14$ **d** $-16x^2 + 6xy$ **e** $-12x + 15$ **f** $-xy - y^3$ **g** $-64x + 72$ **h** $-3m^2 - 8m$ **i** $-2x^2 - 2x$ **3 a** $10x + 35$ **b** $21x - 77$ **c** $-12x + 27$ **d** $-8x - 20$ **e** $48y + 24$ **f** $-15a^2 - 15a$ **g** $-2x^2 - 7x$ **h** $-8m^2 + 5m$ **i** $-6t^2 - 10t$ **4 a** $7x + 6$ **b** $11y + 17$ **c** $2xy - 35$ **d** $p - 70$ **e** $2y^2 + 12y$ **f** $5x + 6$ **g** $12p - 24$ **h** $9m - 16$ **i** $8m^2 + 9m$ **5 a** $6m + 14$ **b** $8x + 12$ **c** $19x + 62$ **d** $8x + 76$ **e** $3y + 37$ **f** $17x + 24$ **g** $5n - 32$ **h** $2x^2 - 3$ **i** $3x + 54$

Page 61 **1 a** 35 **b** 46 **c** 44 **d** 25 **e** 125 **f** 24 **g** 125 **h** 15 **i** 255 **j** 23 **k** 6 **l** $\frac{1}{5}$ **2 a** 6 **b** 17 **c** 4 **d** 40 **e** 36 **f** 31 **g** 320 **h** 34 **i** 12 **j** 6 **k** 7 **l** 12 **3 a** 3, 5, 7, 9, 11 **b** −1 3, 7, 11 **c** −3, 3, 9, 15, 21 **d** 3, 7, 11, 15, 19, 23 **e** 3, 6, 9, 12, 15 **f** −4, −1, 2, 5, 8, 11 **4 a** 5, −1, 6, 4, 9, 23 **b** 10, −2, 24, 16, 36, 46 **c** 9, 1, 20, 25, 16, 40

Answers

d 11, -5, 24, 9, 64, 52 **e** 13, -1, 42, 36, 49, 59 **f** 14, -4, 45, 25, 81, 65 **g** 12, 2, 35, 49, 25, 53 **h** 14, -2, 48, 36, 64, 64 **i** 13, -7, 30, 9, 100, 62

PAGE 62 **1 a** 28 **b** 336 **c** $2\frac{1}{3}$ **d** 784 **e** 4032 **f** 209 **g** 23 **h** 37 **i** -11 **j** 71 **k** 1316 **l** 380 **2 a** 27 **b** 1 **c** 7 **d** 69 **e** 96 **f** 88 **g** 50 **h** 64 **i** 3 **j** 45 **k** 48 **l** -33 **3 a** 10.7 **b** 17.1 **c** 34.2 **d** 23.1 **e** 114.5 **f** 483.8 **g** 425.1 **h** 5.0 **i** 0.2 **j** 4.1 **k** 8.8 **l** 3.9 **4 a** 138 **b** 50 **c** 360 **d** -18 **e** 32 **f** 11 **g** 212 **h** -5 **i** 30 **j** 19

PAGE 63 **1 a** $7(x+1)$ **b** $2(5a-1)$ **c** $7(y+2)$ **d** $4(2x-1)$ **e** $3(a+5)$ **f** $4(x-3y)$ **g** $3(y+1)$ **h** $5(m+6)$ **i** $4(8x-7)$ **j** $5(a-1)$ **k** $2(3a-b)$ **l** $5(a+7)$ **2 a** $y(y+3)$ **b** $3y^2(2y-1)$ **c** $3x^2(2x-1)$ **d** $2m(m+3)$ **e** $4a(2a-1)$ **f** $m(m-9)$ **g** $8a(a+3)$ **h** $3y(5y-2)$ **i** $8a(2a-1)$ **j** $x(x+9)$ **k** $3a^2(3a-2)$ **l** $7x(4x-1)$ **3 a** $-3(x+1)$ **b** $-4(y^2-4)$ **c** $-3(2n+1)$ **d** $-5(x+8)$ **e** $-3(x-9)$ **f** $-7(t+2)$ **g** $-2(m^2+2)$ **h** $-5(m^2-7)$ **i** $-11(x^2+2)$ **j** $-6(a^2-2)$ **k** $-8(a^2+4)$ **l** $-13(y-3)$ **4 a** $-6x(x+2)$ **b** $-4x(4x+1)$ **c** $-3(2x+3)$ **d** $-4y(2x+1)$ **e** $-9x^2(x-3)$ **f** $-5x(3x+y)$ **g** $-3(2mn+1)$ **h** $-5xy(xy-5)$ **i** $-3m(2n-5m)$ **j** $-4xy(2z-1)$ **k** $-3x(3x+5)$ **l** $-x(1+3y)$ **5 a** $8(a+2b)$ **b** $5(a-5)$ **c** $4(2x-y)$ **d** $9(a-b)$ **e** $8(1-t)$ **f** $m^2(m-4)$ **g** $a(16b-a)$ **h** $7(2m-1)$ **i** $8ab(2-a)$ **j** $5xy(3xy-1)$ **k** $-6(m+6)$ **l** $-4(m+3n)$ **6 a** $3(a+b+c)$ **b** $p(16q-7)$ **c** $-x(x+7)$ **d** $8(x+y-2z)$ **e** $x(t-1)$ **f** $5(3a+b-2c)$ **g** $4(x-2y)$ **h** $y^2(y-1)$ **i** $3(6x-3y+2)$ **j** $16(x-2y)$ **k** $3n(5m-1)$ **l** $3(2a^2+a+3)$ **m** $m(m-2n)$ **n** $q(8p-q)$ **o** $-6(x+xy+2)$

PAGE 64 **1 a** $8x+4$ **b** $12x$ **c** $12x+10$ **2 a** $5x^2+7x$ **b** $9x^2$ **c** $6x^2-10x$ **3** $3x+5y$ **4** $40mn$ **5** $5x$ **6** $2x+8$ **7** y^2m^2 **8** $4x+1$ **9** $7x+10$ **10** $8x^3$ cm^3 **11** $3x+5y+2$

PAGE 65 **1** A **2** D **3** D **4** D **5** A **6** C **7** D **8** D **9** C **10** C

PAGE 66 **1** $-2y$ **2** $6a^3$ **3** $7(x+2y)$ **4** $2x^2+7x$ **5** $6x^2-15xy$ **6** 5 **7 a** 1 **b** $-15a$ **c** $2a$ **d** $24m^3$ **e** $16a^2b^2c^2$ **f** $18a^3$ **8** $x(y-5)$ **9 a** $2x+7$ **b** $7a+8b-12$

CHAPTER 6 – Length, mass and time

PAGE 67 **1 a** 300 cm **b** 1500 cm **c** 800 cm **d** 1800 cm **e** 1600 cm **f** 3100 cm **g** 1200 cm **h** 6400 cm **i** 2500 cm **j** 500 cm **k** 600 cm **l** 2000 cm **m** 1 cm **n** 9 cm **o** 95 cm **p** 2.8 cm **2 a** 3 m **b** 14 m **c** 50 m **d** 0.9 m **e** 8 m **f** 21 m **g** 36 m **h** 2.5 m **i** 25 m **j** 8.4 m **k** 16.75 m **l** 12.6 m **m** 3000 m **n** 18 000 m **o** 5650 m **p** 31 870 m **3 a** 2000 mm **b** 3800 mm **c** 5900 mm **d** 16 380 mm **e** 10 000 mm **f** 25 000 mm **g** 63 800 mm **h** 92 600 mm **i** 37 000 mm **j** 53 000 mm **k** 85 900 mm **l** 6750 mm **m** 600 mm **n** 920 mm **o** 356 mm **p** 389 mm **4 a** 800 **b** 6 **c** 9820 **d** 123.6 **e** 12 500 **f** 84.5 **g** 93.6 **h** 3500 **i** 1300 **j** 9000 **k** 800 **l** 9200 **5 a** 8000 **b** 80 **c** 90 **d** 1030 **e** 7 **f** 7 000 000 **g** 1.2 **h** 33 580 **i** 8 **j** 140 **k** 19 500 **l** 630 **6 a** 580 **b** 23 600 **c** 8.64 **d** 85.4 **e** 28 **f** 5600 **g** 4.563 **h** 9350 **i** 15 000 000 **j** 72 560 **k** 8350 **l** 835

PAGE 68 **1 a** $a = 5$ cm **b** $b = 130$ mm **c** $c = 17$ mm **2 a** $a = 3.6$ m **b** $b = 39.7$ cm **c** $c = 3.8$ cm **d** $d = 6.0$ cm **e** $e = 2.5$ m **f** $f = 7.3$ cm **3 a** $x = 4.2$ cm **b** $y = 12$ cm

PAGE 69 **1** $d = 1.3$ m **2 a**

2 m, x, 1.5 m

b The ladder is 2.5 m long. **3 a**

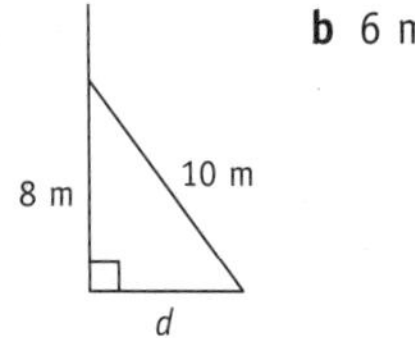

b 6 m

PAGE 70 (All answers are in centimetres.) **1 a** 64 **b** 72 **c** 48 **d** 62 **e** 48 **f** 86 **2 a** 60 **b** 52 **c** 49 **d** 33 **e** 36 **f** 27 **3 a** 40 **b** 48 **c** 24 **d** 24 **e** 38 **f** 48

PAGE 71 **1 a** 50 **b** 61 **c** 54 **d** 50 **2 a** 92 **b** 35 **c** 58 **d** 76 **e** 42 **f** 132 **3 a** 12 m **b** 15.11 m

PAGE 72 **1 a** 8000 kg **b** 6.438 kg **c** 6350 kg **d** 12.67 kg **e** 8160 kg **f** 9135 kg **g** 0.728 kg **h** 9250 kg **i** 7.2 kg **j** 35 000 kg **k** 0.845 kg **l** 430 kg **m** 24.89 kg **n** 0.005 kg **o** 6200 kg **2 a** 9000 g **b** 1 000 000 g **c** 16 450 g **d** 1200 g **e** 13 508 g **f** 4 500 000 g **g** 700 g **h** 87 500 g **i** 3 000 000 g **j** 12 690 g **k** 12 g **l** 52 125 g **m** 680.5 g **n** 930 000 g **o** 4500 g **3 a** 8 kg **b** 63 500 kg **c** 9.607 t **d** 30 kg **e** 4.8 kg **f** 2750 kg **g** 42 t **h** 0.965 kg **i** 67.5 g **j** 0.005 t **k** 9 080 000 mg **l** 70.5 t **m** 12 mg **n** 8.563 kg **o** 3250 kg **4 a** 3 kg **b** 25.006 t **c** 8567 g **d** 2 306 000 g **e** 21 t **f** 3657 g **g** 9.64 kg **h** 8350 kg **i** 12 890 g **j** 25.63 t **k** 0.125 t **l** 198.5 g **m** 8.005 t **n** 37 000 g **o** 500 mg **5 a** 15 mg **b** 32 536 kg **c** 8.36 t **d** 43.25 g **e** 670 000 g **f** 10 t **g** 52 600 mg **h** 53 690 kg **i** 560 000 g **j** 53.467 t **k** 8 700 000 g **l** 4000 g **m** 8.2615 t **n** 3 250 000 g **o** 5.639 kg

PAGE 73 **1 a** 800 **b** 273.6 kg **c** 21 t 480 kg **d** 2.15 kg **2 a** 1.745 kg **b** 45 600 t **c** 31.6 t, 8.65 t, 26.8 kg, 4300 g, 3.6 kg, 460 g **3 a** 5.67 t **b** 3610 g **c** 3.25 kg

PAGE 74 **1 a** minute **b** minute **c** hour **d** hour **e** minute **2 a** 60 **b** 480 **c** 500 **d** 900 **e** 1200 **f** 9 **g** 63 **h** 1800 **i** 8 **j** 84 **k** 6 **l** 366 **m** 12 **n** 84 **o** 1095 **p** 336 **3 a** 60 **b** 180 **c** 720 **d** 1440 **e** 10 080 **f** 30 240 **g** 43 200 **h** 47 520 **4 a** 48 **b** 168 **c** 336 **d** 720 **e** 744 **f** 720 **5 a** 7 **b** 21 **c** 70 **d** 105 **e** 14 **f** 28 **g** 31 **h** 365 **i** 732 **j** 3652

PAGE 75 **1 a** 14 h 37 min **b** 25 h 4 min **c** 19 h 9 min **d** 12 h 13 min **e** 18 h 41 min **f** 10 h 18 min **g** 16 h 16 min **h** 5 h 9 min **2 a** 18 d 22 h 16 min **b** 15 h 12 min **c** 8 **d** 26 **e** 6 **3 a** 9:00 am **b** 7:00 am **c** 4:05 pm **d** 9:00 am e 7:17 pm f 6:57 am g 3:45 pm h 5:56 am i 10:53 am j 2:10 am the next day 4 a 3 h b 4 h 15 min c 4 h d 3 h 28 min e 4 h f 8 h 35 min g 6 h h 9 h i 6 h j 14 h 15 min

PAGE 76 1 a 10:00 am b 11:25 pm c 8:20 pm d 8:00 am e 9:30 am f 8:00 am 2 a 9:00 am b 10:19 pm c 7:32 pm d 6:33 pm e 6:31 am f 8:00 am 3 a 0800 b 1440 c 1720 d 0820 e 0410 f 0935 g 1840 h 1850 4 a ten past ten b five minutes to nine c twenty past nine d ten minutes to four 5 a 3:05 b 8:25 c 6:00 d 12:00 e 5:15 f 6:30 6 a 1:00 pm b 1:00 pm c 11:00 pm d 1:00 pm

PAGE 77 1 a 8:35 am b 9:43 am c 10:25 am d 3:20 pm e 8:36 pm f 11:40 pm 2 a 0900 hours b 1230 hours c 0440 hours d 2310 hours 3 a 0530 hours b 1445 hours c 2320 hours d 0215 hours 4 a 6 h 20 min b 5 h 5 a 1645 hours b 2240 hours 6 9:20 pm 7 1 h 55 min 8 a 4:10 pm b 9:35 am c 11:30 pm 9 a 0530 hours b 1520 hours c 2240 hours 10 2020 hours

PAGE 78 1 1.75 m 2 36 cm 3 12.179 km 4 37.4 cm 5 920 m 6 42 838 7 8.06 t 8 6250 9 576.4 kg 10 1.112 kg 11 3375 12 840

PAGE 79 1 D 2 D 3 C 4 B 5 C 6 C 7 B 8 D 9 B 10 B

PAGE 80 1 36 m 2 86 cm 3 60 m 4 5500 5 $\frac{3}{32}$ 6 1000 7 15.6 8 38 500 9 180 10 192 11 6:50 pm 12 26 min 20 s 13 15 days 15 min 14 750 g 15 5.34 kg

CHAPTER 7 – Area, volume and capacity

PAGE 81 1 a 10 b 100 c 1000 d 10×10; 100 e 100×100; 10 000 f 1000×1000; 1 000 000 g 10 000 h 100 2 a 2000 b 21 500 c 8 d 900 e 0.6 f 200 000 g 2 400 000 h 7.2 i 0.4 j 38 000 k 5 000 000 l 80 000 m 2000 n 3 500 000 o 1280 p 8 q 12 000 r 100 s 36 000 000 t 900 3 a 500 000 000 b 500

PAGE 82 1 a 60 cm^2 b 100 cm^2 c 37.5 cm^2 2 a 126 m^2 b 243 cm^2 c 70 m^2 3 a 24.7 cm^2 b 0.84 m^2 c 42 cm^2 d 550 m^2 e 216 cm^2 f 20.25 m^2

PAGE 83 1 a 126 cm^2 b 240 cm^2 c 336 cm^2 d 96 cm^2 2 a 172.86 cm^2 b 96.6 cm^2 c 172.55 cm^2 3 a 19 m b 5 cm c 216 mm^2 d 480 cm^2 e 12 m f 450 mm^2 g 8 cm h 18 m i 15 cm j 420 cm^2

PAGE 84 1 a 56 cm^2 b 216 cm^2 c 80 cm^2 d 60 cm^2 e 1900 cm^2 f 70 cm^2 3 a 128 m^2 b 12 cm c 120 cm^2 d 16 cm e 11 m f 28 m g 234 cm^2 h 26.86 cm i 30 mm j 42 cm

PAGE 85 1 a 96 cm^2 b 60 cm^2 c 119 cm^2 d 48 cm^2 e 80 cm^2 f 192 cm^2 2 a 1600 $units^2$ b 644 $units^2$ c 48 m^2 3 a 40 m^2 b 24 cm c 110 cm^2 d 34 cm e 20 m f 18 m g 85 cm^2 h 30 cm i 20 mm j 20 cm

PAGE 86 1 a $A = \frac{1}{2}bh$ b $A = lb$ c $A = bh$ d $A = \frac{1}{2}xy$ e $A = \frac{1}{2}h(a + b)$ f $A = \frac{1}{2}xy$ 2 a 15.68 cm^2 b 56.7 cm^2 c 37.21 cm^2 d 70 cm^2 e 105.84 cm^2 f 41.6 cm^2 g 18 cm^2 h 60 cm^2 i 96 cm^2

PAGE 87 1 30 000 cm^2 2 5 000 000 m^2 3 60 000 m^2 4 15.75 ha 5 $52 500 6 a 2100 m^2 b $378 000 7 32 cm 8 23 000 ha 9 a 4 500 000 m^2 b 4.5 km^2 10 a 2000 cm^2 b 1696 cm^2 c 3696 cm^2

PAGE 88 1 a 512 cm^3 b 729 cm^3 c 9261 cm^3 d 1000 m^3 e 2197 m^3 f 15 625 m^3 g 3375 cm^3 h 5832 m^3 i 4913 m^3 2 a 729 m^3 b 274.625 m^3 c 571.787 m^3 d 3652.264 m^3 e 1728 m^3 f 2744 m^3 3 a 21 cm b 16 cm

PAGE 89 1 a 512 cm^3 b 450 cm^3 c 924 cm^3 d 270 cm^3 e 560 cm^3 f 378 cm^3 2 a 616 cm^3 b 1512 cm^3 c 6 cm d 8 cm e 11 cm f 864 cm^3 3 a 720 cm^3 b 1200 cm^3 c 1188 cm^3 d 3960 cm^3 e 4608 cm^3 f 2744 cm^3

PAGE 90 1 a 1600 cm^3 b 3780 cm^3 c 9310 cm^3 d 1080 cm^3 e 576 cm^3 f 3552 cm^3 2 a 816 cm^3 b 869 m^3 c 1680 cm^3 d 1568 cm^3 e 2496 m^3 f 1152 m^3 3 a 3220 m^3 b 68 cm^2 c 504 m^3 d 28 m e 32 cm^2 f 21 cm g 1496 m^3 h 1728 m^3

PAGE 91 1 a 1815.848 cm^3 b 1567.5 cm^3 c 360 cm^3 d 945 cm^3 e 436.8 cm^3 f 1235.4 cm^3

2 a 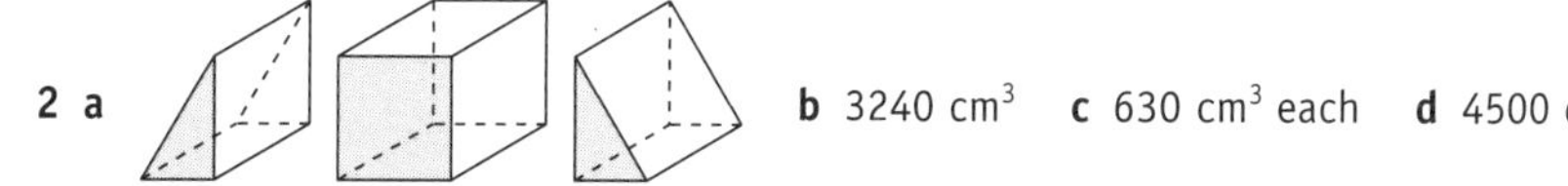 b 3240 cm^3 c 630 cm^3 each d 4500 cm^3 3 a 479 cm^2 b 7185 cm^3

PAGE 92 1 6912 cm^3 2 9 cm 3 a 1 408 000 mm^3 b 1408 cm^3 4 a 8 cm^3 b 400 cm^3 c 8400 5 2744 cm^3 6 18 000 cm^3 7 360 cm^2 8 17 640 cm^3

PAGE 93 1 a 60 b 18 760 c 19 800 d 25 300 e 12 500 f 16 540 g 7000 h 12.865 i 3.486 j 10.58 k 18 700 l 10 700 2 a 10 b 8 c 3.58 d 30 e 7 f 38 g 4 h 26 i 9.635 j 0.08 k 9653 l 8 786 000 3 a 27 b 43 c 875 4 a 176 400 cm^3 b 176.4 L 5 a 1728 cm^3; 1728 mL b 2940 cm^3; 2940 mL c 4050 cm^3; 4050 mL

PAGE 94 1 324.8 L 2 4.8 L 3 21 cm 4 750 mL 5 28 L 6 a 4200 cm^3 b 33 600 cm^3 7 4000 cm^2 8 306.25 cm^2 9 74 cm 10 2500 mL 11 5832 cm^3 12 152 m^2

PAGE 95 1 B 2 A 3 C 4 A 5 B 6 C 7 C 8 B 9 B 10 D

PAGE 96 1 180 cm^3 2 0.375L 3 8 mm 4 343 cm^3 5 7 200 000 mL 6 25 L 7 1 000 000 000 mm^3 8 54.4 m^2 9 2.4 cm^2 10 7 bottles 11 11 doses 12 234 cm^2 13 0.7744 m^3

Answers

Chapter 8 – Circles

Page 97 **1** **a** centre **b** radius **c** diameter **d** arc **e** chord **2** **a** semi-circle **b** minor segment **c** major segment **d** sector **e** tangent **f** secant **g** circumference **h** quadrant **3** **a** $\frac{1}{3}$ **b** $\frac{1}{2}$ **c** $\frac{3}{4}$ **d** $\frac{1}{6}$ **4** **a** 8 cm **b** 15 cm **c** $C = 2\pi r$ **d** $A = \pi r^2$ **e** four

Page 98 **1** **a** $81\frac{5}{7}$ cm **b** $31\frac{3}{7}$ cm **c** $53\frac{3}{7}$ cm **d** $62\frac{6}{7}$ cm **2** **a** 59.69 cm **b** 78.54 cm **c** 87.96 cm **d** 150.80 cm **3** **a** 94.2 cm **b** 133 cm **c** 220 cm **d** 166 cm **4** **a** 169.65 cm **b** 265.78 cm **c** 119.38 cm

Page 99 **1** **a** $18\frac{6}{7}$ cm **b** $25\frac{1}{7}$ m **c** $51\frac{6}{7}$ m **d** $102\frac{16}{35}$ cm **2** **a** 37.70 cm **b** 78.54 m **c** 54.66 cm **d** 260.12 m **3** **a** 141 cm **b** 60.6 cm **c** 69.1 cm **d** 109 m **4** **a** 115.7 cm **b** 28.6 cm **c** 44.8 cm

Page 100 **1** **a** 36.0 cm **b** 94.9 cm **c** 53.2 cm **2** **a** 25.00 cm **b** 74.99 cm **c** 50.74 cm **3** **a** 154 cm **b** 160 cm **c** 129 cm **d** 53.1 cm **e** 25.4 cm **f** 68.6 **g** 54.9 **h** 32.1 cm **i** 105 cm

Page 101 **1** **a** $50\frac{2}{7}$ cm^2 **b** $962\frac{1}{2}$ cm^2 **c** $452\frac{4}{7}$ cm^2 **2** **a** 254.47 cm^2 **b** 55.42 cm^2 **c** 124.69 cm^2 **d** 4.91 cm^2 **e** 196.07 cm^2 **f** 66.48 cm^2 **3** **a** 616 cm^2 **b** 211 cm^2 **c** 1600 cm^2 **4** **a** 115.5 cm^2 **b** 150.8 cm^2

Page 102 **1** **a** 173.2 cm^2 **b** 125.8 cm^2 **c** 131.5 cm^2 **2** **a** 160.61 cm^2 **b** 411.87 cm^2 **c** 1590.43 cm^2 **3** **a** 714.63 cm^2 **b** 298.01 cm^2 **c** 882.47 cm^2 **d** 910.84 cm^2 **e** 125.94 cm^2 **f** 73.06 cm^2 **g** 211.18 cm^2 **h** 139.97 cm^2 **i** 1091.88 cm^2

Page 103 **1** 9.0 cm **2** 182.1 cm **3** **a** 263.9 m **b** 5.7 laps **4** 84.8 cm **5** 992 circles **6** 19.6 m^2 **7** 150.8 m **8** 615.8 cm^2 **9** 53.4 m **10** 69.1 cm **11** 176.7 cm^2

Page 104 **1** C **2** B **3** A **4** C **5** B **6** D **7** A **8** B **9** B **10** D

Page 105 **1** **a** 62.8 cm; 314.2 cm^2 **b** 52.8 cm; 221.7 cm^2 **c** 47.3 cm **2** **a** 52.0 mm **b** 60.1 cm **3** **a** 356.5 cm^2 **b** 69.1 cm^2 **c** 56.5 cm^2 **d** 3396.1 m^2 **4** 1.1 m^2

Chapter 9 – Linear relationships

Page 106 **1** **a** O **b** N **c** Y **d** S **e** E **f** M **g** W **h** V **i** A **j** L **k** K **l** J **m** I **n** P **o** T **p** C **q** X **r** U **s** B **t** Q **u** R **v** D **w** H **x** G **y** F **2** **a** (−2, 6) **b** (−6, 5) **c** (2, 5) **d** (4, 4) **e** (2, 3) **f** (0, 3) **g** (−2, 3) **h** (−5, 3) **i** (−3, 2) **j** (4, 2) **k** (1, 1) **l** (−4, 1) **m** (−6, 1) **n** (−2, 0) **o** (0, 0) **p** (3, 0) **q** (3, −1) **r** (−2, −1) **s** (−5, −1) **t** (−4, −2) **u** (2, −2) **v** (5, −2) **w** (−2, −3) **x** (−3, −4) **y** (1, −4) **z** (1, −5)

Page 107 **1** **a** 3, 4, 5, 6 **b** −1, 1, 3, 5 **c** −8, −5, −2, 1 **d** 1, 2, 3, 4 **e** −4, −3, −2, −1 **f** −5, −3, −1, 1 **2** **a** 0, 2, 4, 6 **b** −2, 0, 2, 4 **c** −7, −4, −1, 2 **d** −3, −1, 1, 3 **e** 3, 4, 5, 6 **f** −4, −3, −2, −1

Page 108 **1** **a**

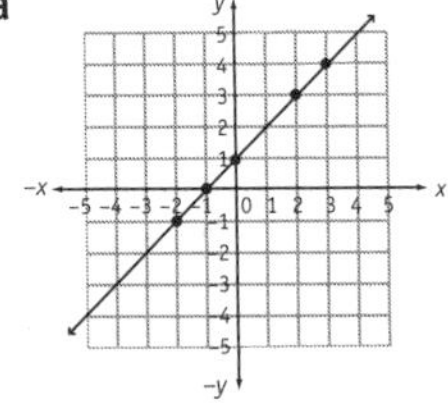

b yes **c** see diagram **d** $y = x + 1$ **2** **a** $y = 2x + 1$

x	−1	0	1	2
y	−1	1	3	5

b (−1, −1), (0, 1), (1, 3), (2, 5) **c**

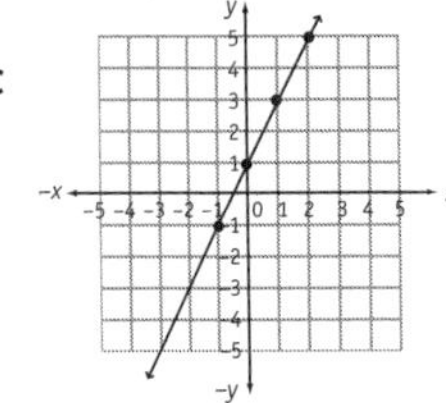

d yes **3** **a**

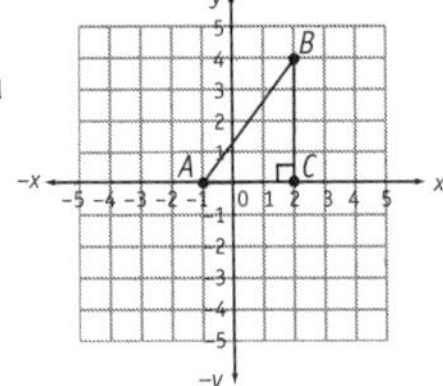

b right-angled triangle **c** 3 units **d** 4 units **e** 5 units **f** 12 units **g** 6 units^2 **4** **a** (0, 5), (1, 6), (2, 7), (3, 8) **b** (−1, 1), (0, 3), (1, 5), (2, 7) **c** (−1, −5), (0, −2), (1, 1), (2, 4)

Page 109 **1** **a**

x	−1	0	1	2
y	1	2	3	4

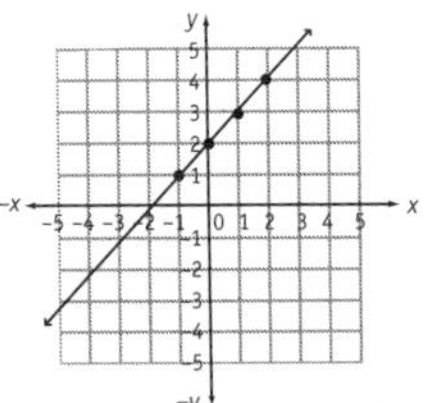

b

x	−1	0	1	2
y	−3	−1	1	3

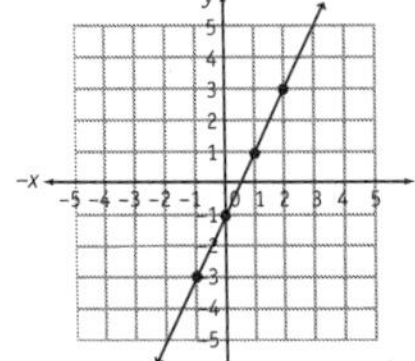

c

x	−1	0	1	2
y	3	1	−1	−3

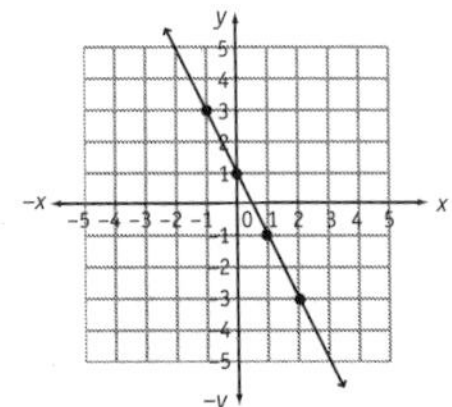

Answers

2 a

x	−1	−1	−1	−1
y	−1	0	1	2

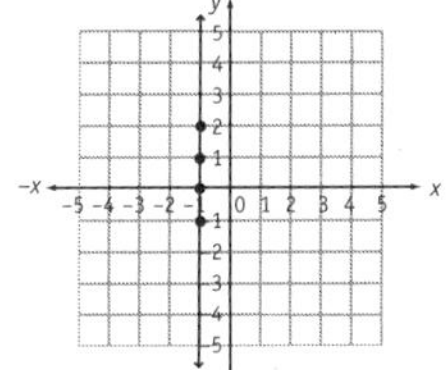

b

x	−1	0	1	2
y	2	2	2	2

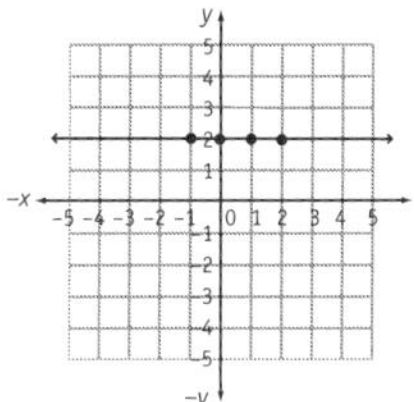

c

x	−4	−4	−4	−4
y	−1	0	1	2

3 a

x	−1	0	1	2
y	−2	0	2	4

x	−1	0	1	2
y	2	0	−2	−4

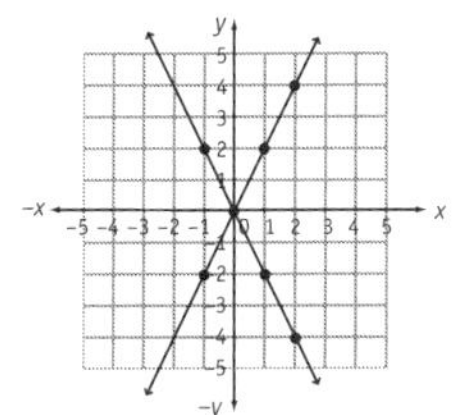

The point of intersection is (0, 0).

b

x	−1	0	1	2
y	0	2	4	6

x	−1	0	1	2
y	1	2	3	4

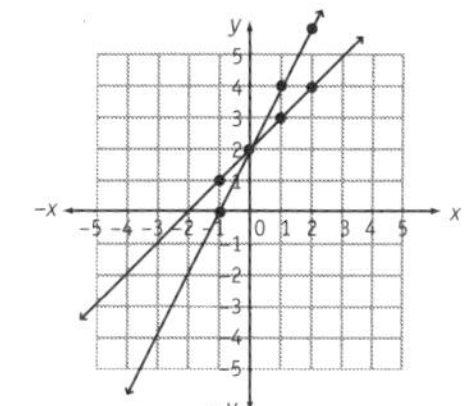

The point of intersection is (0, 2).

Page 110 **1 a** $x = 1$ **b** $x = 2$ **c** $x = 6$ **d** $x = 8$ **e** $x = 1\frac{1}{2}$ **f** $x = 6$ **2 a** $y = -2$ **b** $y = -8$ **c** $y = 6$ **d** $y = -3$ **e** $y = 3$ **f** $y = 3$

3a

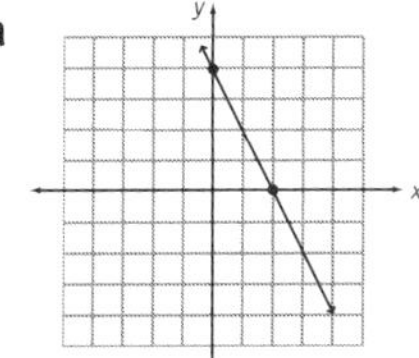

b

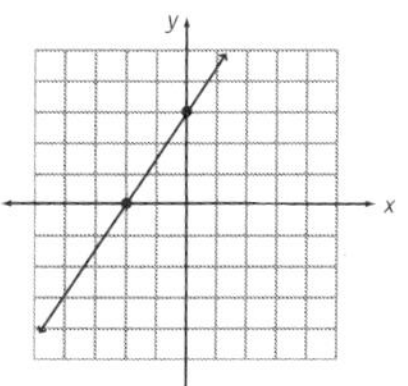

4a

x	0	$\frac{1}{2}$
y	1	0

b

x	0	3
y	3	0

c

x	0	$-\frac{1}{3}$
y	1	0

d

x	0	$\frac{3}{2}$
y	−1	0

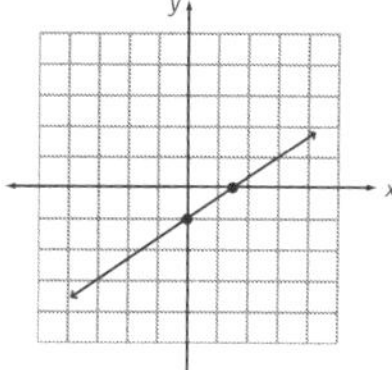

Page 111

1 a

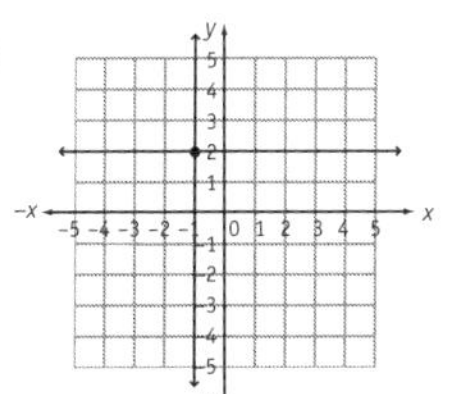

the point of intersection is (−1, 2)

b

the point of intersection is (1, −3)

c

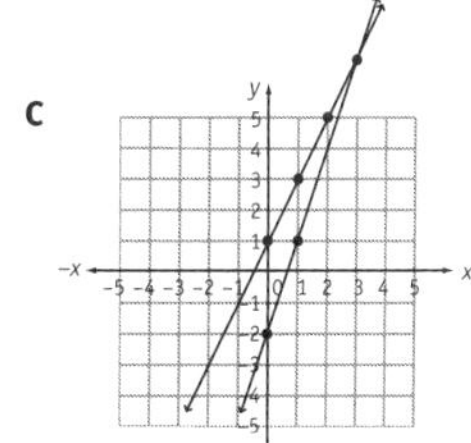

the point of intersection is (3, 7)

d

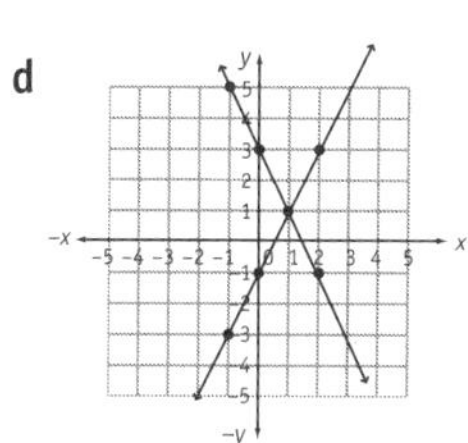

the point of intersection is (1, 1)

Page 112 **1 a** (0, 2) **c** (−3, 4) **d** (3, 0) **f** (6, −2) **2 a** $2x - y = 0$ **b** $y = 3x$ **c** $x - 5y = 0$ **e** $3y = -2x$ **3 a** yes **b** yes **c** yes **d** yes **e** no **f** no **4** $m = 1$ **5** $a = 6$ **6 a** (0, −6) **b** (5, 4) **c** (1, −4) **d** (5, 4) **e** $(\frac{1}{2}, -5)$ **f** (−1, −8)

Page 113 **1 a** $x = 2$ **b** $x = -3$ **c** $x = 4$ **2 a** $x = 2$ **b** $x = -3$ **c** $x = 4$ **3 a** $x = -1$ **b** $x = 1$ **c** $x = -2$ **d** $x = 0$

Answers

4 $y = 2x - 1$

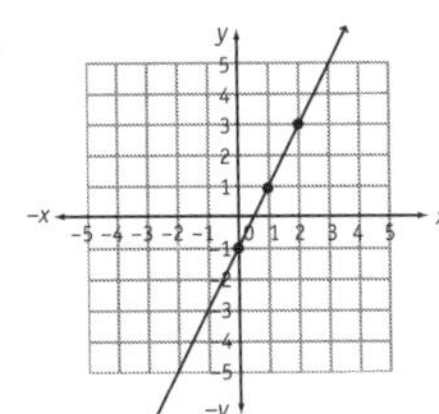

a $x = 2$ **b** $x = -\frac{3}{2}$ **c** $x = 1$ **d** $x = 3$

PAGE 114 **1** A **2** D **3** C **4** B **5** D **6** B **7** C **8** A **9** C **10** D

PAGE 115 **1 a** iv **b** ii **c** vi **d** v **e** i **f** iii

2 a

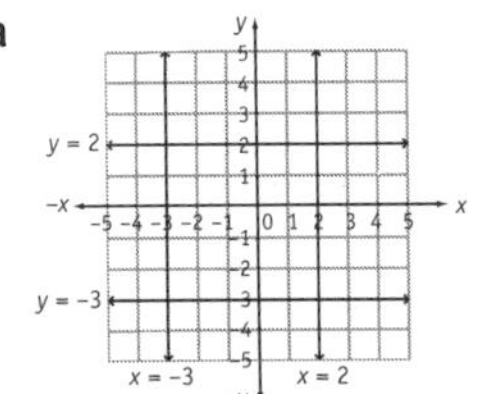

b (−3, 2)
c $y = -3, y = 2$
d $x = 2, x = -3$

3 a $y = 2x + 1$

x	−2	−1	0	1	2
y	−3	−1	1	3	5

b

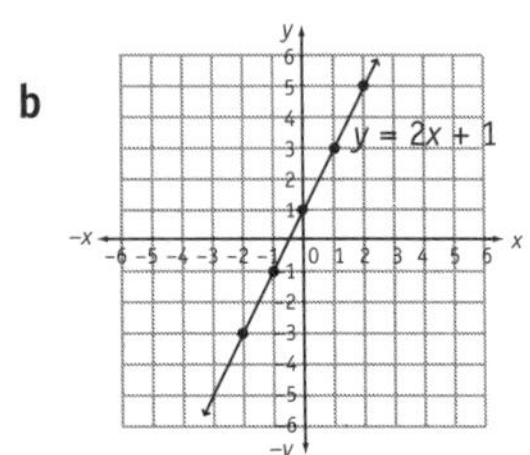

c $x = -\frac{1}{2}$
d $y = 1$
e $x = -3$

CHAPTER 10 – Equations

PAGE 116 **1 a** $x = 7$ **b** $a = 12$ **c** $m = 13$ **d** $b = 16$ **e** $a = 12$ **f** $n = 4$ **g** $x = 14$ **h** $x = 7$ **i** $y = 10$ **j** $p = 34$ **k** $t = 26$ **l** $a = 24$ **m** $y = -1$ **n** $x = -2$ **o** $x = 8$ **2 a** $m = 6$ **b** $x = 10$ **c** $y = 24$ **d** $n = -15$ **e** $x = 24$ **f** $x = 24$ **g** $m = 11$ **h** $y = 30$ **i** $a = 13$ **j** $m = 20$ **k** $n = 18$ **l** $a = 9$ **m** $y = -3$ **n** $t = -4$ **o** $a = -10$

PAGE 117 **1 a** $x = 10$ **b** $a = 21$ **c** $m = 14$ **d** $x = 50$ **e** $y = 27$ **f** $x = -7$ **g** $x = 16$ **h** $x = 5$ **i** $y = 24$ **j** $p = -4$ **k** $y = -32$ **l** $t = 4$ **m** $y = -20$ **n** $a = 4$ **o** $x = 26$ **2 a** $m = 14$ **b** $x = -40$ **c** $y = 10$ **d** $a = -80$ **e** $x = 27$ **f** $b = 36$ **g** $m = 17$ **h** $x = 21$ **i** $a = 2$ **j** $a = 9$ **k** $n = 38$ **l** $x = 10$ **m** $y = -2$ **n** $x = 9$ **o** $a = 17$

PAGE 118 **1 a** $x = 5$ **b** $x = 4$ **c** $y = 6$ **d** $m = 9$ **e** $x = 58$ **f** $a = 7$ **g** $x = 42$ **h** $x = 7$ **i** $a = 27$ **j** $x = 4$ **k** $x = 7$ **l** $t = 4$ **m** $y = 3$ **n** $m = 33$ **o** $k = 5$ **2 a** $x = 4$ **b** $m = 12$ **c** $x = 23$ **d** $x = 23$ **e** $x = 10$ **f** $m = 36$ **g** $m = 9$ **h** $y = 3$ **i** $y = -17\frac{1}{2}$ **j** $x = 3$ **k** $x = 4$ **l** $x = 10$ **m** $a = 10$ **n** $a = 2$ **o** $b = 2.2$

PAGE 119 **1 a** $x = -31$ **b** $x = -4$ **c** $t = 12$ **d** $m = 6$ **e** $m = 11$ **f** $a = 13$ **g** $y = 20$ **h** $x = 15$ **i** $y = 7$ **j** $x = 3$ **k** $a = 7$ **l** $p = 25$ **m** $m = 15$ **n** $y = 7$ **o** $m = 1\frac{1}{3}$ **2 a** $x = 22$ **b** $x = 7$ **c** $x = -5$ **d** $y = 50$ **e** $x = -13$ **f** $x = 5$ **g** $m = 8$ **h** $x = 2$ **i** $y = 4$ **j** $m = -8$ **k** $x = -32$ **l** $y = 19$

PAGE 120 **1 a** $x = 5$ **b** $m = 8$ **c** $m = 5$ **d** $x = 4$ **e** $a = 13$ **f** $n = 12$ **g** $n = 3$ **h** $p = 4$ **i** $x = -4$ **j** $a = 8$ **k** $m = 1$ **l** $x = 5$ **2 a** $a = -46$ **b** $x = 16$ **c** $y = 30$ **d** 86 **e** $x = 6$ **f** $a = -5$ **g** $x = 3$ **h** $a = -17$ **i** $a = 4\frac{3}{13}$ **j** $a = -9$ **k** $a = 27$ **l** $a = -6$ **m** $a = -39$ **n** $a = -71$ **o** $a = 33$

PAGE 121 **1 a** $x = 27$ **b** $a = 40$ **c** $y = 63$ **d** $m = 40$ **e** $m = 42$ **f** $x = 72$ **g** $a = 1$ **h** $x = 6$ **i** $x = 5$ **j** $x = 14$ **k** $x = 52$ **l** $m = 22$ **2 a** $x = 12$ **b** $y = 13\frac{2}{3}$ **c** $t = 35$ **d** $m = 19$ **e** $a = 17$ **f** $p = 9\frac{2}{3}$ **g** $x = 10$ **h** $x = 3\frac{3}{4}$ **i** $x = 0$ **j** $m = 11$ **k** $n = 30$ **l** $m = 10$ **m** −6 **n** $p = 5$ **o** $a = 8$

PAGE 122 **a** 43.2 **b** 121.8 **c** 23.5 **2 a** 60 **b** 14 **c** 19.6 **3 a** −13 **b** 17 **c** −34 **4 a** 33.6 **b** 23.5 **c** 45.1 **5 a** 4 **b** 484 **c** 25 **6 a** 246 **b** 35.2 **c** 12.875

PAGE 123 **1** $x + 15 = 37, x = 22$ **2** $4x - 5 = 23, x = 7$ **3** $x + 21 = 47, x = 26$ **4** $x + 7 = 48, x = 41$ **5** $4x - 9 = 19, x = 7$ **6** $2x = 64, x = 32$ **7** $2x + 3 = 29, x = 13$ **8** $\frac{x}{5} - 7 = 8, x = 75$ **9** $3x - 7 = 23, x = 10$ **10** $x + (x + 10) + x + (x + 10) = 72$, $x = 13$, ∴ width = 13 cm, length = 23 cm **11** $3x = 42, x = 14$, ∴ each side = 14 cm **12** $12 + x + 12 + x = 40, x = 8$, ∴ width = 8 cm **13** $x - 10 = 18, x = 28$ **14** $8 + 12 + 16 + x = 50, x = 14$ **15** $3(x + 6) - 7 = 26, x = 5$

PAGE 124 **1** D **2** C **3** B **4** A **5** D **6** C **7** B **8** D **9** C **10** A

PAGE 125 **1 a** $x = 12$ **b** $y = 20$ **c** $w = 45$ **d** $m = 50$ **e** $n = 14.5$ **f** $a = 77$ **g** $b = 12$ **h** $d = 4\frac{1}{6}$ **i** $e = -5$ **j** $x = 6.5$ **k** $f = 14$ **l** $h = 6$ **2** $S = -8.25$ **3 a** 6.5 **b** 5 cm

Answers

Chapter 11 – Reasoning in geometry

Page 126 1 a three b two c no d three e one f one 2 a equal b equal c 180° d 60° e opposite interior f 360° 3 a four-sided b many c parallel d parallel e equal f right angle 4 a equal, 90° b equal, parallel c 360° d equal e 90° 5 a equal b equal c equal d 180°

Page 127 1 a $x = 110$ b $x = 35$ c $x = 65$ d $x = 65$ e $x = 120$ f $x = 89$ 2 a $x = 110$ (sum of angles) b $x = 54$ (complementary angles) c $m = 60$ (complementary angles, vert. opp. angles) 3 a $a = 70$ (straight angle) b $x = 25$ (straight angle) c $p = 35$ (complementary angles) d $x = 55$ (straight angle) e $x = 35$ (straight angle) f $x = 55$ (sum of angles)

Page 128 1 a $p = 115$ b $x = 70$ c $x = 115$ d $x = 69°$ e $y = 82$ f $m = 70, n = 70$ 2 a $x = 93$ (co-interior angles) b $x = 66$ (alternate angles) c $m = 40$ (co-interior angles, vertically opposite angles) 3 a $y = 143$ (straight angle, corresponding angles) b $t = 35$ (straight angle, corresponding angles) c $x = 130$ (corresponding angles) d $x = 110$ (co-interior angles) e $x = 30$ (alternate angles) f $x = 115, y = 65$ (straight angle, corresponding angles, alternate angles)

Page 129 1 a $m = 45$ b $a = 50$ c $c = 25$ d $x = 18$ e $y = 95$ f $a = 68$ 2 a $x = 70$ b $x = 60$ c $a = 45$ 3 a $x = 139$ b $x = 70, y = 50$ c $m = 46$ d $x = 90$ e $n = 95$ f $x = 75, y = 75, a = 105$

Page 130 1 a $x = 80$ b $m = 70$ c $x = 115$ d $x = 122, y = 80$ e $x = 60$ f $x = 90$ 2 a $x = 215$ b $x = 100$ c $a = 70$ 3 a $x = 95, y = 85$ b $a = 36$ c $x = 150, y = 40$ d $a = 110, y = 70, x = 90$ e $x = 46$ f $m = 36$

Page 131 1 a transformation b sliding c turning d flipping 2 a translation b rotation c reflection d translation e reflection

Page 132 1 a ≡ b same, same c transformation d matching 2 a translation b reflection c rotation 3 a $\triangle ABC \equiv \triangle DEF$ b $\triangle PQR \equiv \triangle LMN$ c $\triangle GHI \equiv \triangle JKL$

Page 133 1 a same, same b i equal ii equal iii equal iv ≡ 2 a *ABCD* and *EFGH*, *ABFE* and *DCGH*, *ADHE* and *BCGF* b *PQR* and *STU* 3 A and M, B and F, C and L, D and I, E and G, H and J, K and N 4 a $\angle A$ and $\angle E$, $\angle B$ and $\angle F$, $\angle C$ and $\angle G$, $\angle D$ and $\angle H$ b $AB = EF$, $BC = FG$, $CD = GH$, $DA = HE$

Page 134 1 a i *AB* and *DE*, *AC* and *DF*, *BC* and *EF* ii $\triangle ABC$ and $\triangle DEF$ iii $\triangle ABC \equiv \triangle DEF$ b i $\angle P$ and $\angle S$; $\angle Q$ and $\angle T$; $\angle R$ and $\angle U$ ii $\triangle PQR$ and $\triangle STU$ iii $\triangle PQR \equiv \triangle STU$
2 i a $AB = CD$, $BC = DA$ b $\angle B = \angle D$; $\angle BAC = \angle DCA$; $\angle BCA = \angle DAC$ c $\triangle ABC$ and $\triangle CDA$ d $\triangle ABC \equiv \triangle CDA$
ii a $EF = GH$; $EH = GF$ b $\angle E = \angle G$; $\angle EFH = \angle GHF$; $\angle EHF = \angle GFH$ c $\triangle EHF$ and $\triangle GFH$ d $\triangle EHF \equiv \triangle GFH$
iii a $SP = SR$; $PQ = RQ$; *SQ* is common b $\angle P = \angle R$; $\angle PSQ = \angle RSQ$; $\angle PQS = \angle RQS$ c $\triangle PSQ$ and $\triangle RSQ$ d $\triangle PSQ \equiv \triangle RSQ$
iv a $TU = VW$; $TW = VU$; *UW* is common b $\angle T = \angle V$; $\angle TUW = \angle VWU$; $\angle TWU = \angle VUW$ c $\triangle TUW$ and $\triangle VWU$ d $\triangle TUW \equiv \triangle VWU$

Page 135 1 a ≡ b 3 sides c two angles and a corresponding side d two sides and the included angle e hypotenuse and one side 2 a RHS b SSS c AAS d SAS 3 a XW b Yes c RHS

Page 136 1 a i $\triangle ABC$ and $\triangle ADC$ ii AAS iii $\triangle ABC \equiv \triangle ADC$ b i $\triangle ADB$ and $\triangle CBD$ ii SAS iii $\triangle ADB \equiv \triangle CBD$ c i $\triangle DEF$ and $\triangle GFE$ ii SSS iii $\triangle DEF \equiv \triangle GFE$ d i $\triangle MTN$ and $\triangle MTP$ ii RHS iii $\triangle MTN \equiv \triangle MTP$ 2 a i $\triangle ACB$ and $\triangle DFE$ ii SSS iii $\triangle ACB \equiv \triangle DFE$ iv $x = 60$ b i $\triangle ABC$ and $\triangle ADC$ ii AAS iii $\triangle ABC \equiv \triangle ADC$ iv $y = 8$ cm c i $\triangle ABE$ and $\triangle CDE$ ii SAS iii $\triangle ABE \equiv \triangle CDE$ iv $x = 12$ cm d i $\triangle LMN$ and $\triangle LPN$ ii RHS iii $\triangle LMN \equiv \triangle LPN$ iv $x = 27$

Page 137 1 a $a = 65$ and $b = 25$ (corresponding ∠s of congruent △s) b $x = 20, y = 70$ (corresponding ∠s of congruent △s) c $x = 12$ cm (corresponding sides of congruent △s) $y = 93$ (corresponding ∠s of congruent △s)
2 a $x = 90$ b $x = 90, y = 45$ c $x = 75, y = 40, z = 65$ 3 a $x = 42, y = 48$ b $y = 15$ cm, $m = 63$

Page 138 1 a $a = 80, b = 100, c = 100, d = 80$ b $x = 36$ c $a = 60, y = 120, x = 60$
2 a $x = 40$, angle sum of a quadrilateral b $a = 110, b = 70, c = 110$, cointerior angles c $x = 30$, angle sum of a quadrilateral
3 a $a = 90, x = 45, y = 45$ b $x = 90, y = 60$ c $a = 80$

Page 139 1 a $x = 50$; 50, 100, 110, 100 b $m = 60$; 60, 60, 120, 120 c $x = 36$; 36, 72, 108, 144
2 a $a = 90, b = 90, c = 90$ b $a = 70, b = 125$ c $m = 18$ d $a = 26, b = 100, c = 54$
e $a = 35, b = 35, c = 55, d = 90, e = 55$ f $x = 25$

Page 140 1 B 2 D 3 C 4 A 5 B 6 C 7 D 8 D 9 C 10 A

Page 141 1 a $PQ = SQ$, $QR = QR$ b $\angle PQR = \angle SQR$ c SAS d $\triangle PQR \equiv \triangle SQR$ 2 a $t = 35$ b $x = 40$ c $y = 42$ d $y = 75$ 3 a AAS b $\triangle DEF \equiv \triangle GFE$ c $x = 9$ cm d $\triangle DEH \equiv \triangle GFH$ 4 a SSS b $\triangle ACB \equiv \triangle CAD$ c $x = 60$

Chapter 12 – Probability

Page 142 1 a $\frac{1}{4}$ b $\frac{1}{2}$ c $\frac{1}{13}$ d $\frac{3}{4}$ e $\frac{1}{26}$ f $\frac{1}{2}$ 2 a $\frac{1}{3}$ b $\frac{2}{3}$ c $\frac{1}{9}$ 3 a $\frac{1}{6}$ b $\frac{1}{2}$ c $\frac{2}{3}$ d 0 e $\frac{1}{2}$ f $\frac{1}{3}$ 4 a $\frac{2}{5}$ b $\frac{1}{4}$ c $\frac{7}{20}$ d $\frac{3}{5}$ e 0 f $\frac{3}{4}$ 5 a $\frac{1}{3}$ b $\frac{2}{3}$ c $\frac{2}{3}$ d 1 e $\frac{1}{3}$ f $\frac{1}{3}$ 6 a $\frac{4}{9}$ b $\frac{5}{9}$ c $\frac{1}{9}$ d 0 e $\frac{4}{9}$ f $\frac{1}{3}$ 7 a $\frac{1}{3}$ b $\frac{2}{3}$ c $\frac{2}{9}$ d $\frac{1}{9}$ e $\frac{5}{9}$ f $\frac{2}{9}$

Answers

PAGE 143 1 a 100 b 0.43 c $\frac{43}{100}$ d 0.57 e $\frac{57}{100}$ f 1 g 100 tails
2 a $\frac{1}{8}, \frac{1}{5}, \frac{1}{4}, \frac{3}{20}, \frac{9}{80}, \frac{13}{80}$ b i $\frac{3}{20}$ ii $\frac{41}{80}$ iii $\frac{9}{20}$ iv $\frac{17}{40}$ v $11\frac{1}{4}\%$ vi $48\frac{3}{4}\%$ vii 0.1625 viii 0.25 ix $\frac{1}{5}$ x $\frac{1}{8}$

PAGE 144 1 a 0 to 1 2 a 0.75 b 0.25 c 0 3 a 10% b 9% c 90% d 20% e 10% f 19% 4 a 20% b 40% c 60% d 40% 5 a $\frac{9}{20}$ b $\frac{7}{20}$ c $\frac{1}{5}$ d $\frac{4}{5}$ e 0 f 1 g $\frac{11}{20}$ h $\frac{13}{20}$ i $\frac{4}{5}$ j 0

PAGE 145 1 a tossing a tail b rolling an even number c not selecting a diamond d losing a game of tennis e spinning a number not less than 10 f complementary events 2 $\frac{1}{3}$ 3 a $\frac{1}{4}$ b $\frac{3}{4}$ c $\frac{1}{2}$ d $\frac{1}{2}$ e $\frac{1}{13}$ f $\frac{12}{13}$ 4 a $\frac{1}{3}$ b $\frac{2}{3}$ c $\frac{1}{9}$ d $\frac{8}{9}$ e $\frac{4}{9}$ f $\frac{5}{9}$ 5 a $\frac{4}{7}$ b $\frac{3}{7}$ c $\frac{4}{7}$ d $\frac{3}{7}$ 6 a $\frac{2}{5}$ b $\frac{3}{5}$ c $\frac{4}{15}$ d $\frac{11}{15}$ e $\frac{1}{3}$ f $\frac{2}{3}$ g $\frac{2}{3}$ h $\frac{11}{15}$

PAGE 146 1 a

2nd die \ 1st die	1	2	3	4	5	6
1	1, 1	2, 1	3, 1	4, 1	5, 1	6, 1
2	1, 2	2, 2	3, 2	4, 2	5, 2	6, 2
3	1, 3	2, 3	3, 3	4, 3	5, 3	6, 3
4	1, 4	2, 4	3, 4	4, 4	5, 4	6, 4
5	1, 5	2, 5	3, 5	4, 5	5, 5	6, 5
6	1, 6	2, 6	3, 6	4, 6	5, 6	6, 6

2 a $\frac{1}{36}$ b $\frac{1}{6}$ c $\frac{5}{36}$ d $\frac{1}{4}$ e $\frac{1}{9}$ f $\frac{1}{12}$ g $\frac{5}{18}$ h $\frac{5}{18}$ i $\frac{1}{6}$ j $\frac{11}{36}$ k 0 3 one die 4 $\frac{1}{4}$

PAGE 147 1 a

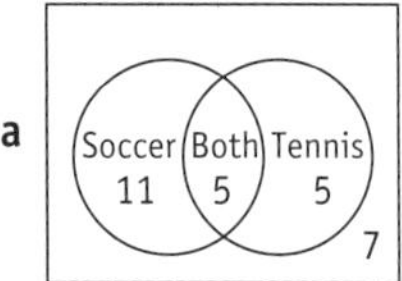

b 5 c $\frac{11}{28}$ d $\frac{4}{7}$ 2 a

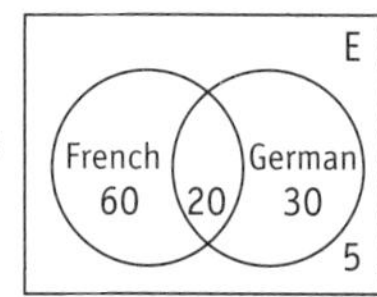

b $\frac{16}{23}$ c $\frac{10}{23}$ d $\frac{1}{23}$
3 a 38 b 15 c 11 d 80

PAGE 148 1 $\frac{7}{8}$ 2 $\frac{15}{16}$ 3 a $\frac{1}{2}$ b $\frac{1}{2}$ c $\frac{3}{10}$ d $\frac{4}{5}$ e $\frac{4}{5}$ f $\frac{1}{5}$ g $\frac{1}{5}$ h 0 i $\frac{1}{2}$ j $\frac{3}{5}$ 4 a $\frac{7}{13}$ b $\frac{3}{26}$ c $\frac{4}{13}$ 5 a $\frac{2}{9}$ b $\frac{1}{6}$

PAGE 149 1 a 575; 150; 400; 900 b i $\frac{23}{36}$ ii $\frac{7}{36}$ iii $\frac{7}{18}$ 2 a $\frac{3}{5}$ b $\frac{2}{5}$ c $\frac{1}{3}$ d $\frac{4}{15}$ e $\frac{13}{15}$ f $\frac{1}{3}$ 3 a $\frac{3}{5}$ b $\frac{8}{25}$ c $\frac{3}{25}$ d $\frac{2}{5}$ e $\frac{2}{25}$ f $\frac{12}{25}$

PAGE 150 1 B 2 D 3 C 4 D 5 C 6 C 7 A 8 D 9 A 10 D

PAGE 151 1 a 67 b 95 c $\frac{5}{19}$ d $\frac{28}{95}$ 2 a $\frac{2}{3}$ b $\frac{1}{3}$ c $\frac{1}{3}$ d $\frac{2}{3}$ e 1 f 0 g $\frac{1}{3}$ 3 a 22 b 36 c $\frac{7}{30}$ d $\frac{32}{47}$

CHAPTER 13 – Statistics

PAGE 152 1 a survey b census c sample 2 a sample b census c sample d sample e census f census 3 a census b sample c census d sample e census f sample

PAGE 153 1 a chance b systematic c strata 2 a stratified b systematic c random 3 a systematic b stratified c stratified d random e systematic

PAGE 154 1 a data b census c sample d categorical, qualitative e quantitative, numerical f discrete g continuous 2 a numerical b numerical c numerical d categorical e numerical f categorical g numerical h numerical 3 a discrete b continuous c discrete d discrete e continuous f discrete g continuous h continuous i discrete j continuous k continuous

Answers

Page 155

1 a

Score (x)	Tally	Frequency (f)
4	||	2
5	|||	3
6	||	2
7	~~||||~~ ||	7
8	~~||||~~ |||	8
9	~~||||~~ ~~||||~~	10
10	~~||||~~ |	6
11	||	2

b, c

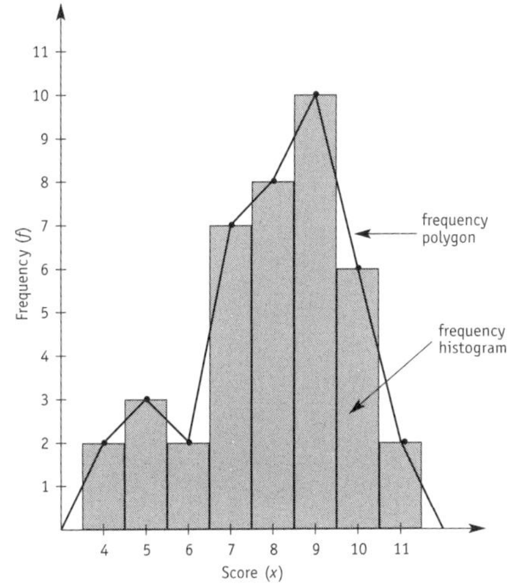

2

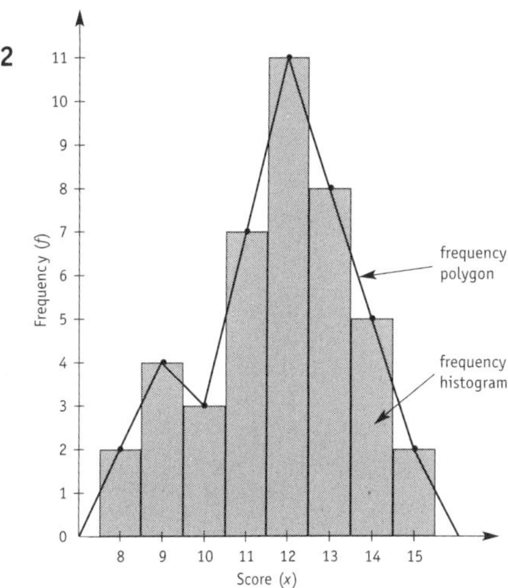

Page 156

1 a

Score (x)	Frequency (f)	Cumulative frequency
5	1	1
6	2	3
7	3	6
8	6	12
9	5	17
10	4	21
11	3	24
12	1	25

b

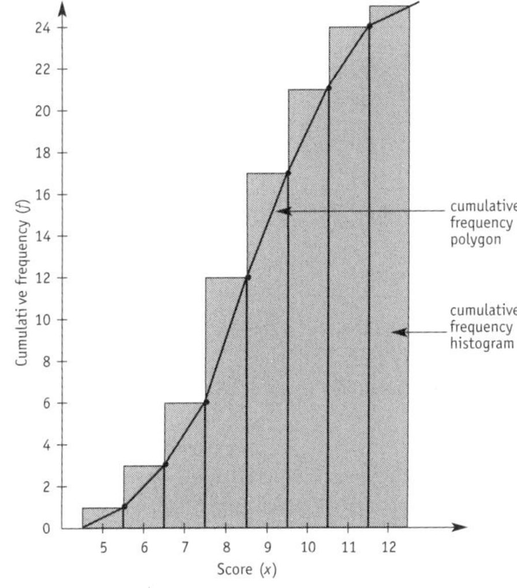

2 a

Score (x)	Tally	Frequency (f)	Cumulative frequency
4	|	1	1
5	||	2	3
6	|||	3	6
7	|	1	7
8	|||	3	10
9	||	2	12
10	~~||||~~	5	17
11	|	1	18
12	||	2	20

b

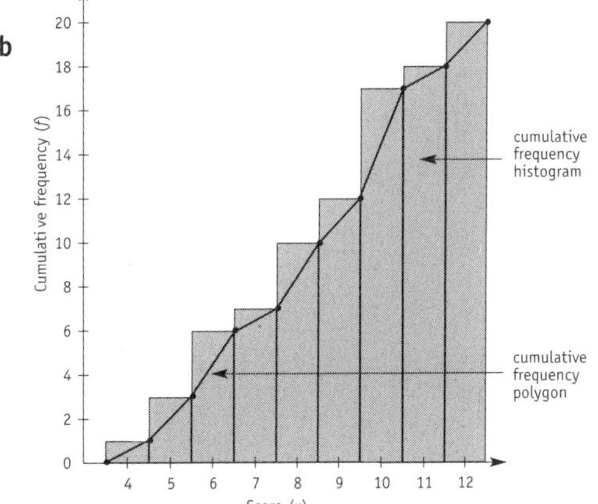

Page 157

1 a

Class	Class centre (c.c.)	Tally	Frequency	Cumulative frequency
57–61	59	~~||||~~	5	5
62–66	64	||||	4	9
67–71	69	~~||||~~ |	6	15
72–76	74	~~||||~~ ||	7	22
77–81	79	~~||||~~ ~~||||~~	10	32
82–86	84	~~||||~~ ~~||||~~ ||	12	44
87–91	89	~~||||~~ |	6	50

b

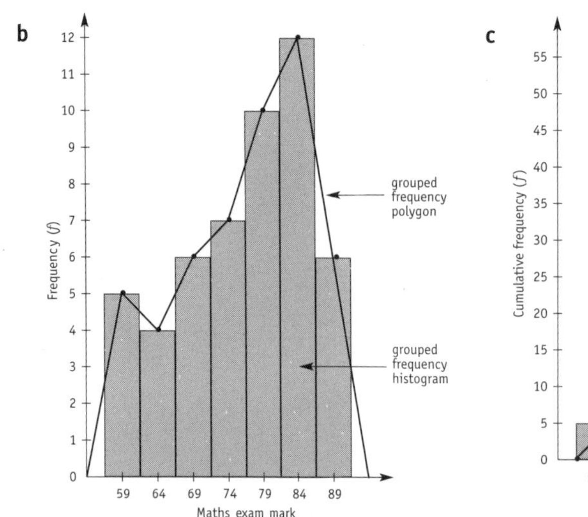

c

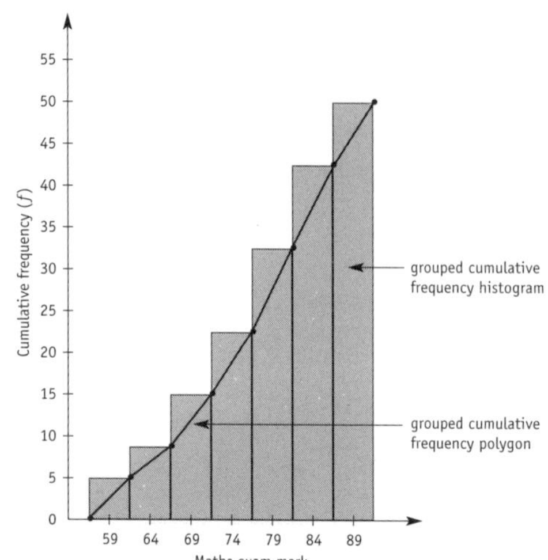

PAGE 158 **1 a** 11 **b** 10.5 **c** 13.5 **d** 17 **e** 12.5 **f** 22 **2 a** 7.4 **b** $11.\dot{3}$ **c** $14.\dot{7}$ **d** 6.9 **e** 9 **f** 15 **3 a** $17.\dot{8}$ **b** 12.375 **c** $8.8\dot{3}$ **d** 9.538 **4 a** 4.2 **b** $5.\dot{3}$ **c** 9.952

PAGE 159 **1 a** 6 **b** 11 **c** 16 **d** 57 **e** 11.5 **f** 9.5 **2 a** 3 **b** 8 **c** 11 **d** 37 **e** 5 **f** 7

3 a

Score	Frequency	Cumulative frequency
1	3	3
2	6	9
3	7	16
4	5	21
5	8	29

Median = 3 Mode = 5

b

Score	Frequency	Cumulative frequency
18	6	6
19	1	7
20	5	12
21	3	15
22	3	18

Median = 20 Mode = 18

c

Score	Frequency	Cumulative frequency
0	5	5
1	2	7
2	6	13
3	9	22
4	2	24
5	1	25

Median = 2 Mode = 3

d

Score	Frequency	Cumulative frequency
10	12	12
11	10	22
12	16	38
13	14	52
14	8	60
15	6	66

Median = 12 Mode = 12

PAGE 160 **1 a i** 94.5 **ii** 83.75 **iii** 96 **b** Edwin was above the average but he did not do well, he was second lowest in the class. When data includes atypical scores, the median should be used as the measure of 'the middle'. **2** Few houses would have identical prices so the mode is not used. If one or several very expensive homes were sold this would significantly increase the mean, the mean would no longer be a good indicator of the price of the majority of houses sold. The median would be unaffected by the few high prices. **3 a** 16.7 **b** 14 **c** 16 **d** The shop owner would sell more of this size and so would need to stock more of the modal size.

PAGE 161 **1 a** 8 **b** 9 **c** 22 **d** 77 **e** 47 **f** 40 **2 a** 7 **b** 10 **3 a** 47 **b** 53 **4 a** 8 **b** 8 **5 a** 15 **b** 75

PAGE 162 **1** C **2** A **3** B **4** B **5** C **6** A **7** D **8** B **9** B **10** B

PAGE 163 **1 a** Census **b** Stratified **c** Categorical **d** mean and range **2 a** 4.4 **b** 3 **c** 5
3 a 4.458 **b** 5 **c** 5 **d & e** **f** 5

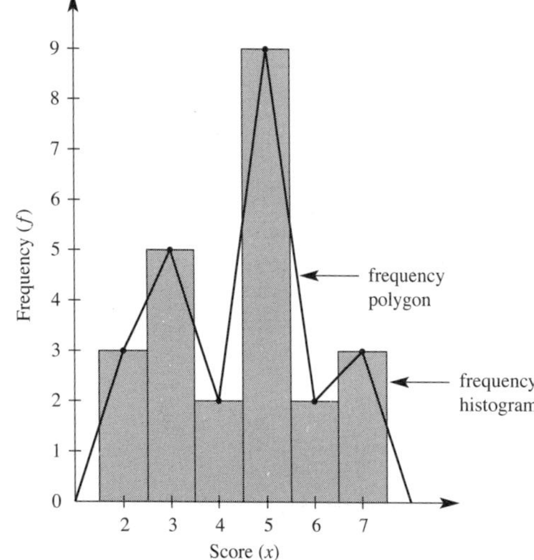

Answers

Exam Papers

PAGE 164 **1** C **2** B **3** B **4** D **5** C **6** D **7** C **8** A **9** C **10** D

PAGE 165 **1** 200 **2** −2 **3** recurring **4** \$200 **5** 2^9 **6** 5:8 **7** 100 000 **8** $m = 15$ **9** $-a - 2b$ **10** 82 **11** $y = 5$ **12** 85° **13** 2 hours 45 min **14** 1:25 **15** 320

PAGE 166 **1 a** $19a$ **b** $21axy + 30xy$ **c** $7x^2$ **d** $9m^3n$ **2 a** 16 **b** \$1080 **3 a** \$120 **b** 80% **4 a** $2m + 2n$; mn **b** $x + 2y + 3z$; xy **5** $P = S - C$ **6** No **7 a** chord **b** radius **8** rotation **9** $a = 108$ (co-interior angles) $b = 72$ (co-interior angles)

PAGE 167 **1** B **2** A **3** A **4** A **5** A **6** D **7** A **8** C **9** A **10** B **11** B

PAGE 168 **1** $4a$ **2** 4 **3** 2.57% **4** 4.5 minutes **5** $9(a - 1)$ **6** $x = 2$ **7** \$3.60 **8** $x = 1\frac{1}{3}$ **9** −24 **10** 85 km/h **11** 0.375 **12** $\frac{3}{4}$ **13** 24 **14** $6 - 6a$ **15** 0.34

PAGE 169 **1 a** 48.8 cm^2 **b** 11.5 m^2 **c** 85.4 cm^2 **d** 2.7 m^2 **2 a** $3a + 12$ **b** $4x^2 - 23x + 3$ **3 a** $x = 48$ (vertically opposite) **b** $x = 25$ (complementary angles) **c** $x = 140$ (angles at a point) **4** \$17 **5** \$54 **6 a** 16 : 3 **b** 9 **c** 13 **d** 5 : 3 **7** $\frac{4}{9}$ **8** 72

PAGE 170 **1** C **2** C **3** B **4** D **5** D **6** A **7** C **8** D **9** D **10** A

PAGE 171 **1 a** 45 **b** $-\frac{4}{21}$ **c** 8.4 **d** 0.15 **2** $1\frac{1}{9}$ **3** $x = 21$ **4** 3.527 **5** $\frac{5}{6}$ **6 a** 11 300 m **b** 200 000 mg **c** 345 kL **d** 62.08 mg **7 a** 2 **b** 2 **c** 1.9

PAGE 172 **1** 30% **2** \$420 **3** 14 **4 a** 15 km **b** 10 km/h **5 a** 4 **b** −2 **c** $\frac{4}{13}$ **6 a** $2a^3$ **b** $-9a - 15b$ **7 a** $c = 50°$ **b** $b = 50°$ **c** a = 80° **8 a** $38\frac{1}{2}$ cm^2 **b** 25 cm **9 a** 1.25 **b** 16 **c** 192 **10 a** $x = 4$ **b** $x = 2$

PAGE 173 **1** C **2** B **3** C **4** A **5** B **6** C **7** A **8** D **9** C **10** D

PAGE 174 **1** 0005 hours **2** 7.52 pm **3** −12 **4** $x = 36$ **5** 147 **6** $2\frac{1}{12}$ **7** 7.761×10^4 **8** 198 **9** 2^{24} **10** true **11** $20ab - 5a^2$ **12** $3(3a + 8)$ **13** \$80, \$120 **14** \$29.40 **15** $v = 11$

PAGES 175–176 **1 a** $\angle BCA = 20°$ (base angles of isosceles triangle equal) **b** SAS **2 a** 66.5 cm^2 **b** 19.5 cm^2 **c** 82.2 cm^2 **d** 89.7 cm^2 **3 a** 19 m^2 **b** 76 m^3 **c** 76 000 L **d** 3.6 kL **4 a** 9 **b** 5 **c** 4 **d** 0 **e** 1 **f** 1.22

5 a [Venn diagram: N only 7, N and E 5, E only 9, outside 3] **b** $\frac{3}{8}$

Notes

Notes

Notes